Advances in Geographic Information Science

Series Editors:

Shivanand Balram, Canada
Suzana Dragicevic, Canada

Zhenjiang Shen

Geospatial Techniques in Urban Planning

Zhenjiang Shen
Kanazawa University
Natural Science and Technology Hall
2C718
Kakuma machi
920-1192 Kanazawa
Japan
shenzhe@t.kanazawa-u.ac.jp

ISBN 978-3-642-13558-3   e-ISBN 978-3-642-13559-0
DOI 10.1007/978-3-642-13559-0
Springer Heidelberg Dordrecht London New York

Library of Congress Control Number: 2011940817

© Springer-Verlag Berlin Heidelberg 2012
This work is subject to copyright. All rights are reserved, whether the whole or part of the material is concerned, specifically the rights of translation, reprinting, reuse of illustrations, recitation, broadcasting, reproduction on microfilm or in any other way, and storage in data banks. Duplication of this publication or parts thereof is permitted only under the provisions of the German Copyright Law of September 9, 1965, in its current version, and permission for use must always be obtained from Springer. Violations are liable to prosecution under the German Copyright Law.
The use of general descriptive names, registered names, trademarks, etc. in this publication does not imply, even in the absence of a specific statement, that such names are exempt from the relevant protective laws and regulations and therefore free for general use.

Printed on acid-free paper

Springer is part of Springer Science+Business Media (www.springer.com)

Editorial Introduction

The book will bring together some of the theoretical and empirical perspectives about application of Geotechnology in the field of urban planning and design. From an integrative viewpoint on Geosimulation, Geovisualization and Geography information system, we give readers a valuable insight into the following subjects:

- Simulation of planning policy impacts on urban land use using cellular automaton (CA) and multi-agent system (MAS) process,
- Visualization of design guideline for improving coordinative design and gaining consensus using virtual reality (VR),
- Development of planning support tools using geography information system (GIS) and integration GIS with CA, MAS and VR.

Currently the three concepts of Geosimulation, Geovisulization and Geography information system are not integrated together so as to build a complete platform of Geotechnology. For example, Google earth is an integrated platform of virtual reality and geography information system. The integrated multi-user environment based on agent-based simulation and virtual reality is also constructed by Secondlife Inc. In this book, we discuss these three concepts and their intended effects on planning support respectively, which imply inspirational possibilities of Geotechnology integration in planning practice. Accordingly, this book will be edited as three parts, which are Geosimulation and land use plan, Geovisualization and urban design, Geography information system and planning support.

For application of CA in urban growth simulation, Chap. 1 discussed about how to predict urban boundary in Beijing and argued about spatial constraints and relevant spatial policies in Beijing. However, the essential thought of constraint CA that spatial constraints decide the urban form, is not true in planning process of a real city. In planning practice, there are more complicated factors that make planners to draw alternative land use plans. The authors propose the concept of "form scenario analysis" (FSA), which is utilized to investigate relationships between planning alternatives and corresponding spatial policies and to adapt spatial policies as necessary for various planning alternatives. Chap. 2 is about a case study in Chuangdong, Guizhou province in China, in which CA is used for simulating urban growth and predicting developed area of a new city with 80 minion populations while considering growth of households in the predicted 20 years in the future. Households are designed as agents in this work who consuming energy and producing waste in their daily life, in which MAS approach is instrumented. Thereby, a low carbon city policy is expected to be conducted in our further research work. In chap. 3, the methodology of CA and MAS are argued with respect to land use planning from viewpoints of farmland conversation quota during urban growth, in which the authors explain the interactions between various driving stakeholders, such as governments, residents, farmers, and industrial enterprises. Furthermore, special attention has been paid to spatio-temporal allocation efficiency of farmland conversion quotas for land use planning. Chap. 4 investigates the implications of introducing planning behaviors into complex organizational systems. As a great progress of Chap. 5 in current CA literatures, alternative plans proposed by urban planners are introduced into CA simulation, while impacts of relevant policies with respect to spatial constraints are retrieved in CA simulation. We suggest that constrained CA simulation can be an approach for negotiation between planners and developers by identifying the required spatial policies for predefined alternative plans in planning practice. For integrating with environmental issues in the CA and MAS approach for urban growth simulation, the household water consumption simulation (HWCSim) model is designed in Chap. 6 to reflect the process of total amount control of household water consumption. As validated by the real data for Kanazawa City, the HWCSim model is found to be applicable for simulating the process of household water consumption regulated by local government in planning practice. The simulation results, including water price and household water consumption, are very similar to the actual data obtained for Kanazawa City.

Regarding Geovisualization, Chap. 7 argued that VR technology has been played very important role in urban planning and managing process, and introduced many case studies of 3D digital city model projects for representing urban design of centre business districts of several metropolitan areas in China. In chap. 8, the authors concentrate on the digital preservation of historic buildings, which is a combination of theoretical methods and real-life measurement and experiment; it offers a detailed description for the process of 3D digitization and virtual simulation. We argued in Chap. 9 about how to use virtual reality for visualizing design guideline, meanwhile, a web-based planning committee is suggested for gaining consensus of design guideline without spatial and temporal limitation. In this Chapter, using some video tapes recorded in deliberation process of a local planning committee, the effectiveness of VR representation is analysed. Some exciting ideas of advanced application of VR in urban planning are introduced in Chaps. 10–12. In Chap. 10, an on-line tool using VRML takes the original icon game as model for design game of a public park in Kanazawa city, Japan, which is used to collect participants' proposals in their community. Participants can use this on-line tool to arrange design elements on a park site, which coordinate information can be saved in a web database for generating a 3D scene of his/her design. In Chap. 11, a VRML tool with prescribed design elements for representing alternatives of townscape design can be changed by users on the internet for gaining consensus of a design guideline in a traditional district. This tool is developed for residents' design learning of traditional style buildings in historical conservation areas. By using this tool, planner can help stakeholders to coordinate alternative designs or review them through multi-user environment, which replace drawings, documents and other traditional media for representing urban planning and design. In chap. 12, we conducted a questionnaire, the results of which verified that virtual representations using Google SketchUp and Google Earth are suitable for improving residents' understanding of the historical landscape. The idea of opening the model to public view on the Internet was evaluated positively by respondents to a questionnaire. Accordingly, Geovisualization can be used to visualize participants' alternatives who are non-experts for representing their proposals of planning and design. Meanwhile, change of those alternatives can be shared between participants on multi-user environment for getting mutual understanding and reaching consensus in deliberation process. Otherwise, visualization is also useful for design review after a consensus of design guideline is reached.

As for GIS that is the most important part of Geotechnology and has longer history than Geosimulation and Geovisualization in planning support, the customized GIS tools with powerful spatial analysis functions can be used for diversified purposes of planning and design. Part III argues about how to customize GIS tool according to different requirements in different urban planning projects. Meanwhile, we also introduced some ideas and tools from an integrative viewpoint on Geosimulation, Geovisualization and Geography information system. In chap. 13, we propose a GIS and CG integrated system that automatically generates 3D building models as 3D city model. In urban planning practice, a 3D city model is quite effective in understanding what will be built, what image of the town will be changed in an urban area for all stakeholders including residents, citizens as well as planners and designers to gain consensus on design alternatives. A urban growth control planning support system (UGC-PSS) tool proposed in Chap. 14 offers a platform for urban growth control analysis. For the Beijing Metropolitan Area, 60 control factors, together with their indicators, were included to reflect urban growth conditions in the study area. The UGC-PSS tool enables complex urban growth control problems to be solved by using multi-disciplinary knowledge, such as flooding control, eco-zone protection, noise prevention, and disaster prevention, which had not previously been possible for a single urban planner with limited education background. In Chap. 15, a planning support system based on ArcGIS is developed to retrieve alternatives of conservation plan for traditional houses in Beijing, which integrates all possible data sources of household and traditional house in the platform of ArcGIS and generates alternatives of the traditional house with a decision tree referred to deducting process using the spatial features' attributes. In Chap. 16, a tool is developed for retrieving ecological network from aero photography and local urban system database, which use not only features' attributes but also raster data for classification of land use types from aero photograph. A decision tree based on features' attributes is also employed as the retrieving rule for generating ecological map. For integration with CA, MAS and

GIS, spatial analysis functions, such as neighbour statistic is employed in Chap. 17 for CA simulation tool in the case of irregular polygons, which are vector dataset of GIS geodatabase. Chap. 17 showed that it is possible to employ CA for simulating land use change at the level of urban partitions using a data set of irregular blocks and parcels, which reflect impacts from urban planning conditions and parcel attributes after an urban project in Kanazawa city, Japan. Finally, approach of MAS that simulating land use policy for regulating locations of large-scale shopping centers is employed to take different scenarios of commercial policies into account in Chap. 18, which shows the possibility of integration of MAS and GIS.

In this book, we try to combine VR, MAS with GIS for Geovisualization and Geosimulation, which imply the prospective integration of virtual reality, simulation and geography information.

Our main objectives as the following:

- Investigating substantial planning problems and visualizing land use plan & policy impacts on land use based on Geosimulation
- Visualizing and coordinating design alternatives using virtual reality and improving deliberation, decision-making and gaining consensus through Geovisulization in cooperative design process and reaching consensus for design guideline and regulations
- Customized tools for planning and design using Geography information system and tools for diversified planning projects through integration of VR, MAS and GIS

Acknowledge

I would like to express my deep appreciation to all the authors for their outstanding contributions to this book, and to series editors of "Advances in Geographic Information Science" Professor Suzana Dragicevic and Professor Shivanand Balram of Simon Fraser University for their kind invitation and encouragement, Dr. Janet Sterritt-Brunner and Dr. Chris Bendall of Springer for their kind editorial work and help to have such diverse topics on "Geospatial Techniques in Urban Planning" published as a book.

I am deeply indebted to my colleague Professor Mitsuhiko KAWAKAMI from the School of Environmental Design, Kanazawa University whose help, stimulating suggestions and encouragement helped me in all the time of research. I also want to thank Professor Jen-Te PAI from Chengchi University and Ms. Yan MA for all their assistance for editing this book.

Especially, I would like to give my special thanks to my wife Fan Huang whose patient love enabled me to complete this work.

Contents

Part I Geosimulation and Land Use Plan

1 A Challenge to Configure Form Scenarios for Urban Growth Simulations Reflecting the Institutional Implications of Land-Use Policy 3
Ying Long, Zhenjiang Shen, Qizhi Mao, and Liqun Du

2 A Planning Tool for Simulating Urban Growth Process and Spatial Strategy of Urban Development in Chuandong, China 27
Yan Ma, Zhenjiang Shen, Dingyou Zhou, and Ke Wang

3 Simulating Spatio-Temporal Allocation of Farmland Conversion Quotas in China Using a Multi-Agent System 49
Zhang Honghui, Zeng Yongnian, Tan Rong, and Shen Zhenjiang

4 Planning in Complex Spatial and Temporal Systems: A Simulation Framework 73
Shih-Kung Lai and Haoying Han

5 Reaching Consensus Among Stakeholders on Planned Urban Form Using Constrained CA 91
Ying Long and Zhenjiang Shen

6 An Agent-Based Approach to Support Decision-Making of Total Amount Control for Household Water Consumption 107
Yan Ma, Zhenjiang Shen, Mitsuhiko Kawakami, Katsunori Suzuki, and Ying Long

Part II GeoVisualization and Urban Design

7 Review of VR Application in Digital Urban Planning and Managing 131
Anrong Dang, Wei Liang, and Wei Chi

8 Virtual Fort San Domingo in Taiwan: A Study on Accurate and High Level of Detail 3D Modelling 155
Shih-Yuan Lin and Sheng-Chih Chen

9 Web-Based Multimedia and Public Participation for Green Corridor Design of an Urban Ecological Network 185
Zhenjiang Shen, Mitsuhiko Kawakami, and Kazuko Kishimoto

10 Online Cooperative Design for the Proposal of Layouts of Street Furniture in a Street Park 205
Zhenjiang Shen, Dingyou Zhou, Mitsuhiko Kawakami, Kazuko Kishimoto, and Seitaro Imai

11 Online Learning Tool for Repair of Traditional Merchant Houses: Machiya 225
Zhenjiang Shen, Mitsuhiko Kawakami, Masayasu Tsunekawa, and Eiichi Nishimoto

12 Historical Landscape Restoration Using Google Technology in a Traditional Temple Area, Kanazawa, Japan 241
Zhenjiang Shen, Mitsuhiko Kawakami, Zheyuang Chen, and Linqian Peng

Part III Geography Information System and Planning Support

13 Automatic Generation of Virtual 3D City Models for Urban Planning 265
Kenichi Sugihara and Zhenjiang Shen

14 An Urban Growth Control Planning Support System for the Beijing Metropolitan Area 285
Ying Long, Zhenjiang Shen, and Qizhi Mao

15 A Planning Support System for Retrieving Planning Alternatives of Historical Conservation Areas from Spatial Data Using GIS 307
Zhenjiang Shen, Mistuhiko Kawakami, Fangfang Lu, Lanchun Bian, Ying Long, Lin Gao, and Dingyou Zhou

16 Visualization of the District Ecological Network Plan at Urban Partitions for Public Involvement 323
Zhenjiang Shen, Mitsuhiko Kawakami, and Satoshi Yamashita

Contents xiii

17 Simulating Land-Use Patterns and Building Types after Land Readjustment at the Urban District Level Using the CAUFN Tool 343
Zhenjiang Shen, Mitsuhiko Kawakami, Takaaki Kushita, and Ippei Kuwamura

18 Integration of MAS and GIS Using Netlogo 369
Zhenjiang Shen, Xiaobai A. Yao, Mitsuhiko Kawakami, Ping Chen, and Masahito Koujin

Index 389

Part I
Geosimulation and Land Use Plan

Investigating urban growth based on computation.
Investigating feasibility of planning alternatives.
Visualizing policies impact and urban activities on urban spaces.

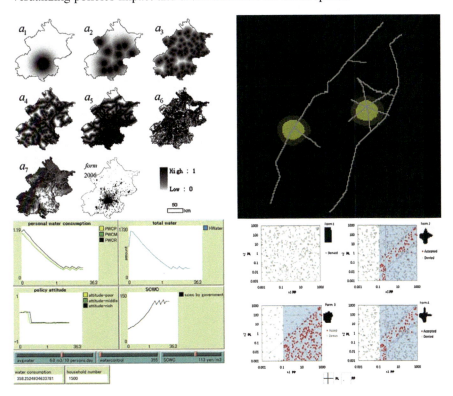

Chapter 1
A Challenge to Configure Form Scenarios for Urban Growth Simulations Reflecting the Institutional Implications of Land-Use Policy

Ying Long, Zhenjiang Shen, Qizhi Mao, and Liqun Du

Introduction

In planning practice, planners and policy makers frequently investigate urban forms, particularly urban growth boundaries (UGBs), using scenario analyses (SA) by regarding development policies as scenario conditions in urban simulations (i.e., Klosterman 1999; Landis 1994, 1995). Couclelis (2005), however, argued that routine land-use modeling has done little in the way of future-oriented research such as investigations of desirable or feared future conditions. This chapter uses planning alternatives, specifically UGBs, as scenarios to identify necessary spatial policies for planners. This is the inverse procedure of traditional urban growth SA. We propose the concept of "form scenario analysis" (FSA), which we employ to investigate relationships between planning alternatives and corresponding spatial policies. This chapter explains an FSA approach using constrained cellular automata (CA), a tool for matching planning alternatives with necessary spatial policies. We look in particular at form scenarios in order to present the institutional implications of different spatial land-use policy options. This novel exploration of FSA can identify necessary policies as well as policy variations required for different planning alternatives. This differs from traditional applications of constrained CA.

Y. Long (✉)
Beijing Institute of City Planning, School of Architecture, Tsinghua University, Beijing, China
e-mail: longying1980@gmail.com

Z. Shen
School of Environmental Design, Kanazawa University, Kanazawa, Japan

Q. Mao
School of Architecture, Tsinghua University, Beijing, China

L. Du
Beijing Institute of City Planning, Beijing, China

Z. Shen, *Geospatial Techniques in Urban Planning*, Advances in Geographic
Information Science, DOI 10.1007/978-3-642-13559-0_1,
© Springer-Verlag Berlin Heidelberg 2012

Unlike in the West, Chinese urban planners, mostly working in government-held planning institutions, hold intrinsic notions from the planned economic system when devising alternatives, resulting in development potential factors not well being taken into account. Planners are accustomed to designing urban structure in master plans without sufficient deliberation on spatial policies that are required to realize the desired urban form. UGBs produced by planners cannot completely conform to the land-use development plan issued by the national land resource bureau, which is dynamically influenced by social, economic, political, technological, and environmental policies such as eco-sensitive land protection, infrastructure development, and other market economy factors. In planning practice, practical urban growth tends to depart from planned UGBs. An Investigation has shown that more than 35% of urban development occurring in Beijing exceeds the urban spaces defined in the original plan (Han et al. 2009). Appropriate policy guidance by planning authorities is necessary to make planned alternatives become reality in the process of urban development. From this point of view, the government is commonly concerned with spatial policies on spatial constraints that are consistent with planned UGBs. Therefore, FSA has significant practical promise as an approach to extract appropriate spatial policies from planning alternatives.

This chapter is organized as follows. In Sect. "What Is Form Scenario Analysis (FSA)?", conventional urban growth simulation and FSA are further introduced. In Sect. "Simulated UGBs Using Conventional Constrained CA", we employ conventional methods to simulate urban growth boundaries. In Sect. "The FSA Approach for Retrieving Policy Parameters that Fit Planned UGBs", we elaborate in detail on our new FSA approach using constrained CA. Section "Experiments in the Beijing Metropolitan Area" describes a case study in which we apply the FSA approach to four planning alternatives in the latest urban master plan for the Beijing Metropolitan Area. We also describe our research materials, including the study area, spatial constraints and planning alternatives. The form scenario analysis results are listed in Sect. "Conclusions and Next Steps". Finally, we give some discussion, a summary, and a description of the next steps of the FSA research.

What is Form Scenario Analysis (FSA)?

Cellular Automata for Simulating Urban Growth Boundaries (UGBs)

Recently, the process of simulating future urban forms using constrained cellular automata (CA) has attracted extensive attention. The results of such simulations, when viewed as alternative future urban forms based on specific assumptions of spatial constraints and policy environment, can be used as a basis for establishing UGBs.

Urban sprawl arising from rapid development is a great challenge in sustainable urban development. It is crucial to design appropriate methods for effective control of urban growth. Among various urban growth management policies, urban containment has been widely adopted to increase urban land-use density and protect open space (Nelson and Duncan 1995). Urban containment policies usually have three components: greenbelts, UGBs, and urban service boundaries (USBs) (Pendall et al. 2002). Through zoning, land development permits and other land-use regulations and methods, UGBs demarcate urban and rural land uses and aim to contain urban development within defined boundaries (Pendall et al. 2002).

In China, concepts resembling UGBs have just started to develop. These ideas are a precondition for land-use planning. The "Urban Planning Compilation Guideline", issued by China's State Ministry of Construction on April 1, 2006, requires that city master plans propose "development exclusion areas", "development control areas", and "suitable development areas". Consequently, development exclusion areas and development control areas have become important references for identifying the boundaries of urban construction and have played a dominant role in controlling urban growth. In response to this guideline, Long et al. (2006) suggested a zoning method for the three types of areas, and established development exclusion and control areas in the Beijing Metropolitan Area (BMA). In addition, planned UGBs for urban growth management have became indispensable because the "People's Republic of China Town and Country Planning Act" introduced on January 1, 2008 gave planned UGBs the legal authority to curb urban growth. The urban planning administrative department issues building permits based on the UGBs. Developments within the planned UGBs are legal, whereas those outside are illegal Therefore, how planners determine the urban growth boundaries is important.

The establishment of UGBs involves comprehensive consideration of various factors related to urban spatial development. In Chinese cities, traditional methods of establishing UGBs, mostly based on planners' experience, often lack adequate scientific basis backed by quantitative analysis, and often fail to adequately control and regulate urban growth. Han et al. (2009) examined the effectiveness of the planned UGBs using multi-temporal remote sensing images and found that more new urban construction lay outside the UGBs than inside them in the case of the area within the sixth ring of Beijing from 1983 to 1993 and from 1993 to 2005. Moreover, Tian et al (2008) and Xu et al (2009) evaluated plan implementation in Guangzhou and Shanghai, respectively. Their results showed that a large number of urban developments were located outside UGBs specified in the urban master plan.

Cellular automata techniques can simulate UGBs and therefore have been widely used in modeling urban growth and as an analytical tool for complex spatio-temporal systems (Tobler 1970; White and Engelen 1993; Xie 1994; White and Engelen 1997; Clark and Gaydos 1998). Because of the complexity of urban growth, simulations of urban growth need to consider various factors that impact the urban growth process. Simple CA models only consider neighborhood effects, without taking factors the effects of planning policies into account. Therefore, many researchers have turned to constrained CA (CCA) models that allow the simulation to provide a more realistic model of urban growth (Engelen et al. 1997; Clark and Gaydos 1998; Wu 1998; Ward et al. 1999, 2000; Li and Yeh 2000; White et al. 2004; Guan et al. 2005). Institutional factors as well as development demands can be regarded as constraints in CCA simulations to model more reasonable future UGBs.

In this work, a constrained CA model incorporating macro socio-economic, locational, neighborhood, and institutional constraints is developed to simulate future urban growth as a basis for establishing UGBs. Because simulated UGBs involve various factors that impact urban growth process, stakeholders will wonder whether planned UGBs are reasonable or whether simulated UGBs are believable. For this, we suggest the use of FSA.

Form Scenario for Planned Urban Growth Boundaries

For finding spatial policy solutions for planning alternatives, particularly UGBs, two important steps are how to configure a scenario in an FSA and how to retrieve policy parameters using CCA. An FSA reflects a particular planning alternative, an urban master plan or land-use plan drawn by urban planners reflecting a planned UGB. To describe a planned UGB as a scenario, a dataset of spatial units and related attributes may be used to define an urban form as shown below:

$$Y = \Psi(U|y_{ij} \in Y, y_{ij} > p_{threshold}) \tag{1.1}$$

Y represents built-up and planned urban areas within the administrative district of city U, which is composed of developed spatial areal units y_{ij} that are assigned values over some constant, which represents a threshold of development potential in the CCA simulation. Ψ stands for the function used to determine places that will be developed based on development probability. y_{ij} stands for the land occupation status at space ij. $y_{ij}=1$ means that space ij is developed (built up), while $y_{ij}=0$ means undeveloped.

FSA, as a vision of future UGB, can be expressed as

$$\text{FSA} = f(X,A)\, x_{ij} \in X, a_{ij} \in A \tag{1.2}$$

In Equation (1.2), x and a are explicit spatial variables with ij subscripts representing spatial locations. A stands for the spatial constraints with respect to factors limiting urban sprawl, such as natural conservation areas, river buffer zones, steep slopes and others features. X, a temporal dynamic variable, is described as a "policy parameter", standing for the intensity of the impact that spatial constraints A exert when the corresponding policy is implemented.

We need affirm three basic premises to apply FSA to urban growth process, considering data availability and calculation time. First, the function f should be based on multi-criteria evaluation (MCE). Second, X and A should remain static from the initial step to the end of the simulation. Third, X is homogenous in the urban space.

In current research on urban growth scenario analysis, there are several competing models. For instance, Klosterman (1999) developed the planning support system "*What if?*" in which *A* represents spatial characteristics like soil conditions, flood control and transportation, and *X* stands for the force with which spatial policies related to *A* are implemented. Landis (1994, 1995) and Landis and Zhang (1998a, 1998b), respectively, developed the CUF (California Urban Future model) and CUF-2, in which *A* stands for spatial characteristics affected by governmental policies on location, environmental conditions, land-use control, zoning, existing development density, and accessibility to development land units (DLUs). *X* stands for the intensity of the effect of implementing related spatial policies. In our current way of thinking about urban growth scenarios, *X* and *A*, input together as one scenario, are used to estimate the corresponding UGB, *Y*.

Our FSA approach can be regarded as the inverse procedure of existing thinking about scenario analysis of urban growth. In FSA, the planned UGB is used as the base scenario. Urban planners are concerned with how to make planned urban forms become reality. From a planner's perspective, FSA is a method with the potential to identify policy solutions that can control urban growth boundaries to match reasonable planned alternatives. The impact of urban policies is their power to shape the future spatial features of an urban area. The FSA we propose can affirm what urban policies can make planning alternatives come to fruition. Before verifying a planned UGB based on FSA, we simulate the process of identifying UGBs in the Beijing Metropolitan Area using a conventional CCA process.

Simulated UGBs Using Conventional Constrained CA

Planners can gain insight on UGBs using the constrained CA model. First, spatial constraints considered by planners when establishing UGBs can be embedded as variables in the constrained CA. A CCA developed in that way will help planners to quantify the influence of these constraints on future urban forms. Second, planners also consider economic demands, which can be represented by land-use demand as a exogenous variable in the CCA simulation. Third, the constrained CA model can be applied for forecasting future urban growth based on historical trends, allowing planners to establish UGBs using the simulated results.

The basic concept of the CCA model as applied to practical urban development can be divided into two general steps. First, the government identifies total land-use demand and allocates that demand to an urban area. The constrained CA then simulates the probability of urban growth for all individual cells in different stages and identifies the spatial distribution of those cells that are predicted to become developed. Elements influencing the urban growth process in CCA models are selected to incorporate a hedonic model. They include the following:

1. Locational constraints: The attractiveness (potential force) of town centers at all levels the central city (*f_tam*), new city areas (*f_city*), towns (*f_town*), the attractiveness of rivers (*f_river*), and the attractiveness of roads (*f_road*). Locational constraints may vary depending on the study area and research purpose.

2. Neighborhood constraint: The development potential affected by the neighborhood (*neighbor*), identified as the ratio of developed cells within the neighborhood. The Moore neighborhood configuration is used in constrained CA, with each cell having nine cells in its neighborhood, including itself.
3. Institutional constraints: agricultural land use (*agri*), and development exclusion areas (*conf*). *agri* indicates suitability for cultivation and can be used to simulate the government's control on farmland to keep it from being encroached on by urban development. *conf* indicates spatial policy, where urban development is forbidden by eco-space protection or disaster prevention regulations.

Based on the above constraints, we use a multi-criteria evaluation (MCE) set of rules for status transition in the CCA as follows:

1. $LandDemand = \sum_t stepNum^t$

2. $s_{ij}^t = w_0$
$$+ w_1{}^*f\_tam_{ij} + w_2{}^*f\_city_{ij} + w_3{}^*f\_town_{ij}$$
$$+ w_4{}^*f\_river_{ij} + w_5{}^*f\_road_{ij}$$
$$+ w_6{}^*conf_{ij} + w_7{}^*agri_{ij}$$
$$+ wN^{**}neighbor_{ij}^t$$

3. $p_g{}^t = \dfrac{1}{1 + e^{-s_{ij}^t}}$

4. $p^t = \exp\left[\alpha\left(\dfrac{p_g{}^t}{p_g{}^t{}_{max}} - 1\right)\right]$

5. for $inStepID = 1$ to $stepNum$
$$if p_{ij}^t = p^t{}_{max} then V_{ij}^{t+1} = 1$$
$$p_{ij}^t = p_{ij}^t - p^t{}_{max}$$
$$p^t{}_{max}\ update$$
next $inStepID$

(1.3)

where *LandDemand* is the total area to be developed in the future; *stepNum* is the number of cells to be developed in each iteration; s_{ij}^t is suitability for development; w is the weighting coefficient for each constraint; $p_g{}^t$ is the initial transition probability; $p_g{}^t{}_{max}$ is the maximum of $p_g{}^t$ in iteration t; α is the dispersion parameter, which ranges from 1 to 10, p^t is the final probability; *inStepID* is the ID of the sub-iteration; V_{ij} is the status of the cell; and $p^t{}_{max}$ is the maximum p^t in iteration t. The cells to be developed in each iteration are those having the greatest development probability based on development suitability, which is determined by constraints and the weights they are assigned.

These status transition rules differ from those used by Wu (2002), which can be described by the equation $P_c{}^t = P_g{}^*con(s_{ij}^t = suitable)^*\Omega_{i,j}{}^t$. In Wu (2002), the user sets the neighborhood parameter Ω by the author's experience, rather than calibrated using historical observed forms. In this chapter, the role of the neighborhood constraint is included in the calculation of the land-use suitability, weighted together with other spatial constraints as in (1.3). This enables the comparison of the role of the neighborhood with other spatial constraints. The simulated urban form will be compared point-by-point with the observed urban form to obtain a *Kappa* index, which serves as a goodness-of-fit indicator. *Kappa* was initially introduced by Cohen (1960) and adapted for accuracy assessment in remote sensing applications by Congalton and Mead (1983). In this chapter, the *Kappa* index is calculated to analyze the similarity between observed and simulated urban forms.

The CCA simulation process shown in Fig. 1.1 is based on established status transition rules. First, the model sets the input variables and calculates the *stepNum* parameter based on the total land-use demand. Then land-use suitability, the initial transition probability, and the final transition probability are calculated. The cell with the maximum final transition probability is identified in space in the allocation process (Fig. 1.1, in grey). The model iterates until the entire simulation process is finally completed.

Constrained CA is used to estimate different future UGBs based on different scenarios devised by planners. The planners' perspectives and preferences can be embedded as constraints in the CCA model to establish UGBs. For instance, expected future urban growth rate and residential standard per capita can be represented as land-use demand in the socio-economic constraints in Fig. 1.1.

The results of the CCA simulation at some time (iteration or step), which are an estimate of that stage of urban growth, can be regarded as UGBs in the corresponding development period. The simulated UGBs, in raster format, can be converted into vector format consisting of many polygons. Small, non-compact polygons (<1 ha) should be first eliminated since they are not feasible for urban development. The remaining polygons can then be regarded as UGBs for different urban hierarchies (e.g., the central city, new cities, or small towns). The areas to be developed can be calculated by subtracting the existing urban construction areas from the UGBs thus established.

The FSA Approach for Retrieving Policy Parameters that Fit Planned UGBs

Policy Parameters and Constraints in Constrained CA

Whether the results of an FSA (a planned UGB) can be regarded as a scenario condition remains unexplored in recent literature on constrained CA. We attempt to follow the transition rules of conventional constrained CA when simulating urban growth with FSA. As usual, historical urban forms (Y) and constrained spatial features (A) are required to identify parameters (X) for the constrained conditions (A). In the routine model calibration process, historical urban forms are employed to

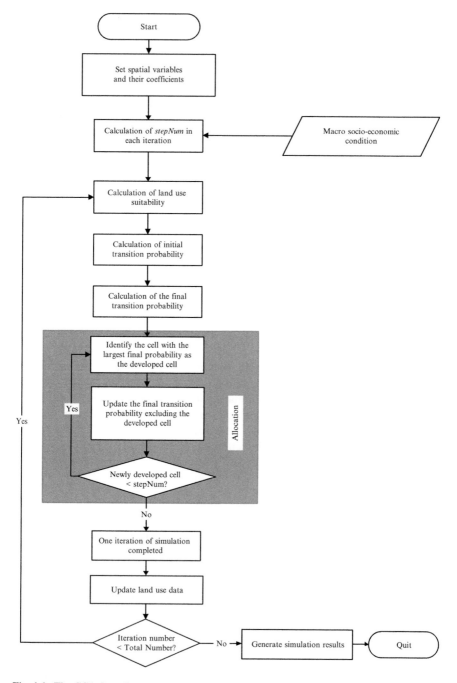

Fig. 1.1 The CCA flow diagram

retrieve X, while in FSA, retrieving X is done using the planned UGB. Therefore, identifying policy parameters (X) through FSA using constrained CA is possible.

Estimated parameters based on an historical urban form can be employed to conduct a simulation, and we can check whether the parameters can replicate the historical form. When the parameter X is positive, the corresponding spatial constrained form A should be encouraged, otherwise A should be rejected. Thus, policy implications can be drawn from the estimated parameters.

Policy Solutions for Planning Alternatives Using FSA

As mentioned above, identification of policy parameters X in an FSA can be used in model calibration for constrained CA simulations, where logistic regression, artificial neural networks (ANN), genetic algorithms (GA), and nested loops are widely employed. The evaluation indicator for the consistency of the form scenario (Y) and current spatial and institution constraints (A) should be established in the FSA. As input to the constrained CA, X is used to get the simulated UGB (Y'). The *Kappa* index, as an evaluation indicator, compares the simulated UGB Y' with the form scenario Y cell by cell and evaluates the goodness-of-fit. A *Kappa* of less than 80% stands for a no-solution condition, meaning that there are no policy parameters that can be used to realize the predefined form. Generally speaking, FSA solutions can be expressed as $\{X|Y' = f(X,A), Kappa(Y,Y') \geq 80\%\}$.

As general constrained CA simulation model, the conceptual constrained CA model for FSA can be shown as

$$y_{ij}^{t+1} = f(y_{ij}^t, G_{mac}, A, X) = f(y_{ij}^t, G_{mac}, A_{loc}, A_{ins}, A_n^t, X) \tag{1.4}$$

in which y_{ij}^{t+1} and y_{ij}^t, respectively, are the cell status at ij at times $t + 1$ and t, and f is the transition rule of constrained CA. There are four types of constraining conditions in the urban growth process. They are total land-use demand G_{mac} (which is a globally constraint, explicit non-spatial variable, with no corresponding policy parameter X), locational constraints A_{loc}, institutional constraint A_{ins}, and neighbor hood constraint A_n^t. Locational and institutional constraints are assumed to remain static during the urban growth process, and the total land-use demand reflects the total number of built-up cells to be developed. The neighbor hood constraint A_n^t, however, changes over model iterations during the constrained CA simulation.

For identifying policy parameters that fit a particular planning alternative, it is possible to retrieve the parameter set from a planned UGB y_e using MCE, just as is done when estimating parameters using historical urban forms in routine constrained CA. FSA, when used as a planned UGB, is a form predicted to develop from the current urban form. The total number of cells to be developed, compared to the current urban form, is defined as *LandDemand*, and *stepNum^t* is the number of cells developed in iteration t reflecting the land-use demand. We assume that *stepNum^t* is kept constant through the entire simulation period as follows.

$$\text{stepNu}m^t = \frac{Y_e - Y_s}{t} \qquad (1.5)$$

In which Y_e and Y_s are, respectively, a scenario form and a current form, and t is the simulated time span. In our constrained CA, we integrate the methods of Wu (2002) and Clark and Gaydos (1998), combining their strengths, to identify the status transition rules. To reduce model estimation computation time, X is divided into two types of parameters, the temporal-dynamic neighbor hood parameter x_n and other temporal-static parameters x_k that represent spatial constraints. We can estimate x_k via logistic regression. Then, x_n can be estimated by finding the form scenario Y_e with the largest *Kappa*, compared with the simulated form Y' using the *MonoLoop* approach (details in Long et al. 2008).

In the following sections, the constrained CA will be applied empirically to the Beijing Master Plan of 2020 to examine the consistency of four planning scenarios and existing spatial policies as well as to identify required policy parameters.

Experiments in the Beijing Metropolitan Area

Study Area

Constrained CA has proven to be suitable to simulate the urban growth of the Pearl River Delta, where growth is quite rapid (Li and Yeh 2000, 2002). The Beijing Metropolitan Area (BMA) has experienced rapid urbanization since the 1978 Reform and Opening Policy was adopted by the central government regulating both GDP and population. The GDP in 2006 was 787 billion CNY, 83.7 times more than in 1976, which was 9.4 billion CNY. The population in 2006 was 15.81 million, 1.9 times the 1976 figure of 8.29 million (Beijing Statistical Yearbook 1987, 2007). Interpreted thematic mapper (TM) images show that the urban built-up area in 2006 was 1,324 km^2 (Fig. 1.2), nearly three times of that in 1976, which was 495 km^2. Moreover, urban growth is predicted to keep increasing for two decades according to the official report of the BMA government. Therefore, stakeholders and decision makers are concerned with how to make the planned UGB become a reality.

It has become necessary to alleviate urban sprawl by establishing UGBs for different city levels. A CCA with cell size of 100 m (a total of 1,640,496 cells in the whole BMA) is used to model the BMA. The model is developed using Python based on the ESRI Geoprocessing module and has a discrete temporal iteration corresponding to 1 month in the real world. The CCA used in the BMA can simulate urban growth based on various urban development spatial policies. The results of

1 A Challenge to Configure Form Scenarios for Urban Growth Simulations Reflecting   13

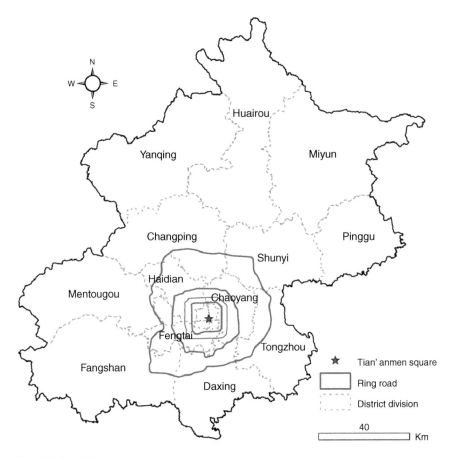

Fig. 1.2 The CCA study area

simulated BMA development can be used as a basis establishing UGBs. In the Beijing City Master Plan (2004–2020) (BCMP), the UGBs were determined by urban planners to meet the land-use demand for development, such as industrial development projects, population growth, and eco-space protection.

Data required for the CCA are shown in Table 1.1 and Fig. 1.3. In the table, the datasets *f_road* and *form* are interpreted from TM images. The locational constraints have been standardized to range from 0 to 1, with higher values representing greater development potential. The locational constraint a_k, which uses the corresponding spatial feature class as a data source, is processed using the "Distance/Straight Line" toolbox of the Spatial Analyst extension in the ESRI ArcGIS package to acquire $dist_k$, which is in turn $dist_k$ used to calculate the corresponding potential attractiveness. For institutional constraints, there are development exclusive policies a_6 and a_7, which describes how suitable an area is for the application of agricultural development policies. For institutional constraints, a value of zero for a development exclusion area means that all urban development is prohibited, and the land

Table 1.1 Statistical descriptions of CCA variables

Type	Variable	Name	Min	Max	Ave.	Std. D	Data source
Locational constraints	a_1	f_tam	0	1	0.037	0.091	Basic data spatial analysis using GIS
	a_2	f_city	0	1	0.214	0.196	
	a_3	f_town	0	1	0.531	0.198	
	a_4	f_river	0	1	0.789	0.162	
	a_5	f_road (1991)	0	1	0.818	0.187	1991-5-16 TM image interpretation
		f_road (2004)	0	1	0.827	0.190	2004-5-29 TM image interpretation
Institutional constraints	a_6	$conf$	0	1	0.570	0.495	Beijing Municipal Planning Commission (2007) (for details see Long et al. 2006)
	a_7	$agri$	0	1	0.418	0.237	Beijing Planning Commission (1988)
		$form$ (1991)	0	1	0.050	0.216	1991-5-16 TM image interpretation
		$form$ (2004)	0	1	0.080	0.266	2004-5-29 TM image interpretation
		$form$ (planned)	0	1	0.150	0.353	Beijing Municipal Planning Commission et al. (2006)
Neighbor hood constraint	a_n^t	$Neighbor^t_{ij}$	0	1			Development intensity in neighborhood; calculated by CA
Macro constraint	$stepNum$	$LandDemand$					Socio-economic development plan

For details of the calculation approach, see Long et al. (2006)

Fig. 1.3 Maps of spatial constraints and the 2006 urban form of the BMA

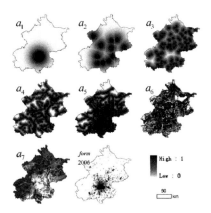

grade identifies how suitable areas are for farming, with larger values indicating lower cultivability. The institutional constraints are extracted from the planning drawings noted in Table 1.1.

Here, we first estimate the UGBs using constrained CA and compare it with the current planned and implemented UGBs in the BCMP. Second, we test the form scenario analysis approach on the BMA to identify policies will affect planning alternatives, in particular the current planned UGBs in the BCMP.

Simulated and Planned UGBs

Parameter Calibration

Historical data is used to retrieve information on urban transition in order to determine the parameter values to use for status transition. We use data from 1991 to 2004 as historical background information for parameter calibration. This requires three steps:

1. For 1991 and 2004 the numbers of developed cells are 80,343 and 125,928 respectively. There are 156 months in this period, and so developed cells is about 292 for each iteration (*stepNum*).
2. The weights of seven spatial variables x_{0-7}, excluding *neighbor* for 1991–2004 are calibrated using logistic regression. From the regression results (see Table 1.2), the significance levels for all variables are less than 0.001 the *Kappa* index is 43.427%.
3. The *MonoLoop* method is used to calibrate the weighting coefficient x_n for *neighbor* with the maximum *Kappa* index. Parameter x_n is set as 13 for the simulations. The *Kappa* index for this weight varies up to 81.419% due to the introduction of the neighborhood effect.

The calibration results are shown in Table 1.2. When the urban form in 1991 is used as the initial status of the model and the parameters in Table 1.2 are input to the

Table 1.2 Calibrated parameters for urban growth in 1991–2004

Name	Value
stepNum	292
x_0	−11.877
x_1	14.266
x_2	2.870
x_3	−2.595
x_4	1.149
x_5	6.743
x_6	1.084
x_7	−0.610
x_n	13.000

CCA, the resulting simulated urban form for 2004 is tested and found to be quite similar to the observed urban form in 2004. The results prove that the CCA is appropriate for simulating future urban growth in the BMA.

Urban Growth Simulation for 2020

The area of planned UGBs in the BCMP is 2388 km^2 expected to be developed till 2020. There are 192 months between 2004 and 2020, and so the *stepNum* value is 588. The baseline urban growth from 2004 to 2020 is continued development at the same rate as from 1991 to 2004. The weightings of all the spatial variables, including *neighbor*, are kept as the same as the calibrated historical values.

According to the simulated UGBs based on the assumption that urban growth will continue as it had up until 2004, UGBs for the central city and 11 new cities are shown in Fig. 1.5. The UGBs of the 142 small towns in the area are not listed here due to limited book space.

We compare the simulated and planned UGBs in the BCMP (see Table 1.3 and Fig. 1.4). The total areas of both versions for the central city and new cities are similar, and both are significantly larger than the 2004 area. If the historical development trend from 1991 to 2004 continues from 2004 to 2020 (the expected implementation period of the BCMP), urban land-use expansion in the southern parts of Beijing (such as Daxing and Yizhuang) will be much larger than planned in the BCMP. Meanwhile, expansion in the north (such as in Changping and Huairou) will be less than projected by the BCMP. The differences occur because the planned form reflects the Beijing government's policy trend toward development in the southern areas. Thus, it is important to take into account future urban development policies to predict the UGB.

The UGB simulated using constrained CA, with the urban form from 2004 as its initial condition and the calibration results from 1991 to 2004 as model inputs is illustrated in Fig. 1.4 (left). The *Kappa* index of the simulation results and planned UGBs is 68.309%, which means there is a relative high degree of difference between the simulation and the plan as can be seen in Fig. 1.4 (right). This also demonstrates that the future urban form and the planned UGBs will be somewhat

1 A Challenge to Configure Form Scenarios for Urban Growth Simulations Reflecting   17

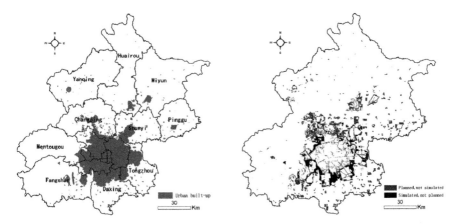

Fig. 1.4 Simulated UGB in 2020 (*left*), comparison with the planned UGB (*right*)

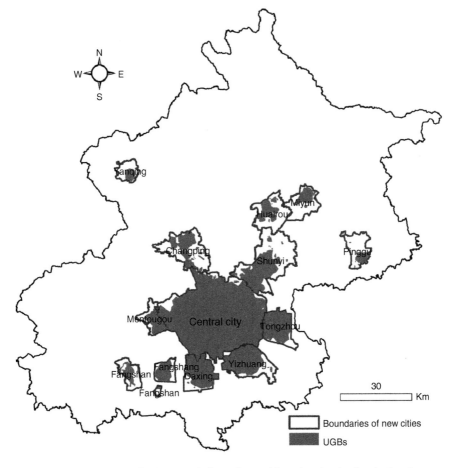

Fig. 1.5 Calculated UGBs for the central city and new cities using the simulated urban form

Table 1.3 UGB areas of the central city and new cities (area unit: km^2)

Name	Area (2004)	Simulated UGBs	Planne UGBs	Simulated-planned	$\frac{\text{Simulated-Planned}}{\text{Planned}}(\%)$
Central city	708.9	1033.0	984.2	48.8	5.0
1. Shunyi	50.2	157.2	180.2	−23	−12.8
2. Yizhuang	25.1	156.8	116.6	40.2	34.5
3. Tongzhou	42.7	133.6	109.5	24.1	22.0
4. Daxing	55.8	132.3	84.7	47.6	56.2
5. Fangshan	52.6	86.4	88.8	−2.4	−2.7
6. Chanping	35.7	82.0	109.8	−27.8	−25.3
7. Mentougou	21.2	52.9	35.5	17.4	49.0
8. Huairou	20.0	48.8	80.0	−31.2	−39.0
9. Miyun	18.9	34.8	46.9	−12.1	−25.8
10. Pinggu	13.1	19.8	31.0	−11.2	−36.1
11. Yanqing	7.3	17.3	20.2	−2.9	−14.4
Sum	1051.7	1954.7	1887.4	67.3	3.6

different, and reconfirms the current observation that partially planned UGBs tend to fail. Moreover, we assume that urban growth trends seen up to 2004 will continue. Thus, the simulated urban form will be different if growth differs during the simulated time period. From this viewpoint, how to set parameters based on future urban growth trends is a key point for identifying corresponding UGBs, for which we suggest employing FSA.

FSA Simulation and Policy Implications

How various constraints are distributed in the BMA influences the pattern of possible future urban growth (see Fig. 1.2). We can easily and directly recognize urban forms corresponding to every single constraint. Considering the combination of constraints and the currently existing urban form, which features central city sprawl, the BMA's future urban form tends to sprawl further around the central city in the plain area, with scattered developed patches in the mountain area. To obtain insight into the BMA's future urban growth pattern, we will set form scenarios (planning alternatives) and identify policy parameters required to produce those forms using constrained CA.

Planning Alternatives – Planned UGB as FSA

There have been five versions of urban master plans for the BMA since 1958. These were issued in 1958, 1973, 1982, 1992, and 2004 (Beijing Municipal Planning Commission et al. 1949). With respect to the 1992 version plan for 1991–2010, Han et al. (2009) argued that up to 51.8% of the urban development from 1991 to 2005 within the sixth ring road area was beyond the planned UGB. The actual urban developments were significantly inconsistent with the planned form in the BMA because urban development policies have been changing and land-use potential has not been taken into account.

The goal of the 2004 version for the BMA is to grow to a metropolis with a population of 18 million, and a developed land area of 2,300 km^2, with an urban spatial structure of "Two axes, two belts, and multi-sub-centers". This plan has four planning alternatives, as shown in Fig. 1.6. The alternatives have almost equal total demand for land-use development but different spatial layouts. During the constitution and approval of the plan, there was no means to efficiently determine the probability that each alternative would prove to be the eventual urban form.

The four planning alternatives (planned UGBs) in the BMA are as follows:

- Alternative A (Y_A), the alternative approved by the State Council of the P.R. of China (see Beijing Municipal Planning Commission et al. 2006), is

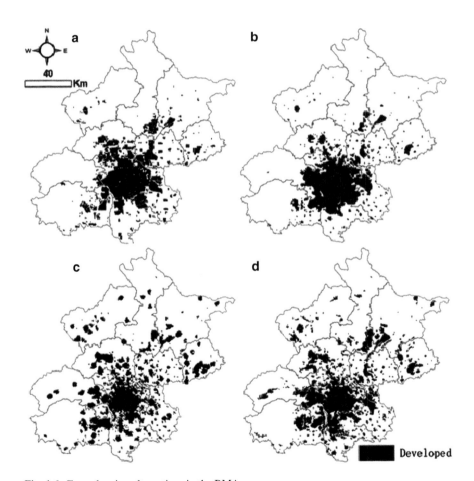

Fig. 1.6 Four planning alternatives in the BMA

characterized by a focus on preventing the central city from over-sprawling and on promoting the development of new cities.

- Alternative B (Y_B), the "sprawl" alternative, termed Tandabing (circle-spread) in Chinese, is characterized by promoting development in the central city and controlling development in new towns.
- Alternative C (Y_C), the "grape-cluster" alternative is characterized by promoting development both along transport corridors and around small towns.
- Alternative D (Y_D), the "sustainable" alternative, is characterized by preventing construction in forbidden areas and protecting high-quality farmland from been encroached upon, resulting in a more dispersed form.

We applied the FSA approach using a constrained CA model to find concealed trends in spatial policies necessary for the planned alternatives. The total number of iterations (*stepNum*) is 168 ([2,020-2,006]*12), with each iteration modeling 1 month. Y_s is equal to 5,011 if the urban form of 2006 (Y_{2006}) is used as the baseline year. In Alternative A, $Y_e = 9,254$ stands for the total number of developed cells in the year 2020, and *stepNum* is equal to 25 ([9,254 − 5,011]/168). The values of *stepNum* identified for the other planning alternatives are listed in Table 1.2. Parameter X is identified by applying an approach that integrates logistic regression and *MonoLoop*. For estimating x_i (x_0 to x_7) through logistic regression, we used the independent variables a_1 to a_7 in Table 1.1, excluding $a_n{}'$. Via logistic regression x_i (x_0 to x_7), the regression coefficients are then identified. Using *MonoLoop*, x_n, the neighborhood effect parameter can be identified, and we can compare the planned UGBs A, B, and C with the simulated UGBs Y_A, Y_B, and Y_C using the *Kappa* index. The simulated urban form Y' can be generated using the constrained CA with the input parameters listed in Table 1.4, including *stepNum*, x_{0-7}, and x_n. When *Kappa* is greater than or equal to 80%, the planning alternative is deemed valid and the parameters identified can be used to realize the planning alternative. When *Kappa* is less than 80%, the planning alternative is deemed not to be validated.

Related Policy Implications

The results of the policy parameter identification, together with the *Kappa* index, are listed in Table 1.4. All the independent variables are significant at the 0.001 level. The *Kappa* indices for Alternatives B, C, and D are all greater than 80%. The cell-by-cell comparison of Y' and Y (the planning alternative) is shown in Fig. 1.7. The confusion matrix is shown in Table 1.5. As a result, the *Kappa* index for Alternative A is as low as 67.5%, indicating that alternative A cannot be achieved using constrained CA in the current policy context because of the effect of existing policies on spatial constraints. The reason alternative A cannot be realized using constrained CA may lie in the spatial distribution of the constraints imposed on the urban space.

Alternatives B, C, and D match relatively better with the simulated UGBs in the cell-by-cell *Kappa* validation. Policy implications for the validated planning

1 A Challenge to Configure Form Scenarios for Urban Growth Simulations Reflecting

Table 1.4 Policy parameters identification results for four planning alternatives in the BMA

Variable	Alternative A	Alternative B	Alternative C	Alternative D	Historical form
Developed cells	9254	9270	9895	10,679	5011
stepNum	25	25	29	34	9
x_0	−8.700	−30.696	−63.599	−55.624	−12.263
x_1	15.268	54.558	15.106	20.849	11.782
x_2	3.575	10.294	10.046	9.701	2.490
x_3	−0.717	5.272	31.639	7.807	−1.872
x_4	4.105	8.765	24.348	11.622	7.574
x_5	1.368	6.027	7.627	8.113	0.917
x_6	1.193	3.672	4.078	23.000	1.535
x_7	−2.396	5.066	6.094	12.003	−1.179
x_n	15	17	9	7	20
Kappa (%)	69.4	91.8	85.0	85.8	67.5
Valid	False	True	True	True	False

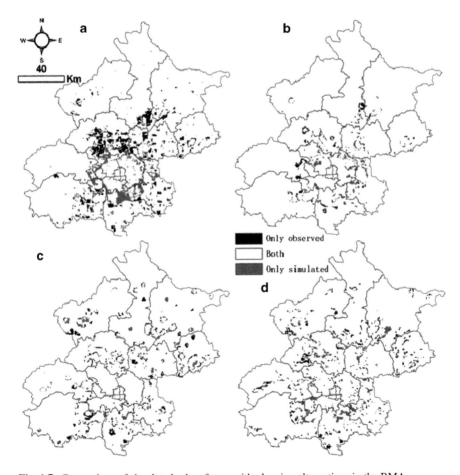

Fig. 1.7 Comparison of simulated urban forms with planning alternatives in the BMA

Table 1.5 Confusion matrix of planning alternatives and simulated urban forms

Alternative	Observed Planned	0	1	SUM	User accuracy (%)
A	0	53,919	2428	56,347	95.7
	1	2440	6826	9266	73.7
	SUM	56,359	9254	65,613	
	Producer accuracy (%)	95.7	73.8	Overall accuracy is 92.6%	
B	0	55,482	620	56,102	98.9
	1	793	8650	9443	91.6
	SUM	56,275	9270	65,545	
	Producer accuracy (%)	98.6	93.31	Overall accuracy is 97.8%	
C	0	54,391	1256	55,647	97.7
	1	1259	8639	9898	87.3
	SUM	55,650	9895	65,545	
	Producer accuracy (%)	97.7	87.3	Overall accuracy is 96.2%	
D	0	53,590	1273	54,863	97.7
	1	1276	9406	10,682	88.1
	SUM	54,866	10,679	65,545	
	Producer accuracy (%)	97.7	88.1	Overall accuracy is 96.1%	

alternatives can be presented based on the results of policy parameter identification. Three aspects of identifying policy implications are as follows:

1. Parallel comparison of parameters within a planning alternative: For Alternative B, built-up land is developed at a rate of $25 \times 12/4 = 75$ km^2/year from 2006 to 2020 according to *stepNum*. Therefore, assuming 100 m^2 of built-up land per capita, the population increase would be 750,000 people per year. Other parameters can also be compared in parallel, and larger parameters will require the implementation of more intense policies to realize the alternative. For instance, Alternative B suggests promoting the development of the central city, new cities, and riverside areas over other spatial policies.
2. Parallel comparison of parameters across planning alternatives: By comparing the same parameter in various planning alternatives, differences in necessary policy implementation of can be conveniently identified. For instance, x_1 of Alternative B is greater than that of Alternative C, meaning that to realize Alternative B, policies regarding the development of the central city will need to be implemented much more strenuously than to realize Alternative C. In the same way, much more intense protection against construction in prohibited areas will be needed to realize Alternative D than Alternative C since parameter x_6 in Alternative D is significantly greater than in Alternative C. The expansion of neighboring development should also be controlled to realize planning alternatives, especially Alternatives C and D.

The results of calibrating Alternative A, however, demonstrate that no policy parameter can be used to make this planned UGB become a reality. For this condition, either Y_A or A should be adjusted to align development policies with the predefined plan. By adjusting the spatial distribution of constraints affected by development policies such as the transportation network, or eco-space distribution, planners can experiment with the constrained CA model until the planned UGBs can be achieved.

This case study shows that a CCA simulation can be employed with FSA to determine required spatial policies for planned urban forms. In urban planning practices, UGBs tend to reflect the planners' preferences, knowledge, and possible urban development policy changes that may occur in the future, but lacking of consideration of land-use potential that geospatial policy analysis. Inconsistencies between planned urban growth patterns and practical development often imply that it is impossible to achieve the planned UGBs. FSA enables us to detect precautions that should be taken in planning practice.

Conclusions and Next Steps

In this chapter, we attempted to show how planning alternatives could be used as scenario conditions based on which discussions of whether UGBs developed by planners are feasible within the current framework of development policies regarding land-use, geographical conditions, and institutional controls. First, we proposed a method to establish UGBs using constrained CA. After calibrating the model using historical data, we simulated future urban forms based on current urban growth trends, and recommended new UGBs based on the simulations. The spatial distribution and UGBs estimated in the CCA simulation differ significantly from the UGBs specified in the Beijing City Master Plan (2004–2020). Our results reveal shortcomings in the CCA process of simulating urban form if it does not consider possible policy changes and planners' thinking about future urban development policies. Second, we incorporated constrained CA into the form scenario analysis approach, and successfully used four planning alternatives from the Beijing Metropolitan Area master plan as a case study to demonstrate our approach.

Combining FSA with constrained CA is a breakthrough in the application of CA. It can be employed to solve problems faced by planning departments and planners in China. Using FSA, planners can also evaluate how well planned UGBs and development policies fit together. Nowadays in China, planning implementation evaluation (PIE) is required in urban planning practice to examine how well real urban spatial developments agree with planned forms after several years of plan implementation. Existing PIE reports are not optimistic because development has sometimes broken out of the planned forms. Rather than PIE, FSA can be conducted at the very beginning of the planning process to examine the potential success of plan implementation under existing urban development policies and thus increase the likelihood that the planned form will become a reality.

We bring up three ways to simplify the FSA process in this chapter, including not taking into account spatial-temporal policy dynamics in the model parameters, as well as not separating development into various urban land uses, such as residential, commercial, and industrial types. To simulate urban growth much more factually, we suggest that further research focus on the current simplifications. The spatial and institutional constraints A, as well as the policy parameter X, can be included in FSA to identify not only how intense policy implementation must be, but how policies may need to vary in different areas. On the other hand, spatial heterogeneity of the policy parameter needs to be investigated since various studies show that the forces driving urban growth in China vary by sub-region in the metropolitan area (Liu et al. 2005; Li et al. 2008). Furthermore, we suggest that agent based modeling (ABM) be applied in FSA to represent planners' and other decision makers' preferences in another FSA research stream (Ligtenberg et al. 2001; Saarloos et al. 2005).

We did not investigate the sensitivity of policy parameters in FSA of this chapter. As our further research, we will undertake a global sensitivity analysis for FSA incorporating a regionalized sensitivity analysis approach called the HSY algorithm (Spear et al. 1994) that is more able to detect the parameters' global dynamics than local one-at-a-time (OAT) sensitivity analysis. Policy implications can meanwhile be identified in the global sensitivity analysis. Furthermore, with global sensitivity analysis it is possible to identify all the policy parameters that meet *Kappa* values over 80% instead of just finding the unique optimal solution described here.

Acknowledgments We would like to thank the National Natural Science Foundation of China (No. 51078213), the Grants-in-Aid for Scientific Research (No. 23404022B), Japan Society of the Promotion of Science for the financial support.

References

Beijing Municipal Planning Commission, Beijing Institute of City Planning, and Beijing Academy of Urban Planning. Beijing City Planning Atlas (1949–2005), 2006. [in Chinese]

Beijing Municipal Planning Commission. Ecologically Limited Land-use Planning in Beijing (2006–2020), Unpublished report, 2007. [in Chinese]

Beijing Planning Commission, (1988) Beijing Land Resource. Beijing: Beijing SciTech Press, [in Chinese]

Beijing Statistical Yearbook. (1987) Beijing Statistics Press, Beijing.

Beijing Statistical Yearbook. (2007) Beijing Statistics Press, Beijing.

Clark KC, and Gaydos L.J (1998), Loose-coupling a cellular automation model and GIS: Long-term urban growth prediction For San Francisco and Washington/Baltimore DOI:dx.doi.org. International Journal of Geographical Information Science, 12(7): 699–714.

Cohen, J. (1960) A coefficient of agreement for nominal scales. Educational and Psychological Measurement, 20: 37–46.

Congalton RG, and Mead R.A., (1983) A quantitative method to test for consistency and correctness in photo interpretation, Photogrammetric Engineering and Remote Sensing, 49(1): 69–74.

Couclelis, H. (2005) "Where has the future gone?" Rethinking the role of integrated land-use models in spatial planning, Environment and Planning A, Vol. 37, No. 8, 1353–1371.

Engelen, G., White, R., and Uljee, I., (1997) Integrating constrained cellular automata models, GIS and decision support tools for urban and regional planning and policy making.

Timmermans, H. (ed): Decision Support Systems in Urban Planning, London: E&FN Spon. pp. 125–155.

Guan G, Wang, L., and Clark, K.C. (2005), An artificial-neural-network-based, constrained CA model for simulating urban growth. Cartography and Geographic Information Science, 32(4): 369–380.

Han, H., Lai, S., Dang, A., Tan, Z., and Wu, C., (2009) Effectiveness of urban construction boundaries in Beijing: An assessment DOI:dx.doi.org. Journal of Zhejiang University SCIENCE A, 10(9): 1285–1295.

Klosterman, R.E. (1999) The What if? Collaborative planning support system, Environment and Planning B: Planning and Design, Vol. 26, No. 3, 393–408.

Landis, J.D. (1994) The California Urban Future model: A new generation of metropolitan simulation models, Environment and Planning B: Planning and Design, Vol. 21, No. 4, 399–420.

Landis, J.D. (1995) Imaging land use futures: Applying the California Urban Future model, Journal of American Planning Association, Vol. 61, No. 4, 438–457.

Landis, J.D., and Zhang, M. (1998a) The second generation of the California Urban Future model, Part1: Model logic and theory, Environment and Planning B: Planning and Design, Vol. 25, No. 5, 657–666.

Landis, J.D., and Zhang, M. (1998b) The second generation of the California Urban Future model, Part2: Specification and calibration results of the land-use change submodel, Environment and Planning B: Planning and Design, Vol. 25, No. 6, 795–824.

Li, X., Yang, Q., and Liu, X. (2008) Discovering and evaluating urban signatures for simulating compact development using cellular automata, Landscape and Urban Planning Vol. 86, 177–186.

Li, X., and Yeh, A. G. O. (2000) Modeling sustainable urban development by the integration of constrained cellular automata and GIS DOI:dx.doi.org. International Journal @@of Geographical Information Science, 14(2): 131–152.

Li, X., and Yeh, A.G.O. (2002) Neural-network-based cellular automata for simulating multiple land use changes using GIS, International Journal of Geographical Information Science, Vol. 16, No. 4, 323–343.

Ligtenberg, A., Bregt, A.K. (2001) Multi-actor-based land use modelling: Spatial planning using agents, Landscape and Urban Planning, Vol. 56, No. 1–2, 21–33.

Liu, X., Wang, J., Liu, M., and Meng, B. (2005) Spatial heterogeneity of the driving forces of cropland change in China, Science in China Series D-Earth Sciences, Vol. 48, 2231–2240.

Long, Y., He, Y., Liu, X., and Du, L. (2006) Planning of the controlled-construction area in Beijing: Establishing urban expansion boundary (in Chinese), City Planning Review, Vol. 30, No.12, 20–26.

Long, Y., Shen, Z., Du, L., Mao, Q., and Gao, Z., (2008) BUDEM: An urban growth simulation model using CA for Beijing Metropolitan Area DOI:dx.doi.org. Proc. SPIE, Vol. 7143, 71431D.

Nelson, A. C., and Duncan, J. B., (1995) Growth Management Principles and Practices. Chicago, IL; Washington D.C.: Planners Press; American Planning Association.

Pendall, R., Martin, J., and Fulton, W., (2002), Holding the Line: Urban Containment in the United States. Washington, D C: The Brookings Institution Center on Urban and Metropolitan Policy.

Saarloos, D., Arentze, T., Borgers, A., and Timmermans, H. (2005) A multiagent model for alternative plan generation, Environment and Planning B: Planning and Design, Vol. 32, 505–522.

Spear, R.C., Grieb, T.M., and Shang, N. (1994) Parameter uncertainty and interaction in complex environment models, Water Resources Research, Vol. 30, No. 11, 3159–3169.

Tian, L., Lv, C., and Shen, T., (2008) Theoretical and empirical research on implementation evaluation of city master plan: A case of Guangzhou City Master Plan (2001–2010), Urban Planning Forum, (5): 90–96. [in Chinese]

Tobler, W. R. (1970) A computer movie simulating population growth in the Detroit region. Economic Geography, 42: 234–240.

Ward, D. P., Murray, A. T., and Phinn, S. R. (1999) An optimized cellular automata approach for sustainable urban development in rapidly urbanizing regions. GeoComputation99, 4th International Conference on GeoComputation, Virgina, USA.

Ward, D. P., Murray, A. T., and Phinn, S. R., (2000) A stochastically constrained cellular model of urban growth. Computers, Environment and Urban Systems, 24(6): 539–558.

White, R. W., and Engelen, G. (1993) Cellular automata and fractal urban form: A cellular modeling approach to the evolution of urban land use patterns DOI:dx.doi.org. Environment and Planning A, 25(8), 1175–1193.

White, R. W., and Engelen, G. (1997) Cellular automaton as the basis of integrated dynamic regional modeling Environment and Planning B: Planning and Design, 24(2), 235–246.

White, R., Straatman B., and Engelen, G. (2004), Planning scenario visualization and assessment - A cellular automata based integrated spatial decision support system. In: Goodchild, M.F., Janelle, D.G., Shrore, Z.G. (Eds.), Spatially Integrated Social Science, Oxford University Press. 420–442.

Wu, F. (1998), Simland: A prototype to simulate land conversion through the integrated GIS and CA with AHP-derived transition rules. International Journal of Geographical Information Science, 12(1): 63–82.

Wu, F. (2002) Calibration of stochastic cellular automata: The application to rural-urban land conversions, International journal of Geographical Information Science, Vol. 16, No. 8, 795–818.

Xie, Y. (1994) Analytical Models and Algorithms for Cellular Urban Dynamics. Unpublished Ph. D. dissertation, State University of New York at Buffalo, Buffalo, N.Y.

Xu, Y., Shi, S., and Fan, Y., (1992) Methodology of Shanghai City Master Plan in new position, Urban Planning Forum, 2009, (2). [in Chinese]

Chapter 2
A Planning Tool for Simulating Urban Growth Process and Spatial Strategy of Urban Development in Chuandong, China

Yan Ma, Zhenjiang Shen, Dingyou Zhou, and Ke Wang

Introduction

This century is marked by rapid urbanization. Some statistics predict that the world's urban population will reach five billion by 2030 (United Nations 2007). This rapid urbanization not only has resulted in expanded land-use, but also has led to population increases in urban areas and development-related environmental impacts. Researchers, therefore, have begun to be concerned about the relationships between urbanization, population dynamics, and development-related environmental impacts (Lambin and Geist 2001; Fontaine and Rounsevell 2009). In this chapter, we introduce a simulation tool, spatial strategic plan support system (SSP-SS), to support local decision-making of urban development, which includes two separate parts for urban growth simulation and total amount of household energy consumption and waste discharge. To represent the process of urban growth, the constraint cellular automata (CA) approach is utilized, and an agent-based model is employed for calculating the total amount of household energy consumption and waste discharge.

Nowadays, urbanization is taking place at an unprecedented pace in China, and it will continue over the next decades. According to some statistics, the total migration into urban areas during 1995–2000 amounted to 128 million and the amount of

Y. Ma (✉) • Z. Shen
School of Environmental Design, Kanazawa University, Kanazawa, Japan
e-mail: shenzhe@t.kanazawa-u.ac.jp

D. Zhou
School of Architecture, Dongnan University, Nanjing, China

K. Wang
VPN, Beijing Office, Beijing, China

Z. Shen, *Geospatial Techniques in Urban Planning*, Advances in Geographic
Information Science, DOI 10.1007/978-3-642-13559-0_2,
© Springer-Verlag Berlin Heidelberg 2012

built-up urban area almost doubled from the 1990s to the 2000s (National Bureau of Statistics et al. 1996–2005). This rapid urbanization will lead to the problems of landscape deterioration and unexpected urban spatial change. Therefore, some prior researchers have represented the process of urban growth in the past in order to predict its possible spatial changes in the future (Catalan et al. 2008; Han et al. 2009; He et al. 2008). Meanwhile, some researchers began to focus on simulating the phenomenon of segregation and suburbanization in order to better analysis the spatial pattern of urban areas (Jayaprakash et al. 2009). Additionally, some other researchers concentrated on revealing the relationship between landscape change and household residential location (Fontaine and Rounsevell 2009; Brown et al. 2008). Meanwhile, it is a challenge to develop practical and effective planning tools to support local governments in making decisions about planning policies. In this *chapter*, we propose a planning support tool to investigate the impacts of strategic spatial plans on urban growth.

In China, local governments attempt to determine future urban characteristics, spatial development, and industry distribution by implementing strategic spatial plans aimed at achieving objectives such as sustainable economic growth, land reclamation, infrastructure construction, and urban development. Making a strategic spatial plan will involve at least three local government planning departments, namely the local economic and social development planning department, the land-use planning department and the urban planning department. The planning work they do is implemented independently by the economic and social development commission, the land-use planning and management bureau and the urban planning and management bureau. Local governments should have an integrated process for making strategic spatial plans that integrates all three planning functions. Additionally, when considering environmental issues arising from urban development, work done by the environmental protection department cannot be ignored. Thus, how to combine the work proposed by the four departments into a sensible overall plan is a key factor in supporting local government decision-making.

In the Chuandong area of China, the local strategic spatial plan includes development of a new city. The proposed new system, *SSP-SS*, is intended to assist local government decision makers by enabling them to produce a visible, pellucid model of the effects of local strategic spatial plans. As mentioned above, this system simulates urban growth and integrates the economic and social development plan with the land-use plan and urban plans while taking into consideration environmental issues, such as the total amounts of natural resources used and waste discharged. In this *chapter*, we use the cellular automata (CA) approach to represent the process of urban growth. Moreover, we employ an agent-based model to simulate population change, resource use, and waste discharge. Much research has demonstrated that the CA approach has advantages over other methods and can simulate urban growth in a very realistic manner (Batty et al. 1997, 1999; Clarke and Gaydos 1998; White and Engelen 1997, 2000; Wu and Webster 1998; Sui and Zeng 2001; Li and Yeh 2002b; Fang et al. 2005; Shen and Kawakami 2008). Urban growth, however, is a complex process influenced by many macro-scale driving forces. The pure CA model is inadequate to take these political, economic, and cultural forces into account (White and Engelen 1997, 2000). To improve the CA model so that it can represent the spatial complexity of the urban growth process, a constrained CA model has been developed (Li and Yeh 2000). It can simulate urban growth as influenced by local, regional, and global sustainability constraints. Despite its advantages in spatial representation, the CA approach also has its own drawbacks. CA models are arrangements of individual automata over tessellated space, where automata in neighboring cells influence each individual automaton. Neighbors transfer information by diffusion (Torrens and Benenson 2005). This results in one shortfall in the CA approach; cells cannot move and make their own decisions (Torrens 2007). Therefore, urban growth simulations tend to depend too much on how cells are assigned using predefined neighborhood rules. Although CA-based models are good for representing and simulating the spatial processes of urban growth that underlie decisions related to neighborhood interactions (Batty et al. 1997), agent-based models (ABMs) are much better at handling individual decisions.

For these reasons, we employ both constrained CA and ABM methods in the SSP-SS simulation of urban growth and in calculating the total amount of resources used and waste discharged in the Chuandong area. Moreover, work proposed by different planning departments can be integrated in the SSP-SS, aiding local governments in their strategic spatial planning and decision-making processes.

The whole *chapter* comprises five sections. In Sect. "A Systematic Approach for Supporting Local Government Decision Making", we introduce a systematic approach for supporting local governments' decision making. In Sect. "Description of SSP-SS", we give details about the models used to build the SSP-SS. In Sect. "SSP-SS Implementation", we describe SSP-SS implementation, and then an analysis of the experimental results and conclusions are presented in "Conclusions".

A Systematic Approach for Supporting Local Government Decision Making

As mentioned above, in China, a strategic spatial plan will involve the work of urban planning, land-use planning, and local economy and social development planning departments. Moreover, related environmental assessments conducted by the department of environmental protection should also be considered to take into account the environmental consequences of implementing the plan. In planning practice, however, these four departments are independent from each other, and their work is always carried out under different frameworks.

In China, local development and reform commissions authorize large-scale investment projects in urban areas in accord with local economic and social development plans. The department of land management devises land-use plans to be implemented in both urban and rural areas based on the land use demanded by these large-scale investment projects. Then, the department of urban planning takes its turn, planning and managing urban development. Obviously, this is a top-down relationship between the local economic and social development planners, land-use planners, and urban planners. Ideally, these three planning divisions should cooperate with each other instead of working in sequence. Moreover, to achieve local sustainable development targets, the total amount of resources used and waste discharged should not exceed the local resource supply and environmental carrying capacities determined by the environmental planning department. Thus, a practical strategic spatial plan should bring factors from across all four functions together. The current situation, in which the four departments operate independently, results in difficulty in coordinating planning work and analysis of the impacts the plans have on urban development. Hence, it is necessary to for local governments to develop an integrated planning system for making decisions about their strategic spatial plans.

To find a solution for the problems noted above, we have developed the SSPSS decision-making tool, which targets full integration of land-use planning, urban planning, and economic and social development planning on the same platform. To achieve this, we employed a constrained CA model to simulate the process of urban growth driven by land development potential in the Chuandong area of China. Economic investment is the direct driving force for urban growth, and we therefore chose the locations of investment projects as the seeds (initial areas) for our urban growth simulation. According to Chuandong's local economic and social development plan, there are two types of investment projects for new urban development in the area. One is for highway construction and the other is for new urban center construction. Based on this division, we generate two scenarios corresponding to the two projects in order to reveal the different urban patterns resulting from the types of investment projects recognized in the local economic and social development plan. Evaluation of whether developments are a suitable use of land is a precondition for local government's production of a land supply plan. In this *chapter*, we assume that local governments make land supply plans based on land-use suitability evaluation. The spatial process of urban growth, then, can be evaluated using the probability of land development using CA transition rules. Meanwhile, urban planning controls related to each cell are employed to regulate development projects in urban spaces. The factors of land-use types, population density, and landscape that are determined by urban planning are regarded as institutional constraints for our urban growth simulation. Hence, the three types of planning work conducted by the different departments are integrated into the SSPSS to allow visualization of the spatial process of urban growth based on the strategic plan.

Finally, numbers of households or business owner are assumed to be consistent with the population density as controlled by urban plans for the Chuandong area. Correspondingly, the total resources used and waste discharged can be simulated through an agent-based model. In this model, the interactions between local government agents and other agents are defined by rules that adjust individual resource use and waste discharge to satisfy the total amount of resources used and waste discharged. These total control demands are predefined as proposed by the department of environmental protection.

As mentioned above, the urban growth simulation can be used to predict resource use and waste discharge amounts that reflect the impact of implementing a spatial plan. It is possible for local governments to use urban growth simulation a tool to assist in making decisions about strategic spatial plans.

Description of SSP-SS

Simulation of Urban Growth Using Constrained CA

As mentioned above, we assume that areas that have high development potential, based on land-use suitability evaluations, urban plans, and economic and social development plans, will be developed. In order to visualize the possible spatial process of urban growth in the Chuandong area, reflecting a dynamic balance between land supply and land-use demand that results from implementing strategic spatial plan, we applied the constrained CA model built by Wu (1998) and Li and Yeh (2000). In this model, the constraints include spatial constraint A_s reflecting the land-use suitability evaluation, macro-social economic constraint A_{mac}, reflecting urban development projects outlined in the economic and social development plan, institutional constraint A_{ins}, and neighborhood constraint A_{nei} (Long et al. 2008, 2009). In this simulation, the institutional constraint can be understood to be urban planning. Investment projects for urban development determined by local economic and social development plan can be taken as the macro-social economic constraint. Additionally, the Moor Neighborhood scheme is utilized for this simulation and each cell is defined as having four neighbors. The four spatial constraints can be as expressed as

$$V_{i,j}{}^{t+1} = f\left(V_{i,j}{}^t, A_{mac}, A_{spa}, A_{ins}, A_{nei}{}^t\right)$$

$V_{i,j}{}^t$: cell status at time t and in location ij

$V_{i,j}{}^{t+1}$: cell status at time $t+1$ and in location ij

f : transition rule of cell status

$$(2.1)$$

Equation 2.2 details the rule governing transitions of cell status $V_{i,j}$ as follows:

$$s_{ij}{}^t = x_0 + \sum_{k=1}^{n-1} x_k \times a_k + x_n \times a_n{}^t = s_0 + x_n \times a_n{}^t$$

$$p_g{}^t = \frac{1}{1 + e^{-s_{ij}{}^t}}$$

$$p_{ij}{}^t = \exp\left[\alpha\left(\frac{p_g{}^t}{p_g{}^t{}_{max}} - 1\right)\right]$$

if $p_{ij}{}^t \geq p_{threshold}(p_{ij}{}^t, Development\text{stage}^{t+1})S^{t+i}(ij) = $ Developed

otherwise $S^{t+i}(ij) = $ Un Developed

$$(2.2)$$

Among the variables, $s_{ij}{}^t$ reflects the development potential of cell (i, j) at time t as influenced by the land-use suitability evaluation and the urban, economic, and social development plans. In addition, x_0 is a constant and s_0 is the static part of development potential during each simulation step. $S^{t+i}(ij)$ is the status of cell (i, j) at time t + 1. x_k is a parameter set reflecting institutional policies other than neighborhood rule, such as land-use suitability. x_n is a parameter reflecting neighborhood impacts, a_n represents the spatial features of the neighborhood, a_k reflects spatial features excluding the neighborhood, and p_g^t is the global probability for transition. $p_{g\max}^t$ is the maximum global probability during each step. α is a diffusion coefficient ranging from 1 to 10, $p_{ij}{}^t$ is the final probability, $p_{threshold}(p_{ij}{}^t, Development\mathrm{stage}^{t+1})$ is the neighborhood of the cell and controls development speed, which can change depending on $p_{ij}{}^t$ and $Dvelopment\mathrm{stage}^{t+1}$ in order to make the development speed in $Dvelopment\mathrm{stage}^{t+1}$ match the settings.

The basic principle of this transition rule is as follows. First, simulation initial areas are determined. Then, the final probability of one cell will be employed to decide if the cell can be developed or not by comparing with the threshold, which is expressed as $p_{threshold}(p_{ij}{}^t, Development\mathrm{stage}^{t+1})$.

As determined in the strategic spatial plan, the development periods in the Chuandong area are divided into initial, mid-development, and mature stages.

Resource Use and Waste Discharge Amount Calculations

Simulation of Agent Generation

Urban growth inevitably leads to population centralization in the urban area. Using the constrained CA approach, urban growth can be visualized to aid local governments' strategic decision-making. It is consider how the spatial changes will impact population growth in the Chuandong area. To answer this question, we assumed that the total population that will become concentrated in the new urban area will come from the surrounding countryside and that the urbanization objectives of the local socio-economic plan will be achieved. However, we do not simulate the process of moving the population from the surrounding countryside and only generate agent numbers based on the density proposed by urban plan. Therefore, if the simulation shows that the total population scale can meet government urbanization objectives while satisfying density requirements and the total inhabitable space in the Chuandong area, the local government can gain insight on how implementation of the strategic spatial plan can impact urban development. Simulating the movement of the population from the countryside to the city in the urbanization process remains as a further task that should be carefully investigated in the future.

Recently, the agent-based modeling approach has been applied extensively to mimic individual behaviors in virtual societies. We use our SSP-SS tool, which employs an agent-based approach, to generate agents and simulate resource use and waste discharge. Agents, including households and business owners, are assumed to migrate to each developed residential and commercial cell. Their numbers are controlled by the density proposed in the urban plan. According to the Chuandong area urban plan, there is only one small industrial area where a single industrial agent is located. Among the formulas in (2.3)–(2.5), N_r, N_c, N_i, and N_h, respectively, stand for the number of households in residential, commercial, and industrial areas and the number of households all developed areas. D_{rn} and D_{cn} are the population densities in residential and commercial areas, respectively. ε is the error term (a random value with a mean of 1 and a standard deviation of 0.5) used to generate each area's population density. S is the area of each different land type, and S_{cell} gives the real area of each cell. $\sum NumCell$ represents the total number of developed cells for each land-use type.

Based on the predefined population densities for the different types of land use zones, the number of households in residential areas is calculated using (2.3) as follows:

$$Nr = \sum_{n=1} (Drn^* \varepsilon rn) \times Srn$$

$$Srn = Scell \times \sum NumCellr$$

$$Nr = \sum_{n=1} (Drn^* \varepsilon rn) \times Scell \times \sum NumCellr \qquad (2.3)$$

The number of company owners in commercial areas is expressed as

$$Nc = \sum_{n=1}^{2} (Dcn^* \varepsilon cn) \times Sin$$

$$Scn = Scell \times \sum NumCellc$$

$$Nc = \sum_{n=1}^{2} (Dcn^* \varepsilon cn) \times Scell \times \sum NumCellc \qquad (2.4)$$

Consequently, the total number of agents in all developed areas is expressed as

$$Nh = \sum_{x=r,i,c} Nx \qquad (2.5)$$

Residential Resource Use and Waste Discharge

Local governments attach great importance to the principle of sustainable development and environmental protection. The implementation of a strategic spatial plan will lead not only to urban growth but also to population concentration in developed urban areas. Thus, the great amounts of resources used and waste discharged by agents should not be neglected when making strategic decisions. We utilize the SSP-SS tool to reveal agents' resource use and waste discharge while taking into account the interaction between local government agents and other agents under the assumption that all agents achieve a consensus and cooperate to comply with the planned total amounts of resources used and wastes discharged.

Some basic principles governing these calculations are: First, the initial values of agents' resource use and waste discharge are set for each agent based on the different land-use types proposed by the local environmental department. Second, limits on total resource use and waste produced proposed by the environmental department are used in the initial step. Accordingly, the total amount of agents' resource use and waste discharge will be calculated in each simulation step, and adjustments will be made by each agent when the total resource use and/or waste discharge amounts exceed control values proposed by the local government. A flow diagram for this calculation is shown in Fig. 2.1. The model for calculating resource

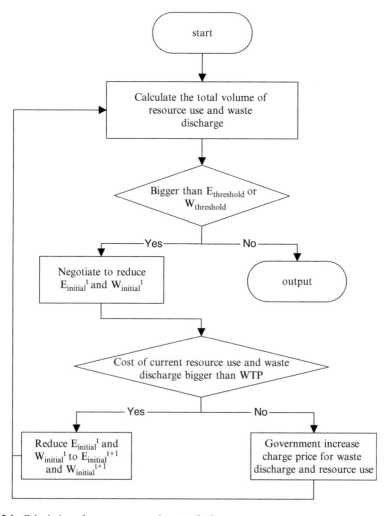

Fig. 2.1 Calculation of resource use and waste discharge

use is shown in (2.6). E_i^t stands for the resources consumed in industrial areas in step t, E_r^t stands for resources used by households in residential areas in step t and E_c^t stands for resources used in commercial areas in step t. E^t is the total amount of resources used in step t; $E_{initial}^t$ is the initial value of resources used by each agent in step t; $E_{initial}^{t+1}$ is the updated value for each agent's resource use, which has been adjusted by agents' interactions and governed by the control policy; and $E_{threshold}$ is the value to which the total amount is limited.

$$E^t = E^t i + E^t r + E^t c = N^t i \times E^t iinitial \times \varepsilon i$$
$$+ N^t r \times E^t initial \times \varepsilon r + N^t c \times E^t iinitial \times \varepsilon c$$
$$= E^t initial \times (N^t i^* \varepsilon i + N^t r^* \varepsilon r + N^t c^* \varepsilon c)$$
$$= E^t initial \times \sum^{x=i,r,c} N^t x^* \varepsilon x$$

$$\text{if } E^t \geq E_{threshold}$$

$$E^{t+1} = E^{t+1} initial \times \sum^{x=i,r,c} N^t x^* \varepsilon x (E^{t+1} initial - E^t initial < 0)$$

$$\text{otherwise } E^t = E^t initial \times \sum^{x=i,r,c} N^t x^* \varepsilon x \tag{2.6}$$

The model for calculating waste discharge is shown in (2.7). W_i^t stands for the waste discharged by industrial agents in step t, W_r^t is household waste discharged from residential areas in step t, and W_c^t stands for waste discharged from commercial areas in step t. W^t stands for the total amount of waste discharged in step t, $W_{initial}^t$ is the initial value for waste discharged by each agent in step t, and $W_{initial}^{t+1}$ is the updated value for each agent's waste discharge, which has been altered by agents' interaction based on agreement that the total waste to be discharged will be controlled. $W_{threshold}$ is the value to which the total amount is limited.

$$W^t = W^t i + W^t r + W^t c = N^t i \times W^t iinitial \times \varepsilon i$$
$$+ N^t r \times W^t initial \times \varepsilon r + N^t c \times W^t iinitial \times \varepsilon c$$
$$= W^t initial \times (N^t i^* \varepsilon i + N^t r^* \varepsilon r + N^t c^* \varepsilon c)$$
$$= W^t initial \times \sum^{x=i,r,c} N^t x^* \varepsilon x$$

$$\text{if } W^t \geq Wthreshold \tag{2.7}$$

$$W^{t+1} = W^{t+1} initial \times \sum^{x=i,r,c} N^t x^* \varepsilon x (W^{t+1} initial - W^t initial < 0)$$

$$\text{otherwise } W^t = W^t initial \times \sum^{x=i,r,c} N^t x^* \varepsilon x$$

SSP-SS Implementation

Implementation Area

The Chuandong area belongs to Tongren City, which is in northeast Guizhou Province, southwest China. The whole area is located in the middle of the Lingwu Mountains next to Hunan Province. According to Tongren City's latest local economic and social development plan, it is going to develop a new urban area in Chuandong. The new urban area will be located in the vicinity of 108° 23′ to 108° 26′ east and 30° 80′ to 30° 83′ north. The whole planning area covers about 18.7 km². It is just 5 km from the local airport, Daxing Airport, and 10 km away from the local railway station. The route for Provincial highway N201 crosses the planning area. The spatial pattern of this new urban area was designed by the local government's department of urban planning. Its configuration is as shown in Fig. 2.2 with the whole urban area to be divided into industrial, landscape, residential, and commercial zones. Meanwhile, the planned highway will play an important role in the new urban development. Traffic conditions provide Chuandong area with special development opportunities and challenges.

According to the local economic and social development plan, the total population is planned to be 800,000 in the Chuandong area, and development will cover about 20 km². There are two development proposals for this new urban area that have been permitted by the local government. One proposal is intended to accelerate new urban development through investment in specific projects in Chuandong's new business district, as well as construction of two university campuses in the urban center. As shown in Fig. 2.2a, the new urban area comprises the locations of educational organs (T4), residential areas (T5), and commercial areas (T6). The other development proposal is targeted to accelerate land-use development through construction of the provincial highway, which already under way. The site of the highway is shown in grey in Fig. 2.2. We will, therefore, introduce two urban growth scenarios for the Chuandong area corresponding to these development proposals. The first will take the urban centers in Fig. 2.2b as the initial area for

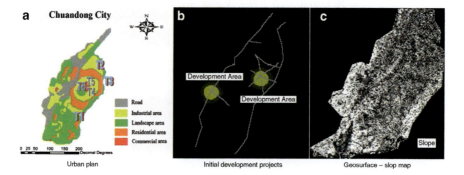

Fig. 2.2 Spatial pattern of land-use zoning in urban plan of Chuandong area

development, and the second will take the highway construction site as its initial land-use development area.

SSP-SS Framework

As shown in Fig. 2.3, SSP-SS fits into a three-part framework of data importing, data processing, and output.

In the first phase of the urban growth simulation, spatial patterns, geo-surface information, and initial development zones within the Chuandong area must be imported from GIS data. Among these data, the spatial patterns are plan drawings as shown Fig. 2.2a, which is a plan proposed by VBN design Inc. but not the final approved urban plan for the Chuandong area. Geo-surface information includes the

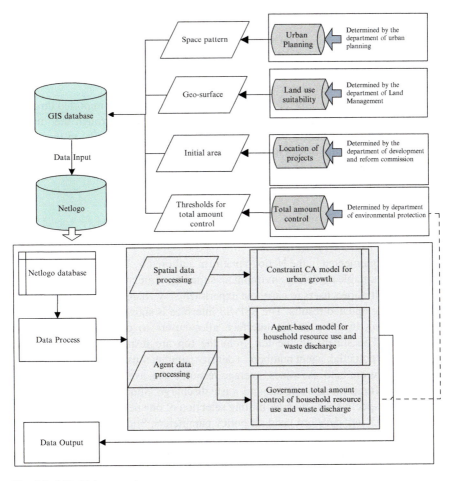

Fig. 2.3 SSP-SS framework

results of land-use suitability evaluations, which we generated using remote sensing data (Fig. 2.2c). The initial development areas (Fig. 2.2b) for our simulation are based on the locations of the investment projects discussed above, for which consensus was gained through deliberation among local officers, and abiding by the local economic and social development plan. We believe that all the data can be prepared by local urban planning, land-use management, and development departments. As a case study, we prepared these data ourselves in order to be consistent with planning practices in China for possible future application of this model by local governments.

There will be two choices of how to launch investment projects in the initial development phase in the Chuandong area, and therefore two urban growth scenarios can be simulated, based on the different locations of areas to be developed in that initial phase. One of the scenarios is that urban growth will be spurred by construction of the highway (the gray in Fig. 2.2a). In the other scenario, growth is promoted by developing the city centers, as shown in Fig. 2.2b.

As shown in Fig. 2.4, the constrained CA model can simulate Chuandong's urban growth process. Based on the simulated urban growth, resource use and waste discharge can be calculated. Moreover, threshold values for controlling resource use and waste discharge should be based on values determined by the local environmental protection department.

Figure 2.4 details the flow of the simulation. First, data can be imported through the "setup" procedure. Then, the simulation is implemented through the "urban growth" and "environmental assessment" processes. Next, the results of the simulation can be saved and output in the "display statistics" procedure. Users can then analyze the results displayed on the system interface.

SSP-SS Implementation and Results

SSP-SS is implemented in the Netlogo platform authored by Uri Wilensky in 1999. It is a programmable modeling environment suited for simulating natural and social phenomena. It allows users to carry out experiments by changing parameter values and observing updated results. The SSP-SS interface is shown in Fig. 2.5.

Sliders and switches in the interface allow users to change the values of parameters. The sliders on the left from the top are respectively threshold of development potential, agent numbers, a density parameter used to generate density values in different land-use zones, average resource use, average waste discharge, and the total limits on resource use and waste discharge. Sliders in the middle from the top are respectively a slider permitting selection of one new initial development area (*developX and developY*) and set policy parameters. Sliders from the top right are global demand (developed cell numbers), development years, and other policy parameter settings. There are some settings that allow the user to choose different results that they want to display. These settings, as shown in Fig. 2.5, include space potential, geosurface, and environment assessment.

2 A Planning Tool for Simulating Urban Growth Process and Spatial Strategy

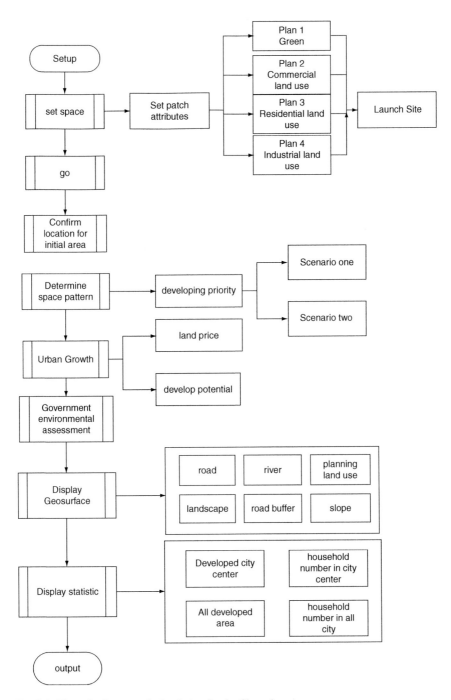

Fig. 2.4 Flow of urban growth simulation for the Chuandong area

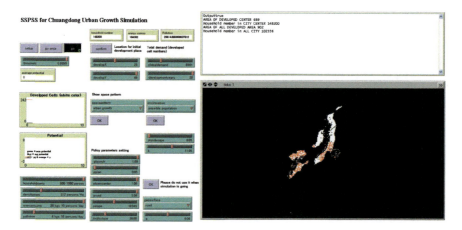

Fig. 2.5 SSP-SS interface in Netlogo

Results of First Scenario Simulation

Under the urban growth scenario characterized by urban center development, the region formed by T4, T5, and T6 land-use categories is determined to be the area where initial development will occur. According to the strategic spatial plan, investment in urban center development will be divided into three stages, and we therefore implemented the simulation in three stages as well: initial development, mid-stage development and the mature stage where development is complete.

According to local social-economic conditions given in the strategic spatial plan, there will be a population demand of 800,000 people in the Chuandong area. However, because the strategic plan does not explicitly describe economic demand with regard to industrial structure or commercial use at this stage, we took only simulated household resource use and waste discharge estimates for our analysis. The standard daily household waste discharge is about 14.6 kg/10 persons according to the statistics from the city of Statistic Bureau of Shanghai (2000). We assumed a standard daily household resource use of 20 kg/10 persons. Due to a lack evaluation data judging the environmental carrying capacity and energy supply capability in the Chuandong area, we assumed that the total resource use and waste discharge thresholds are respectively 1,470,000 kg and 30,000 kg. If the total resources used or wastes discharged exceed these thresholds, household will reduce their daily resource use and/or waste discharge in order to meet the total limit prescribed in the simulation. The results of the initial stage simulation are shown in Fig. 2.6.

As shown in Fig. 2.6, promotion of city center development induced rapid growth in the number of households, which increased from 30,000 to 200,000 over the total urban area. This occurs even though, except for the development of university campuses in the urban center, there is no explicit development potential determined by the strategic spatial plan. Thus, the developed cells tend to be

Fig. 2.6 Simulated urban growth in the initial stage under the first scenario

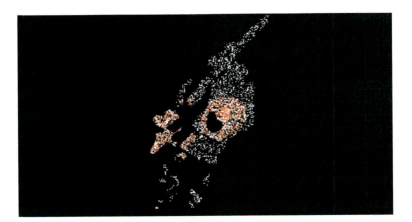

Fig. 2.7 Simulated urban growth at mid-stage under the first scenario

centralized in the urban center area. In Fig. 2.6, areas shown in red and orange are the regions with high population density and large resource use and waste discharge.

As the simulation continues we assumed that a population of 300,000 would be needed at the beginning of mid-stage development. As shown in Fig. 2.7, formation of the new urban center has been basically achieved. The more colorful areas in the T4, T5, and T6 zones show that the population is higher there than in the areas promoted for T2 and T3 land uses. This means that households are mainly centralized in the urban center.

When the simulation reached the mature stage, the number of households rapidly increased to more than 800,000 over the whole area. As shown in Fig. 2.8, urban development in the urban center area has improved even more and the population density has increased to 600 persons/ha. Moreover, the city center is connected with

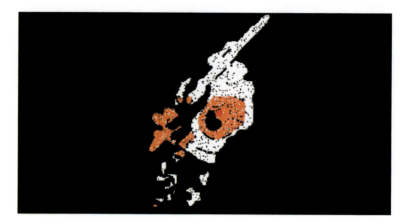

Fig. 2.8 Simulated urban growth in the mature stage under the first scenario

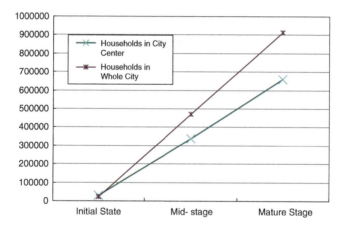

Fig. 2.9 Household numbers in the first scenario

Daxing Airport via the highway. At this stage, the development of the new urban area in Chuandong has been achieved. Compared with Fig. 2.2, it can be seen that the simulated space patterns of Chuandong's new urban area basically match the planned space pattern.

The simulation also shows that if the developed area in Chuandong were increased to 24 km^2 the population density would increase to 500 persons/ha in the T4, T5, and T6 land-use areas, and would reach 400 persons/ha in the T2 and T3 areas. The population would reach 650,000 in the city center and more than 800,000 over the entire area (Fig. 2.9). As shown in Fig. 2.10, the total daily household resources used and daily waste discharged are respectively more than 1,000,000 and 200,000 kg, which do not exceed the total control limits.

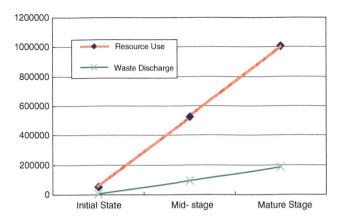

Fig. 2.10 Total daily household resource use and waste discharge at each simulation stage

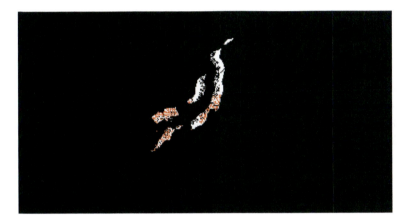

Fig. 2.11 Simulated urban growth in the initial stage under the second scenerio

Results of Second Scenario Simulation

Under the urban growth scenario characterized by the development of the highway, we took the areas shown in gray in Fig. 2.2 as locations where initial development would occur. Except for the initial area, all other simulation conditions and parameters were the same as in the first scenario.

In the simulation's initial stage, areas around the highway developed first, with people attracted to the area because of the road. Meanwhile, households gradually centralized into areas along the highway as shown in Fig. 2.11. Population density increased to 600 persons/ha in the urban center and 350 persons/ha in the surrounding areas. Although the number of households around the highway nearly doubled by the end of the initial stage, the spatial pattern in the urban center is still not clear. Thus,

Fig. 2.12 Simulated urban growth at mid-stage under the second scenario

we see that construction of the city centers, should be strenuously promoted to induce formation of the urban center and growth of the entire urban area.

By stressing the city center construction, we were able to induce the spatial pattern of the urban center to become clearer by the end of the simulation's mid-stage. Meanwhile, since urban growth was promoted by both highway development and the city centers construction, urban development sprawled across the entire area. Figure 2.12 shows that most of the areas near the highway and the urban center have been developed. Moreover, in the mature stage, the colored areas became even more aggregated, showing that the number of households, as well as the amount of resources they use and waste they discharge, will have increased even further there. The local government should be sensitive to urban development near the highway and should separate the urban area from the highway with a green belt.

Figure 2.13 shows that, compared with Fig. 2.11, Chuandong's spatial pattern becomes more compact as development proceeds. Displaying aggregation of developed urban areas, the colored areas are more intense than in the previous two stages. This indicates a boom in household mobility and increased household resource use and waste discharge. Urban growth under the second scenario can be summarized by noting that the spatial pattern tends to be different (as shown in Fig. 2.11) from the pattern devised for the strategic urban plan if development of the urban center is not strenuously promoted.

Through last stages of the simulation, the purpose of urban development in Chuandong area, namely meeting population and land-use demands, have been basically achieved through the promotion of highway development and campus construction. In order to make the spatial pattern of the Chuandong area more compact, development around the highway should be restricted in a rational way, inducing newly developing areas to aggregate around the city center (Fig. 2.14).

The simulation also showed that if the developed area in Chuandong reached 20 km^2 the population density would increase to 600 persons/ha in the T5 and T6 areas, while in the surrounding T2, T3, and T4 areas, densities of 350 persons/ha

Fig. 2.13 Simulated urban growth in the mature stage under the second scenerio

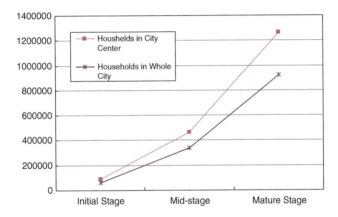

Fig. 2.14 Numbers of households in the second scenario

could be achieved. As Fig. 2.15 shows, total household resource use and waste discharge are nearly 1,400,000 kg/day and 260,000 kg/day, respectively, which do not exceed the control limits.

Conclusions

The SSP-SS tool, which integrates constrained CA and ABM models, is able to describe potential spatial urban growth processes. Using this tool, the planning work proposed by local urban planning, land management, and environmental protection departments and development and reform commissions can been combined in the same platform to visualize possible spatial patterns of urban growth.

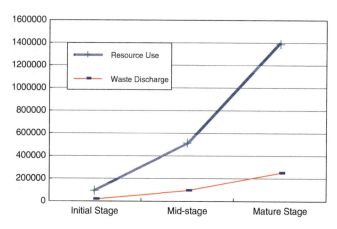

Fig. 2.15 Total daily household resource use and waste discharge at each simulation stage

In this *chapter*, we have focused on how to apply the agent-based model as a tool to aid local governments in investigating growth possibilities through urban modeling in planning practice rather than in model building and validation. We employed the SSP-SS tool to integrate urban and land-use plans with an economic and social development plan on the same platform in order to study the dynamic, interactive influences between the three plans. These influences are reflected in the different outcomes for various spatial strategy plan scenarios in the urban growth process. As a result, it is possible to bring population and land-use demand into an economic and social development plan. This allows urban regulation planning and land-use suitability evaluations found in land-use plans to be used together to gain insight into the impact urban policies laid out in strategic spatial plans can have on urban growth. We also integrated resource supply capability and waste discharge carrying capacity drawn from Chuandong's environmental plan into the Netlogo platform for investigating environmental issues in the urban growth process.

In this research, all the data were prepared by the research team to test whether an accurate dataset could be imported into the system. With the success of this attempt, we can try to validate the model in planning practice. For this purpose, historical datasets for estimating parameters and a planning dataset including urban, land-use, and socio-economic development plans are necessary. This remains as further work in this research stream. Finally, we did not present modeled behaviors in this *chapter* nor was interaction between agents validated in the estimation of the total control amounts of resources used and wastes discharged. These should be considered carefully in future work.

Acknowledgment This research was supported by Minister of Environment of Japan (S-3 project), Grants-in-Aid for Scientific Research (No. 23404022B), Japan Society of the Promotion of Science.

References

Batty, M., Couclelis, H., Eichen, M. (1997), "Urban systems as cellular automata," Environment and Planning B: Planning and Design, vol. 24, 159–164.

Batty, M., Xie, Y., Sun, Z. (1999), "Modeling urban dynamics through GIS-based cellular automata," Comput. Environ. Urban Syst, vol. 23, 205–233.

Brown Daniel G. et al., (2008), "Exurbia from the bottom-up: Confronting empirical challenges to characterizing a complex system," Geoforum, vol. 39, 805–818.

Clarke, K.C., Gaydos, L.J. (1998), "Loose-coupling of a cellular automaton model and GIS: long-term growth prediction for the San Francisco and Washington/Baltimore," Int. J. Geogr. Inform. Sci, vol. 12, 699–714.

Catalan B., Sauŕl D., Serra P. (2008), "Urban sprawl in the Mediterranean? Patterns of growth and change in the Barcelona Metropolitan Region 1993–2000," Landscape and Urban Planning, vol. 85, 174–184.

Fontaine Corentin M. and Rounsevell Mark D. A. (2009), "An agent-based approach to model future residential pressure on a regional landscape," Landscape Ecol, vol. 24, 1237–1254.

Fang, S., Gertner, G., Sun, Z., Anderson, A. (2005), "The impact of interactions in spatial simulation of the dynamics of urban sprawl," Landscape Urban Plan, vol. 73, 294–306.

Han J., Hayashi Y., Cao X., Imura H. (2009), "Application of an integrated system dynamics and cellular automata model for urban growth assessment: A case study of Shanghai, China," Landscape and Urban Planning, vol. 91, 133–141.

He Ch, Okada N., Zhang Q., Shi P., Li J. (2008), "Modelling dynamic urban expansion processes incorporating," Landscape and Urban Planning, vol. 86, 79–91.

Jayaprakash C. et al. (2009), "The interaction of segregation and suburbanization in an agent-based model of residential," Environment and Planning B: Planning and Design, vol. 36, 989–1007.

Li, X., Yeh, A.G-O. (2000), "Modeling sustainable urban development by the integration of constrained cellular automata and GIS" Int. J. Geogr. Inf. Sci, vol. 14, 131–152.

Li, X., Yeh, A., G.-O. (2002b), "Neural-network-based cellular automata for simulating multiple land use changes using GIS," International Journal of Geographical Information Science, vol. 16(4), 323–343.

Long Y, Shen, Z, Du L, Mao Q, Gao Z, (2008), "BUDEM: an urban growth simulation model using CA for Beijing metropolitan area," Proc. SPIE, Vol. 7143, 71431D.

Long Y, Shen Z, Mao Q, Dang A, (2009), "Form scenario analysis using constrained CA," Proceedings of Computers in Urban Planning and Urban Management.

Lambin, E. F. and Geist H. J. (2001), "Global land use and land cover change: What have we learned so far?," Global Change News Letter, vol. 46, 27–30.

National Bureau of Statistics, (1996–2005) China Statistical Yearbook, Beijing: China Statistics Press, [中国统计年鉴].

Peccol, E., Bird, A.C., Brewer, T.R. (1996), "GIS as a tool for assessing the influence of countryside designations and planning policies on landscape change," J.Environ. Manage, vol. 47, 355–367.

Shen Z, Kawakami M (2008), "Geo-simulation model using geographic automata for simulating land use patterns in urban partitions," Environment and planning B: planning and design, vol. 36, 802–823.

Sui, D. Z., Zeng, H. (2001), "Modeling the dynamics of landscape structure in Asia's emerging desakota regions: a case study in Shenzhen," Landscape and Urban Planning, vol. 53, 37–52.

Statistic Bureau of Shanghai (2000), Statistical Yearbook of Shanghai city, Beijing: China Statistics Press, [上海市统计年鉴].

Torrens Paul M., Itzhak Benenson, (2005), "Geographic Automata Systems," Int. J. Geogr. Inform. Sci, vol. 19 (04), 385–412.

Torrens Paul M. (2007), "A Geographic automata model of residential mobility," Environment and Planning B: Planning and Design, Vol. 34, 200–222.

Wu, F. (1998), "Simland: a prototype to simulate land conversion through the integrated GIS and CA with AHP-derived transition rules," Int. J. Geogr. Inform. Sci, vol. 12(1), 63–82.

White, R. and Engelen, G. (1997), "Cellular Automata as the Basis of Integrated Dynamic Regional Modelling," Environment and Planning B: Planning and Design, vol. 24(2), 235–246.

White, R. and Engelen, G. (2000), "High-resolution integrated modelling of the spatial dynamics of urban and regional systems," Computers, Environment and Urban Systems, vol. 24(5), 383–400.

Wilensky U NetLogo Center for Connected Learning and Computer-Based Modelling, Northwestern University, Evanson, IL, http://ccl.northwestern.edu/netlogo

Chapter 3
Simulating Spatio-Temporal Allocation of Farmland Conversion Quotas in China Using a Multi-Agent System

Zhang Honghui, Zeng Yongnian, Tan Rong, and Shen Zhenjiang

Introduction

Farmland conversion, i.e., converting farmland into non-agricultural uses, is an unavoidable trend in global economic development (Fazal 2001; Tan et al. 2005; Bugri 2008). China is facing a relative shortage of farmland resources in its accelerating urbanization. Farmland conversion is closely related to national food security and ecological security, as well as sustainable land use. Therefore, it is of great importance to allocate farmland conversion quotas in a rational manner conforming to the principle of maximal spatial-temporal allocation efficiency, in which the total revenue derived from farmland conversion is maximized in terms of spatial and temporal allocation of land resources.

Although there are few special research papers on the spatio-temporal allocation of farmland conversion quotas, it has been touched on in studies on land-use and land-cover change (LUCC) as well as on urbanization (Yeh and Li 1998; Li and Yeh 2000; Ryan 2002; Liu et al. 2005, 2006, 2008; Manson 2006; Lichtenberg and Ding 2008; Li and Liu 2008; Thompson and Prokopy 2009; Zhang et al. 2010). In these studies, LUCC or urbanization models can simulate and predict spatial patterns of farmland conversion with high precision, but they cannot explain the interactions between various driving stakeholders, such as governments, residents, farmers, and industrial enterprises (Li and Liu 2008). Furthermore, little attention

---

Z. Honghui (✉) • Z. Yongnian
School of Geoscience and Geomatics, Research Center of Space Info-Technique and Sustainable Development, Central South University; Changsha Planning Information Service Center, Changsha, China
e-mail: zhhgis@163.com

T. Rong
School of Public Administration, Zhejiang University, Hangzhou, China

S. Zhenjiang
School of Environment Design, Kanazawa University Kakuma CHO, Kanazawa City, Japan

Z. Shen, *Geospatial Techniques in Urban Planning*, Advances in Geographic Information Science, DOI 10.1007/978-3-642-13559-0_3,
© Springer-Verlag Berlin Heidelberg 2012

has been paid in these models to spatio-temporal allocation efficiency of farmland conversion quotas. Consequently, these models can neither explain the driving mechanisms of farmland conversion nor be used to achieve a scientific spatial-temporal method of allocating farmland conversion quotas.

The multi-agent system (MAS) method is an integration of the theories and technologies of complex adaptive systems, artificial life, and distributed artificial intelligence. Because of its bottom-up ideas, powerful complex computations, and spatio-temporal dynamic features, MAS has been very advantageous in simulating a complex spatial system (Chebeane and Echalier 1999); and there have been numerous successful applications to spatio-temporal allocation of land use (Ryan 2002; Manson 2006; Liu et al. 2006; Li and Liu 2008; Zhang et al. 2010). Moreover, compared with traditional methods such as empirical statistics, systems dynamics, and cellular automaton (CA) models, MAS, which integrates macroscopic phenomena and micro decision-making, is more able to interpret the driving mechanisms of LUCC considering the interactions between driving stakeholders that act as agents in MAS (Manson 2006). Since MAS has advantages in the study of LUCC, this research aims to explore the interactions and negotiations of different stakeholders. Based on the decision-making framework of multi-agents and integrating a resource economics model, a spatio-temporal model for allocating farmland conversion quotas was developed which conforms to the principle of spatial-temporal allocation efficiency.

The proposed model is used to simulate scenarios regarding the allocation of farmland conversion quotas in Changsha City, which is in the National Comprehensive Reforms Test Area for building a resource-saving and environment-friendly society in China. We also use the loss of farmland's ecosystem service value to measure the reasonableness of farmland conversion under various scenarios. We expect the model to contribute to a deeper understanding of the mechanisms driving farmland conversion and provide useful reference points for reasonable spatial-temporal allocation of farmland conversion quotas.

Modeling Spatio-Temporal Allocation of Farmland Conversion Quotas

Overview of the Model

In this *chapter*, the model simulating the spatio-temporal allocation of farmland conversion quotas consists of four parts, i.e., the environment, a multi-agent system, GIS data and the core element, a decision-making framework. The environment includes land use, soil, land prices, transportation and public facilities. Multi-agent systems include various agents, such as governments, residents, farmers, and industrial enterprises. Agents may be affected by the environment and in turn may also affect the environment and change its status. Decision-making behaviors

3 Simulating Spatio-Temporal Allocation of Farmland Conversion Quotas in China 51

of multi-agent systems are defined in a decision-making framework, and decision-making behaviors integrated into GIS information can be input to the model. Land already built up in this model is assumed to be unchanged.

Defining the Decision-Making Behaviors of Multi-Agents

Government Agents

As a kind of special agent, governments do not have spatial attributes, but they allocate farmland conversion quotas through formulating land use plans. In the course of planning, the overall objective is spatio-temporally optimal allocation of farmland conversion quotas. For this, the government agent will offer some scenarios for spatio-temporally optimal allocation of farmland conversion quotas at the beginning of the simulation.

In modeling farmland conversion, the government agent will interact with other agents in the simulation process to determine farmland development potential. Meanwhile, the government agent needs to search for farmland with high development potential using cellular automata (CA) methods.

Drawing on the idea of cellular automata, ${}^{t}P_{x,y}^{K}$, namely, farmland cell (x, y)'s land development potential at time t, can be considered to be the result of the joint effects of ${}^{t}S_{x,y}^{K}$, ${}^{t}N_{x,y}^{K}$, and V, which are individually cell (x, y)'s suitability for land use type K at time t, cell (x, y)'s neighborhood's effect on its conversion into land use K at time t, and stochastic chaotic factors. Therefore, ${}^{t}P_{x,y}^{K}$ can be represented as below:

$$ {}^{t}P_{x,y}^{K} = f({}^{t}S_{x,y}^{K}, {}^{t}N_{x,y}^{K}, {}^{t}V). \tag{3.1}$$

Referring to Barredo et al. (2003), (3.1) can be further transformed for convenience of calculation:

$$ {}^{t}P_{x,y}^{K} = ((1 + {}^{t}S_{x,y}^{K}) \times (1 + {}^{t}N_{x,y}^{K}))^{t}V \tag{3.2}$$

According to (3.2), we can calculate each farmland cell's farmland development potential and non-agricultural (construction) development potential, and consequently determine each cell's farmland conversion potential (Zhang et al. 2010).

In order to calculate ${}^{t}P_{x,y}^{K}$, cell (x,y)'s potential for non-agricultural development as well as farmland development should be predetermined. Cell (x, y)'s suitability for land-use type K at time t can be represented by

$$ {}^{t}S_{x,y}^{K} = \sum_{i=1}^{m} W_{i,x,y}^{K} Z_{i,x,y}^{K}, \tag{3.3}$$

where, i represents factors for evaluating suitability for land-use K, such as terrain and transportation, $Z_{i,x,y}^K$ represents cell (x,y)'s standardized evaluation of its suitability to land-use K; $W_{i,x,y}^K$ represents its weight, including evaluation factors for non-agricultural construction land and farmland and their weights (Xu 2002). $^tN_{x,y}^K$, the neighborhood effect on cell (x, y) at time t, is represented as the proportion of the total area in land-use K within cell (x, y)'s 3 × 3 neighborhood to the total area of all land uses within the window. Stochastic chaotic factor V can be represented as

$$V = [-\ln(rand)]^a, \tag{3.4}$$

where $rand$ is a stochastic number ranging between 0 and 1 and a is a constant indicating the degree of disturbance.

After the government agent identifies the farmland with high development potential in the initial step of the simulation, the development potential is converted by interactions between resident, industrial, and farmer agents.

Resident Agents, Industrial Enterprise Agents, and Farmer Agents

Different from government agent, resident agents, industrial enterprise agents and farmer agents are characterized as mobile, adjustable, and durative agents. Moreover, due to their different behavior modes, they operate under different decision-making rules in the process of farmland conversion.

Resident Agents

Those residents who are in better economic conditions or who are unsatisfied with their present living environment will hope to relocate to new, more suitable residences. As existing urban land is limited, farmland is always occupied in the course of developing new residences; it is a natural process of farmland conversion. There are three types of residents: those moving into city from outside, existing residents relocating into new places and new residents produced by existing residents. Resident agents in this study exhibit such behaviors as moving into the city, relocating into new residences, reproducing, and dying. We define attributes like education level (E), income (S), payout (P), property (W) (difference between payout and income), age (A), and distance between working place and residence (D) to model these agents.

We assume that resident agents will die out from the model when their economic conditions cannot support the elementary needs of urban life. Also, when residents are old enough and their economic conditions can support the cost of reproducing, new resident agents will be produced. Resident agents' decision-making behaviors are described in detail in the following sub-sections.

3 Simulating Spatio-Temporal Allocation of Farmland Conversion Quotas in China

Relocation of Resident Agents

Following the model of Li and Liu (2008), $Res_{x',y'}$, the relocation utility of resident agents located at cell (x', y'), can be defined as follows:

$$Res_{x',y'} = a_1 e^{(1-E_{nei})} + a_2 e^{(1-RD)} + a_3 e^{ID} + a_4 e^{D_{x'y'}} + a_5 e^{H_{x'y'}} + a_6 e^{(1-K_{x'y'})}$$
$$+ a_7 e^{(1-T_{x'y'})} + a_8 e^{(1-F_{x'y'})}. \tag{3.5}$$

In (3.5), E_{nei} represents the proportion of residents of the same education level as the resident agent i within the neighborhood of cell (x', y'); RD and ID separately represent the density of resident agents and industrial enterprise agents in cell (x', y'); $D_{x'y'}$ represents the standardized distance from cell (x', y') to its workplace; $H_{x'y'}$, $K_{x'y'}$, $T_{x'y'}$, and $F_{x'y'}$ represent respectively the standardized housing price, environmental comfort level, transportation accessibility, and community facility convenience of cell (x', y'); $a_i (i = 1,2,3,4,5,6,7,8)$ represents preference weights, where $\sum_{i=1}^{8} a_i = 1$. Residents moving into city from outside and residents newly produced from existing residents (A = 0) will search directly for a suitable residence, and so no consideration is needed to decide whether to relocate.

In the model, a resident agent can choose a certain number of suitable optional cells within its spatial access. Then, after the relocation cell set RCS_0 of resident agent i is obtained, each member in RCS_0 is evaluated, when $t = 0$, as to whether the property W of resident agent i satisfies

$$W > \beta L_{x,y}; \tag{3.6}$$

And when t > 0, whether it satisfies

$$S - P > \beta (L_{x,y} - L_{x',y'}). \tag{3.7}$$

In (3.6) and (3.7), $L_{x',y'}$ represents the standardized house price of cell (x', y'); $L_{x,y}$ represents that of a relocation cell candidate, and β is a constant. Then, Attraction_res$_{x,y}$, the utility of each cell in RCS to resident agent i, can be calculated using

$$Attraction\_res_{x,y} = a_1 e^{E_{nei}} + a_2 e^{RD} + a_3 e^{(1-ID)} + a_4 e^{(1-D_{x,y})} + a_5 e^{(1-H_{x,y})}$$
$$+ a_6 e^{K_{x,y}} + a_7 e^{T_{x,y}} + a_8 e^{F_{x,y}}. \tag{3.8}$$

In (3.8), according to McFadden's theory (McFadden 1974), the probability of cell (x, y) being selected by resident agent i as a residence is equal to the probability that the attraction of cell (x, y) is more than or equal to any other cell's attraction. $^i P_{x,y}^R$, resident agent i's probability to select cell (x, y) as a residence is

$$^i P_{x,y}^R = Pr(Attraction\_res_{x,y} \geq Attraction\_res_{x',y'}) = \frac{\exp(Attraction\_res_{x,y})}{\sum_i \exp(Attraction\_res_{x,y})},$$

$$\tag{3.9}$$

where $\Pr(\text{Attraction\_res}_{x,y} \geq \text{Attraction\_res}_{x',y'})$ is expressed as the probability that the utility of cell (x, y) is more than or equal to any other cell's utility, and $\sum_i \exp(\text{Attraction\_res}_{x,y})$ is the sum of all the exponential attractions of cells resident agent i could select. The final selection of cell (x, y) as the residence is decided by the Monte Carlo method in order to implement a more realistic decision (Wu and Webster 1998).

Renewal of Property

If a resident agent relocates to new cell at time $t + 1$, renewal of the property can be displayed as

$$W_{t+1} = W_t + S - P + \beta(L_{x,y} - L_{x',y'}), \tag{3.10}$$

where W_{t+1}, W_t are, separately, the property of resident agent i at times $t + 1$ and t.

Determining Whether a Resident Agent Will Die Out from the Model

A resident agent dies when $W < 0$.

Determining Whether a New Resident Agent Will be Generated

When either a resident agents' age (A) or property (W) is higher than the value of age and property set in the model, new resident agents will be generated.

Industrial Enterprise Agents

An enterprise agent's behavior is driven by the pursuit of more profit and broader development space. Industrial enterprise agents display an assembling feature, as they tend to move to industrial zones around urban areas to improve their competitiveness. Industrial enterprises in the model have such properties as scale (Z), assets (U), sales (S), and consumption (P), while exhibiting a behavior flow quite similar to resident agents, but with a specific difference in behavior motivation. When an industrial enterprise's assets cannot maintain its operation, it has to die out of the model. As industrial enterprises' reproductive behavior is rather complex and involves such factors as the society and economy, we do not take it into consideration in this *chapter*. Based on this, an industrial enterprise's decision-making behaviors can be described as follows:

Industrial Enterprise Agent Relocation

Similar to that of resident agents, $Ind_{x',y'}$, the relocation utility of industrial enterprise agents located at cell (x', y'), can be defined as follows:

$$Ind_{x',y'} = a_1 e^{(1-ID)} + a_2 e^{RD} + a_3 e^{PD_{x'y'}} + a_4 e^{LP_{x'y'}} + a_5 e^{1-EV_{x'y'}}$$
$$+ a_6 e^{1-PC_{x'y'}} + a_7 e^{1-T_{x'y'}} + a_8 e^{1-I_{x'y'}}. \tag{3.11}$$

where, ID and RD represent the densities of industrial enterprises and resident agents within the neighborhood of cell (x', y'); $PD_{x',y'}$, $LP_{x',y'}$, $EV_{x',y'}$, $PC_{x',y'}$, $T_{x',y'}$, $I_{x',y'}$ respectively represent the standardized slope, land price, environment value, planning completeness level, transportation accessibility and industrial agglomeration level of cell (x', y'); $a_i(i = 1,2,3,4,5,6,7,8)$ represents preference weights with $\sum_{i=1}^{8} a_i = 1$.

The relocation cell set RCS_0 for industrial enterprise agent i is obtained through a local space search and each member in the set is evaluated, while t = 0, as to whether the assets U of the industrial enterprise agent i satisfy

$$U > \beta L_{x,y}. \tag{3.12}$$

When t > 0, whether they satisfy

$$S - P > \beta(L_{x,y} - L_{x',y'}). \tag{3.13}$$

In (3.12) and (3.13), $L_{x',y'}$ represents the standardized land price of cell (x', y'); $L_{x,y}$ represents the standardized land price of relocation cell candidates; and β is a constant. Based on these equations, relocation cell sets RCS can be obtained and the utility of each cell in the RCS to industrial enterprise agent i, Attraction_ind$_{x,y}$, can be calculated using

$$Attraction\_ind_{x,y} = a_1 e^{ID} + a_2 e^{(1-RD)} + a_3 e^{(1-S_{x,y})} + a_4 e^{(1-L_{x,y})} + a_5 e^{E_{x,y}}$$
$$+ a_6 e^{P_{x,y}} + a_7 e^{T_{x,y}} + a_8 e^{I_{x,y}}. \tag{3.14}$$

Moreover, the method of calculating the probability that cell (x, y) will be selected as industrial land use by industrial enterprise agent i at time t $^tP^1_{x,y}$ is the same as given in (3.9). The final selection of cell (x, y) for industrial land use is decided by the Monte Carlo method in order to implement a more realistic decision (Wu and Webster 1998).

Renewal of Property

If an industrial enterprise agent relocates to new cell at time $t + 1$, renewal of property can be described by

$$U_{t+1} = U_t + S - P + \beta(L_{x,y} - L_{x',y'}), \tag{3.15}$$

where U_{t+1}, U_t are respectively the property of industrial enterprise agent i at times $t + 1$ and t.

Determining Whether an Industrial Enterprise Agent Will Die Out from the Model

A industrial enterprise agent will die out from the model due to economic problems resulting from competition when $U < 0.5U_0$ (U_0 represents the assets of an industrial enterprise agent at $t = 0$).

Farmer Agents

In the course of farmland conversion, farmer agents' behavior is controversial, as they hope to be closer to an urban area in order to live a more convenient life on one hand, but on the other hand, they want to keep their farmland, which is their livelihood. They have such attributes as scale (Z), income (S), and payout (P), and they die out from the model when S is below P. Slope, slope direction, soil productivity, and land price are selected as feature variables, and distance to the town center, distance to a government administration center (such as the location of municipal or district government offices), distance to a railway, distance to urban roads, and distance to an expressway are selected as location feature variables. Whether the land is within a farmland preservation or an urban planning area, and the permitted development densities are selected as planning restriction variables. Neighborhood densities for farmland preservation and construction land are selected as neighborhood variables, and population density of the country, as well as the density of the gross output values of agriculture and industry in the country, are selected as socio-economic statistic variables. These variables jointly influence farmers' decisions on farmland conversion. Different from resident agents' decision-making behaviors, the decision behaviors of farmer agents have the characteristic of bivalence (they either accept or reject their own farmland being converted into non-agricultural construction land). Therefore, a binary logistic regression model can be used to simulate farmer agents' decision-making behaviors (Kleinbaum 1994). The general form of the logistic regression formulation is represented as follows:

$$Z = a + b_1 h_1 + b_2 h_2 +, ..., + b_n h_n,$$ (3.16)

and

$$^t\mathbf{P}^F_{x,y} = \frac{1}{1 + e^{-Z}},$$ (3.17)

where $h_1, h_2, ..., h_n$ are independent variables which separately represent various decision-making variables affecting farmer agents' decision-making as depicted above; n is the number of independent variables; Z is a linear function of independent variables representing the state of land use with its value being 1 for conversion from agricultural land to non-agricultural construction land or 0 for cases

3 Simulating Spatio-Temporal Allocation of Farmland Conversion Quotas in China

where no conversion happens. The constant a and parameters $b_1, b_2, \ldots, b_n$ are regression coefficients to be calculated, representing the contribution each decision-making variable makes to probability ${}^t P^F_{x,y}$; and ${}^t P^F_{x,y}$ represents the probability of farmer agents' acceptance of cell (x,y) being converted into non-agricultural construction land from farmland after decision-making takes place.

Interactions Between Government Agents and Resident, Industrial Enterprise, and Farmer Agents

The interaction between the government agent and other agents is assumed to be a process of psychological expectations that other agents will ask the government agent to accept their candidate cells as non-agricultural construction land. For this, we suggest two types of interaction in this simulation, which are the government at the macro-level, and the residents, the industrial enterprises, and the farmers at the micro-level. In the set of potential land use cells searched by government agents, the more cells government agents allow to be converted into non-agricultural construction land, the bigger the potential is that farmland cell (x, y) will be applied for or accepted as non-agricultural construction land by a resident, industrial enterprise, or farmer agent and vice versa. The interaction at the micro-level reflecting each agent's expectation can be represented as follows:

$$
{}^{new,t}P^j_{x,y} = \begin{cases} {}^t P^j_{x,y}(1+f_{j,x,y}) & \Delta s > 0 \\ {}^t P^j_{x,y} & \Delta s = 0 \\ {}^t P^j_{x,y}(1-f_{j,x,y}) & \Delta s < 0 \end{cases}, \qquad (3.18)
$$

where Δs is the difference of between actual expectations and psychological expectations of other agents.

At the micro-level, ${}^t P^j_{x,y}$ represents the initial potential for farmland cell (x, y) to be applied for or accepted as non-agricultural construction land by resident, industrial enterprise, or farmer agents, and j is the sequence number of the resident, industrial enterprise, or farmer agent. The agent set of resident, industrial enterprise, and farmer agents will have different ${}^t P^j_{x,y}$ values.

In (3.18), ${}^{new,t}P^j_{x,y}$ represents the potential for farmland cell (x, y) to be applied for or accepted as non-agricultural construction land in consideration of the government agent's influence; $f_{j,x,y}$ is an adjustment parameter reflecting agents' psychological expectations for applying for or accepting candidate cells as non-agricultural construction land, which can be calculated using

$$
f_{j,x,y} = |\Delta s|/n = \left| PE_{x,y} - PE^{min}_{x,y} \right|/n, \qquad (3.19)
$$

where n represents the number of candidate cells applied for or accepted as non-agricultural construction land by resident, industrial enterprise, and farmer agents. $PE^{min}_{x,y}$ represents resident, industrial enterprise, and farmer agents' minimal psychological expectations for applying for or accepting candidate cells as non-agricultural construction land, which varies with the contrast between simulated and real situations and has an initial value of 0.05; $PE_{x,y}$ represents the actual expectations of resident, industrial enterprise, and farmer agents considering the results of government agents' allocations, which can be obtained from

$$PE_{x,y} = \sum_{i=1}^{n} {}^{t}P_{x,y}^{g}\, {}^{t}P_{x,y}^{j} \bigg/ \sum_{i=1}^{n} {}^{t}P_{x,y}^{g}. \tag{3.20}$$

From the viewpoint of the government agent at the macro-level, the more frequent a farmland cell (x,y) is applied for or accepted as non-agricultural construction land by resident, industrial enterprise, and farmer agents, the higher the probability is that government agents will allocate it as non-agricultural construction land, as shown in the following:

$$^{new,t}P_{x,y}^{g} = {}^{t}P_{x,y}^{g} + \Delta p \cdot N, \tag{3.21}$$

where $^{new,t}P_{x,y}^{g}$ represents government agents' decision-making probability considering resident, industrial enterprise, and farmer agents' influence; N represents frequency that cell (x, y) is applied for or accepted as non-agricultural construction land by resident, industrial enterprise, and farmer agents; Δp represents the increased probability for government agents to allow cell (x, y) to be converted into non-agricultural land when cell (x, y) is applied for or accepted as non-agricultural construction land by resident, industrial enterprise, and farmer agents, which has initial value in the simulation model set as 0.005 and varies depending on the contrast between simulated results and the actual situation.

Transition Rule for Farmland Conversion with Agent Interaction

Conversion of a farmland cell is realized under the influence of a multi-agent system. Adapting a weighted summation model, the potential of farmland conversion is

$$P = (\theta C_{x,y}^{g} + (1 - \theta)C_{x,y}^{j})\, \forall \theta \in [0, 1], \tag{3.22}$$

where θ represents the weighting parameter; P represents the potential for farmland conversion of farmland cell (x, y) under multi-agent influence; $C_{x,y}^{g}$ represents the probability of farmland conversion under the influence of government agents (formula 3.2); $C_{x,y}^{j}$ represents probability of farmland conversion under the influence of resident, industrial enterprise, and farmer agents. It can be obtained using following, which is based a discrete choice model:

$$C_{x,y}^{j} = \sum_{j=1}^{L} w_{j}\,{}^{new,t}P_{x,y}^{j}. \tag{3.23}$$

Here, j represents the agent type sequence number, i.e., resident agents, industrial enterprise agents, and farmer agents; $^{new,t}P^j_{x,y}$ represents the potential for farmland conversion of cell (x, y) by an individual agent. Farmland conversion quotas are allocated in the simulation according to the numerical value of P, from high to low.

Simulation Process and System Framework

We assume that there is a farmer agent in each farmland cell and an industrial enterprise agent or a resident agent in each newly converted non-agricultural land cell. An agent here does not represent a farmer, resident, or industrial enterprise, but indicates certain proportional relationship, i.e., its actual meaning in this research is the total number of residents, farmers, or industrial enterprises in a 100 × 100 m cell. The flow of the simulation implementation and the system framework are represented in Figs. 3.1 and 3.2.

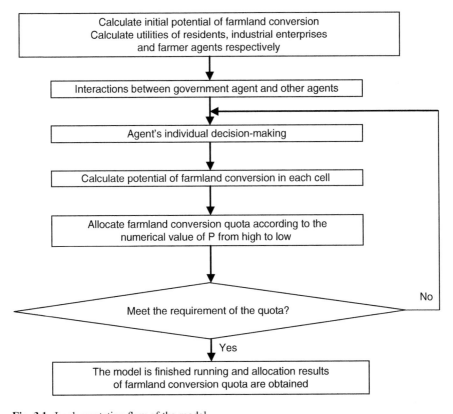

Fig. 3.1 Implementation flow of the model

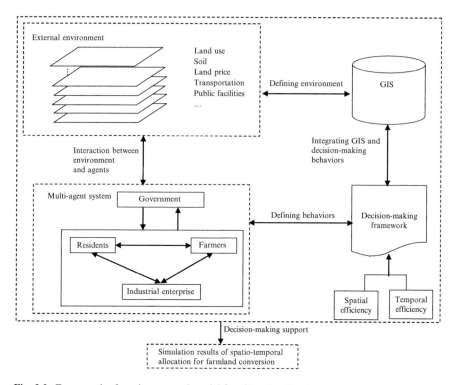

Fig. 3.2 Framework of spatio-temporal model for allocating farmland conversion quotas

Application of the Model

Simulation Implementation

Study Area and Dataset

We selected Changsha City, the core of the Changsha-Zhuzhou-Xiangtan city cluster, which is a national comprehensive reform test area for building a resource saving and environmentally friendly society, as our study area. The data for the application includes remote sensing data, GIS data, and socio-economical statistics. The data details are listed in Table 3.1.

According to land use data for Changsha City in 2005, its total urban area was 556.33 km^2, of which 254.61 km^2 is non-agricultural land, making up 45.77% of the total study area. According to the general land use plan for Changsha City (2006–2020), non-agricultural land will, by 2020, have taken 75% of the total study area, increasing to 417.25 km^2. Consequently, the total farmland to be converted in Changsha City from 2006 to 2020 is 162.64 km^2.

Table 3.1 Data types and content for simulation application

Type	Content
Remote sensing data	TM data for years 1993, 1996, 1999, 2002, and 2005
GIS data	Land use in 2001, 2005
	General land use planning (2006–2020)
	General urban planning (2003–2020)
	Soil type in 2005
	Transportation in 2005
	Electronic map containing information about schools, hospitals, banks, residential areas, and commercial and industrial enterprises in 2005
	Land price in 2005
	Digital elevation model in 2005
Socio-economic statistical data	Distribution of commercial housing development projects and their prices in 2005
	Population statistics in 2005
	Output value of industry and agriculture in 2005
	Income statistics of residents and farmers in 2005

Simulation Process

Model implementation can be summed up in the following steps:

1. Scenario configuration of the demand for farmland conversion in Changsha City from 2006 to 2020.
2. Based on the total area of farmland (260.69 km^2) in Changsha City in 2005, the number of farmer agents is set as 26,069 in the initial phase, and decreases as farmland cells are gradually replaced by non-agricultural cells. The total number of resident and industrial enterprise agents can be obtained using the predicted farmland conversion quota. Changes in the ratio of residential land area to industrial land area from 1996 to 2005 can be used to estimate the ratios in 2010, 2015, and 2020. Based on these, we can estimate the number of resident and industrial enterprise agents in the three phases: 2006–2010, 2011–2015, and 2016–2020. According to the ratio of farmland conversion quotas allocated to the Furong, Tianxin, Yuhua, Kaifu, and Yuelu districts by the scenario provided by the government agent, the number of industrial enterprise agents and residential agents to be allocated to the five sub-areas in three phases can be obtained.
3. Generate the resident agents and industrial enterprise agents required by the model with a Monte Carlo approach.
4. Calculate the farmland development potential and non-agricultural land development potential of a cell based on the cellular automata model.
5. Determine the parameters of variables that influence the decision-making behaviors of resident agents and calculate the probability of a cell selected as residential land by resident agents based on the simulated distribution of resident agents.

6. Determine the parameters of variables that influence the industrial enterprise agents' decision-making behaviors and calculate the probability that cells will be selected as industrial land by industrial enterprise agents based on the simulated distribution of industrial enterprise agents.
7. Determine the parameters of variables that influence decision-making behaviors of farmer agents and calculate probability of transformation from farmland into non-agricultural by means of a logistic regression model.
8. Simulate the interaction behaviors between government agents and resident, industrial enterprise, and farmer agents.
9. Calculate the farmland conversion potential of each cell according to (3.22) and (3.23) by combining the results of the multi-agent decision-making.
10. Simulate the spatial distribution of farmland conversion over a 5-year interval according to the numerical value of P, proceeding from high to low. If the allocation results meet the quota requirement, the simulation process ends. Otherwise it returns to the multi-agent comprehensive decision-making phase and iterates until the results meet the quota requirement.

Determining Model Parameters

Government Agents

Given the total farmland conversion constraint in Changsha City from 2006 to 2020, we made each of the five sub-regions have equal net marginal farmland conversion revenues and obtained each sub-region's farmland conversion quota (Zhang et al. 2010), which were 19.52 km^2 for the Furong district, 32.31 km^2 for the Tianxin district, 26.91 km^2 for the Yuhua district, 48.79 km^2 for the Kaifu district and 35.41 km^2 for the Yuelu district.

Using population statistics for Changsha City from 1993 to 2005, we can predict the population in Changsha City from 2006 to 2020 and calculate the population increases in each period (Deng 1998). On the basis of that data, the government's temporal allocation of farmland conversion quotas from 2006 to 2020 can be determined in a fashion consistent with maximal temporal allocation efficiency (Zhang et al. 2010) (Table 3.2).

Table 3.3 was used to estimate the initial development potential of each cell.

Table 3.2 Population increases and allocation of farmland conversion quotas in each period from 2006 to 2020

	2006–2010	2011–2015	2016–2020
Increased population/10^4 people	30.75	36.31	42.88
Farmland conversion quota/km^2	45.56	53.73	63.35

3 Simulating Spatio-Temporal Allocation of Farmland Conversion Quotas in China

Table 3.3 Weights of evaluation factors in land use suitability evaluation

Factor	Non-agricultural land			Farmland		
	Traffic accessibility	Land price	Slope	Soil type	Soil productivity	Slope
Weight	0.462	0.297	0.241	0.214	0.549	0.237

Table 3.4 Decision-making parameters for different categories of resident agents

	a_1	a_2	a_3	a_4	a_5	a_6	a_7	a_8
Primary school education and below	0.165	0.108	0.147	0.097	0.22	0.051	0.129	0.083
Middle school education	0.158	0.123	0.135	0.084	0.196	0.061	0.118	0.125
Higher education	0.171	0.124	0.145	0.059	0.167	0.121	0.092	0.121

Table 3.5 Decision-making parameters for different categories of industrial enterprise agents

	a_1	a_2	a_3	a_4	a_5	a_6	a_7	a_8
Large-scale enterprises	0.135	0.096	0.078	0.159	0.078	0.098	0.185	0.171
Middle-scale enterprises	0.129	0.089	0.093	0.177	0.073	0.093	0.182	0.164
Small-scale enterprises	0.119	0.085	0.126	0.193	0.067	0.084	0.169	0.157

Resident, Industrial Enterprise, and Farmer Agents

As in Li and Liu (2008), agents should be classified into categories so that their properties can be heuristically defined. Agents' attributes are determined using social and economic data. This study takes into account such attributes as education level, enterprise scale, and agricultural scale for resident agents, industrial enterprise agents, and farmer agents. Accordingly, resident agents can be divided into those having only a primary school education or below, a middle school education, and a higher education. Industrial enterprise agents can be divided into large-scale enterprises, middle-scale enterprises, and small-scale enterprises. Farmer agents can be divided into large-scale farmer households, middle-scale farmer households, and small-scale farmer households. Xu (2002) has necessary decision-making parameters for different categories of resident agents and industrial enterprise agents as shown in Tables 3.4 and 3.5.

After testing the series of farmer agents' decision-making variables described in Sect. "Resident Agents, Industrial Enterprise Agents, and Farmer Agents" for collinearity and removing insignificant variables, nine variables remained, including land price, distance to urban roads, distance to an expressway, distance to a town center, whether the land is within farmland preservation area, whether it is within an urban planning area, the neighborhood density of protected farmland, the neighborhood density of construction land, and the density of the gross output value of agriculture and industry in the town. After standardizing each independent variable, the contributions made by the various farmer agents' decision-making variables can be determined using a logistic regression model as shown in Table 3.6.

Table 3.6 Coefficients of decision-making variables for different categories of farmer agents

	Decision-making variable	Contribution		
		Large-scale farmer household	Middle-scale farmer household	Small-scale farmer household
Micro feature variables	Land price	2.101	2.123	2.447
Location feature variables	Distance to urban roads	−2.614	−2.515	−2.193
	Distance to expressway	−1.932	−1.774	−1.321
	Distance to town center	−4.554	−4.236	−3.877
Planning restriction variables	Within farmland preservation area?	−1.521	−1.129	−0.845
	Within the urban planning area?	1.434	1.578	1.853
Neighborhood variables	Neighborhood density of protected farmland	−1.323	−1.271	−1.199
	Neighborhood density of construction land	1.753	1.804	1.896
Socio-economic statistical variables	Density of gross output value of agriculture and industry in town	0.884	0.673	0.519
Constant		3.142	2.976	1.815

Simulation of a Spatio-Temporal Farmland Conversion Allocation Scenario

Confronted with a more and more serious farmland conversion situation, local authorities in Changsha City have instituted some farmland policies, including a requisition-compensation balance for arable land, land use regulations, and dynamic monitoring farmland protection, to minimize farmland conversion. However, little concern has been paid to these policies' spatio-temporal allocation efficiency. Since there is no policy for the spatio-temporal allocation of farmland conversion quotas, we need to design different allocation scenarios and integrate them into the model to simulate their influence on the spatio-temporal allocation of farmland conversion. We carried out simulations assuming the four scenarios below.

Scenario A: farmland protection intentions are similar to the simulation benchmark period. The government continues its previous farmland conversion trends with no consideration of the principle of maximal spatio-temporal allocation efficiency.

Scenario B: based on previous farmland protection policies, farmland is more strenuously protected, with 10% increase in protection. The government allocates farmland conversion quotas in accordance with the principle of maximal spatio-temporal allocation efficiency and sustainable development.

Scenario C: farmland will be protected in a manner similar to the previous farmland protection policy. At the same time demand for industrial zones, which contribute greatly to the regional economy such as the National High and New Tech Zone, the Yuhua Environmental Protection Industry Zone, Tianxin Ecological Town, and the Great Hexi Prior Region should be satisfied first. The principle of maximal spatio-temporal allocation efficiency is not taken into account in the government's allocation of farmland conversion quotas.

Scenario D: Based on previous farmland protection policies, the overall level of farmland protection level will be improved, with cultivated land increasing by 5%, orchard land by 5% and forest land by 5%. The government allocates farmland conversion quotas in accordance with the principle of maximal spatio-temporal allocation efficiency and sustainable development.

For the scenarios above, results of spatio-temporal allocation of farmland conversion quota simulations in Changsha City from 2006 to 2020 are displayed in Figs. 3.3 and 3.4 and the amounts of farmland lost under different allocation scenarios are given in Table 3.7.

After comprehensive analysis of the various modeled scenarios, we can observe the following: Under scenario A, present farmland conversion trends in the study area will continue until 2020; large quantities of non-agricultural construction land will expand along the present construction land area to the suburbs; cultivated land, forest, orchard land, and water areas will be continuously converted into non-agricultural land, and the conversion will come mainly from cultivated and forest land. Under scenario B, as the principle of maximal spatio-temporal allocation efficiency is incorporated in the government's allocation of farmland conversion quotas and cultivated land is protected more strenuously, farmland conversion will be restricted to a certain degree. Loss of cultivated land, forest, and orchard land, as well as water area, is correspondingly less than in scenario A. This is more obvious in the northern and southeastern parts of the study area. Under scenario C, where priority is placed on satisfying demand for industrial lands in the National High and New Tech Zone, the Yuhua Environmental Protection Industry Zone, the Tianxin Ecological Town, and the Great Hexi Prior Region, non-agricultural land use expands greatly around these areas. The main source of these lands is cultivated and forest land. The expansion will occur mainly in the National High and New Tech Zone and the Great Hexi Prior Region in the west and the Yuhua Environmental Protection Industry Zone and Tianxin Ecological Town in the south. In the north, however, there is less farmland loss because of a lack of industrial zones which demand large amounts of land. Under scenario D, because of the double constraints of the principle of maximal spatio-temporal allocation efficiency and increased overall protection of farmland, farmland conversion is more strictly controlled and cultivated land, forest, and orchards, as well as water areas, are better protected, though non-agricultural construction land is still expanding.

For further evaluation of farmland conversion allocation efficiency under various allocation scenarios, loss of farmland's ecosystem service value is used to measure the reasonableness of various allocation scenarios (Costanza et al. 1997). Xie et al. (2003) have established Chinese ecosystem services values for different types of ecosystems.

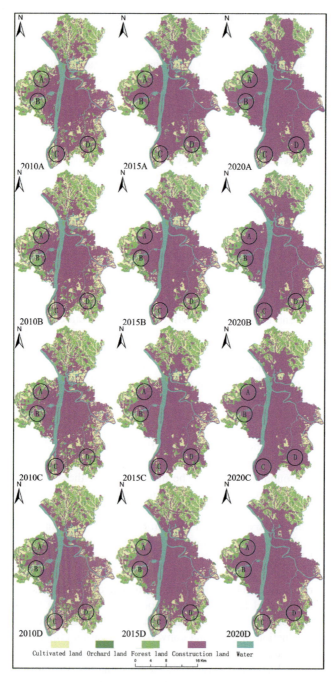

Fig. 3.3 Simulations of farmland conversion development trends under different scenarios in Changsha City from 2006 to 2020 (A, B, C, and D represent, respectively, the High and New Tech Zone, the Great Hexi Prior Region, Tianxin Ecological Town, and the Yuhua Environmental Protection Industry Zone)

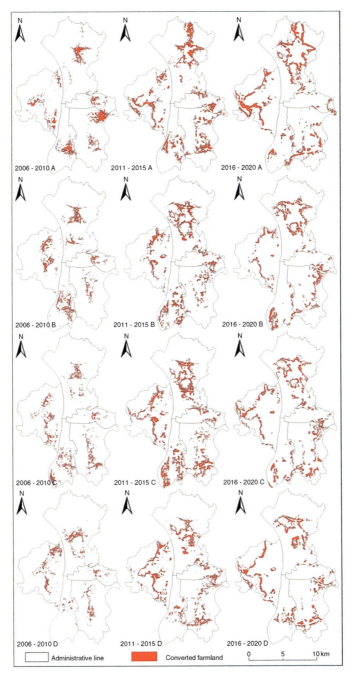

Fig. 3.4 Simulations of spatio-temporal allocation for farmland conversion under different allocation scenarios in Changsha from 2006 to 2020

Table 3.7 Farmland loss under different farmland allocation scenarios (km^2)

	A	B	C	D
Cultivated land	92.93	76.54	84.45	82.31
Orchard land	9.44	4.67	7.08	3.07
Forest land	77.16	62.09	74.38	58.16
Water	20.19	19.34	20.41	19.10
Sum	199.72	162.64	186.32	162.64

Table 3.8 Predicted loss of ecosystem service value under different allocation scenarios in Changsha City from 2006 to 2020

	Cultivated land	Orchard land	Forest land	Water	Total loss
Scenario A	5682.02	577.19	14918.11	8212.57	29389.89
Scenario B	4679.89	285.54	12004.48	7866.82	24836.72
Scenario C	5163.53	432.89	14380.63	8302.05	28279.10
Scenario D	5032.68	187.71	11244.65	7769.19	24234.24

Ecosystem service value can be calculated using

$$Val = \sum_{i=1}^{n} Price_i.Area_i, \tag{3.24}$$

where Val represents the total loss of ecosystem services value in the study area; $Price_i$ represents the value of ecosystem services per unit area for converted farmland of type i, and $Area_i$ represents the area of farmland type i lost within study area.

Table 3.8 shows the total loss of ecosystem service value under different allocation scenarios in Changsha City from 2006 to 2020, calculated using (3.24).

From Table 3.8, we can see that scenario A has the highest total loss, followed by scenario C, scenario B, and scenario D. This indicates the following:

1. Compared with scenario A, which follows present developmental trend and scenario C, which focuses on economic development, scenarios B and D, which conform to the principle of maximal spatio-temporal allocation efficiency and sustainable development, are preferable with respect to the spatio-temporal allocation efficiency of farmland conversion.
2. Because of its improvement in farmland protection, scenario D has a lower total lost ecosystem services value than scenario B.
3. Scenario A has the largest loss of ecosystem services value, which indicates that present farmland protection policies need to be improved. Otherwise, farmland conversion in Changsha City will become more serious and regional ecological and food security will be severely endangered.
4. Because land demand for industrial zones is fully satisfied for economic development under scenario C, its total loss of ecosystem services value is second only to scenario A. This indicates that it is not preferable or sustainable to sacrifice farmland for regional economic development and such land development policies should be avoided as much as possible.

Conclusions and Discussion

At present, when China's rapid economic increase depends on farmland conversion and its farmland resources have become precious, the country needs a reasonable method for allocating farmland conversion quotas. However, because the present administrative system for farmland conversion quota is inefficient, revenue has been lost in the spatio-temporal allocation of farmland conversion quotas. Therefore, if maximal spatio-temporal allocation efficiency is achieved with other conditions unchanged, farmland conversion allocation efficiency will increase. With this background, we have developed an agent-based model for spatio-temporal allocation of farmland conversion quotas on the basis of our former work (Zhang et al. 2010) regarding the principal of maximal spatio-temporal allocation efficiency and sustainable development. Through integrating a multi-agent system and a resource economics model, we have defined decision-making rules for the different agents, established comprehensive decision-making rules, and obtained explicit spatial and temporal representation of various agents' decision-making behaviors in the course of farmland conversion.

We applied the proposed model to scenarios simulating the allocation of farmland conversion quotas in Changsha, at the core of the Changsha-Zhuzhou-Xiangttan city cluster, which lies in the national comprehensive reforms test area for building a resource-saving and environmentally friendly society in China. We calculated the loss of farmland area and ecological system service value resulting from farmland conversion under various allocation scenarios, which will provide valuable decision-making support for more reasonable allocation of farmland conversion quotas.

Though we have made some contribution to development of the proposed model, further exploration is still needed. First, while all kinds of agents' farmland conversion decision-making rules are set using a rational decision-making framework in this chapter, they are not completely rational in actual situations. Therefore, further effort is needed to establish agents' farmland conversion decision-making rules in a bounded rationality framework. Second, while we have designed four kinds of agents' decision-making rules in farmland conversion, there is a larger system of agents pushing for farmland conversion and a more complicated series of decision-making rules exists in reality; therefore, further effort should be made to improve the multi-agent system and gain a more precise representation of decision-making rules.

Acknowledgments This study was supported by the National Natural Science Foundation of China (Grant no. 40771198), and the Natural Science Foundation of Hunan province, China (Grant no. 08JJ6023).

References

Barredo, J. I., Kasanko, M., McCormick, N. & Lavalle, C. (2003). Modelling dynamic spatial processes: simulation of urban future scenarios through cellular automata. *Landscape and Urban Planning*, 64(3), 145–160.

Bugri, J.T. (2008). The dynamics of tenure security, agricultural production and environmental degradation in Africa: Evidence from stakeholders in north-east Ghana. *Land Use Policy*, 25 (2), 271–285.

Chebeane, H. & Echalier, F. (1999). Towards the use of a multi-agents event based design to improve reactivity of production systems. *Computers & Industrial Engineering*, 37(1–2), 9–13.

Costanza, R., d'Arge, R., de Groot, R., Farber, S., Grasso, M., Hannon, B., Limburg, K., Naeem, S., O'Neill, R. V., Paruelo, J., Raskin, R.G., Sutton, P. & van den Belt, M. (1997). The value of the world's ecosystem services and natural capital. *Nature*, 387(6630), 253–260.

Deng, J. L. (1998). Introduction to grey system theory. *The Journal of Grey System*, 1(1), 1–24.

Fazal, S. (2001). The need for preserving farmland: A case study from a predominantly agrarian economy (India). Landscape and Urban Planning, 55(1), 1–13.

Kleinbaum, D.G. (1994). *Logistic regression: A self-learning text*. New York: Springer.

Li, X. & Liu, X. (2008). Embedding sustainable development strategies in agent-based models for use as a planning tool. *International Journal of Geographical Information Science*, 22(1), 21–45.

Li, X. & Yeh, A.G.-O. (2000). Modelling sustainable urban development by the integration of constrained cellular automata and GIS. *International Journal of Geographical Information Science*, 14(2), 131–152.

Lichtenberg, E. & Ding, C. (2008). Assessing farmland protection policy in China. *Land Use Policy*, 25(1), 59–68.

Liu, J., Liu, M., Tian, H., Zhuang, D., Zhang, Z., Zhang, W., Tang, X. & Deng, X. (2005). Spatial and temporal patterns of China's cropland during 1990–2000: An analysis based on Landsat TM data. *Remote Sensing of Environment*, 98(4), 442–456.

Liu, X., Li, X., Liu, L., He, J. & Ai, B. (2008). A bottom-up approach to discover transition rules of cellular automata using ant intelligence. International Journal of Geographical Information Science, 22(11), 1247–1269.

Liu, X., Li, X. & Anthony, G.-O. (2006). Multi-agent systems for simulating spatial decision behaviors and land-use dynamics. Science in China Series D: Earth Sciences, 49(11), 1184–1194.

Manson, S. (2006). Land use in the southern Yucatán peninsular region of Mexico: Scenarios of population and institutional change. *Computers, Environment and Urban Systems*, 30(3), 230–253.

McFadden, D. (1974). *Conditional logit analysis of qualitative choice behavior*. In P. Zarembka (Eds.), Frontiers in Econometrics (pp. 105–142). New York: Academic Press.

Ryan, R.L. (2002). Preserving rural character in New England: local residents' perceptions of alternative dential development. *Landscape and Urban Planning*, 61(1), 19–35.

Tan, M., Li, X., Xie, H. & Lu, C. (2005). Urban land expansion and arable land loss in China–a case study of Beijing-Tianjin-Hebei region. *Land Use Policy*, 22(3), 187–196.

Thompson, A.W. & Prokopy, L.S. (2009). Tracking urban sprawl: Using spatial data to inform farmland preservation policy. *Land Use Policy*, 26(2), 194–202.

Wu, F. & Webster, C.J. (1998). Simulation of land development through the integration of cellular automata and multicriteria evaluation. *Environment and Planning B: Planning and Design*, 25 (1), 103–126.

Xie, G., Lu, C., Leng, Y., Zhen, D., Li, S. (2003). Ecological assets valuation of the Tibetan Plateau. *Journal of natural resources*, 18(2), 189–195. (in Chinese)

Xu, J. (2002). *Mathematical Method in Modern Geography*. Beijing: China Education & Culture Publishing Company.(in Chinese)

Yeh, A.G.O., Li, X. (1998). Sustainable land development model for rapid growth areas using GIS. *International Journal of Geographical Information Science*, 12(2), 169–189.

Zhang, H., Zeng, Y., Bian, L. & Yu, X. (2010). Modelling urban expansion using a multi agent-based model in the city of Changsha. *Journal of Geographical Sciences*, 20(4), 540–556.

Chapter 4
Planning in Complex Spatial and Temporal Systems: A Simulation Framework

Shih-Kung Lai and Haoying Han

Introduction

Planners are confident that their planning affects not only behaviors in organizations, but also outcomes. There is, however, little backing for this confidence. Surprisingly little is known about planning processes and how they affect organizations. One approach to gaining understanding of planning behaviors in organizations is to develop and analyze simulation models. The framework presented here builds on two streams of previous work: the garbage can models of organizational behavior presented by Cohen et al. (1972) and the spatial evolution models of Nowak and May (1993). Our objective is to develop a framework sufficient to investigate the implications of introducing planning behaviors into complex organizational systems evolving in space and time. Our primary focus for this *chapter* is on devising simulations from which we might discover general principles about the effects of planning the behavior of organizations. Additional work will be necessary to determine the external validity of these simulations, that is, to interpret concrete situations in terms of such principles.

We focus on the planning activities of considering related choices (Hopkins 2001), setting aside for this *chapter* planning with respect to uncertainty about planning objectives, environments, and available alternatives. Information that reduces uncertainty arises from some regularity about observed phenomena that permits prediction across actors, space, or time. The level of planning investment can be measured by the number of comparisons and judgments made in gathering

---

S.-K. Lai
College of Public Administration, Zhejiang University, Hangzhou, China

Department of Real Estate and Built Environment, National Taipei University, Taipei, Taiwan
e-mail: lai@mail.ntpu.edu.tw

H. Han (✉)
Department of Land Resource Management, Zhejiang University, Hangzhou, China
e-mail: hanhaoying@zju.edu.cn

Z. Shen, *Geospatial Techniques in Urban Planning*, Advances in Geographic
Information Science, DOI 10.1007/978-3-642-13559-0_4,
© Springer-Verlag Berlin Heidelberg 2012

information about related choices while making a plan, as manifested by the manipulation of these choices. Plans are sets of decisions, which are contingent on outcomes resulting from prior decisions and system behavior based on exogenous parameters. Plans persist in time and space. As decisions become actions yielding outcomes, further contingent decisions can be enacted. These contingent decisions are, however, part of the persistent plan. Revising a plan thus implies changing the contingencies on which ensuing decisions are based.

We construct the elements of this simulation framework in sequence. Planning is understood here as gathering information to reduce uncertainty (e.g., Friend and Hickling 2005; Schaeffer and Hopkins 1987). Section "Planning Behaviors in the Planning Process" explains the garbage can model and develops one definition of planning as manipulation of decision situations within that framework. Section "Incorporation of Spatial Relationships" explains the spatial process of urban modeling and the spatial framework in terms of evolutionary planning behavior in a prisoner's dilemma game. Section "Idea for Integrating the Garbage Can Model with a Spatial Evolution Game for Planning Simulation" introduces an idea of how these two types of models can be integrated and the questions that might be addressed by analysis of such simulations. Section "Conclusions" concludes.

Planning Behaviors in the Planning Process

As everyone knows, scientific technology is developing day and night. Since computer models have emerged, there are various models developed for scientific research. While one of them, namely agent-based modeling (ABM) has been proven to be an effective way to simulate activities in which entities participate (Torrens 2007). This kind of simulation is expected to provide a valuable tool for exploring the effectiveness of policy measures in complex environments (Jager and Mosler, 2007; Jager 2007). Before discussing planning the behavior of organizations and its effects on a complex urban society, we should take a retrospective glance reviewing the current research on ABM with respect to urban social systems.

Agent-Based Modeling

ABM for Planning Support Systems

In the last decades, influenced by rapid urbanization, the relationships between policies, the location and intensity of urban activities, and related urban environmental problems have become a hot topic for planners and researchers (Chin 2002; Ewing 1994, 1997; Neuman 2005). This research is always carried out using statistical analysis or investigated in ways such that the variability of entities' activities and the influences between different entities cannot been represented particularly well. When computer models were first constructed for urban systems, they were built for testing the impacts of urban plans and policies rather than for scientific understanding purposes (Batty 2008). The basic argument was that given a good theory, a model would be constructed based on it, which would then be validated and, if acceptable, used in policy making (Batty 1976). Topical examples can be gleaned from urban growth simulations, in which the spatial process of urban growth can be visualized and represented in a very realistic way using cellular automata (CA) models. These can be used to support decision-making (Batty et al. 1997, 1999; Clarke and Gaydos 1998; Wu 2002; Li and Yeh 2000; Fang et al. 2005).

Now ABM is becoming the dominant paradigm in social simulations due primarily to its priority on reflecting agents' choices in complex systems. Researchers employ ABM for planning support and decision-making on urban policies. One example is a role-playing approach introduced by Ligtenberg, in which a complex spatial system including a multi-actor spatial planning process can be simulated for spatial planning support (Ligtenberg et al. 2010). Furthermore, in regard to the highly complex process of making urban policy decisions, a multi-agent paradigm has been built to develop an intelligent and flexible planning support system, within which three types of agents, including interface agents who improve the user–system interaction, tool agents to support the use and management of models, and domain agents to provide access to specialized knowledge, were created (Saarloos et al. 2008). Researchers also utilize ABM to simulate urban development processes. As described by a CityDve model, the economic activities of agents (e.g., family, industrial firms, and developers) that produce goods by using other goods and trade their goods on the markets have been simulated to visualize urban development processes resulting from urban policies (Semboloni et al. 2004).

In China, one of the countries around the world whose urbanization is taking place at an unprecedented rate, the conflict between human activities and urban environments is very serious. Agent-based simulation has easily gained much attention from Chinese researchers. Some of these researchers have improved traditional urban growth models by building up a set of spatial-temporal land resource allocation rules and developing a dynamic urban expansion model based on a multi-agent system (MAS) to simulate the interactions among different agents, such as residents, farmers, and governments (Zhang et al. 2010). This work is able to reflect basic urban growth characteristics, explain the reasons for the urban growth process and explain the effects of agents' behavior on urban growth. MAS simulations have shown a higher precision than cellular automata models, which suggests that these models could provide land use decision-making support to government and urban planners. Meanwhile, other researchers have focused on solving urban transportation problems using ABM. One example is a qualitative model of a multi-lane environment that has been built to simulate several cars acting in a multi-lane circuit (Claramunt and Jiang 2001). This work is an illustrative example of a constrained frame of reference. The potential of this model is that it was illustrated and calibrated using an agent-based prototype, within which the modeling objects were individual cars.

Residential Motility Simulation

In a study of ABM challenges for geo-spatial simulation, seven challenges for ABM work were illustrated, including the purpose of model building, the independent theory of model rooting-in, and interactions among agents (Crooks et al. 2008). Within this work ABMs have been utilized to model different urban systems problems, such as residential location, urban emergence evaluation, and residential segregation. Similar studies have been carried out on this topic such as a MASUS model (Multi-Agent Simulator for Urban Segregation) which provides a virtual laboratory for exploring the impacts of different contextual mechanisms on the emergence of segregation patterns (Feitosa et al. 2011). A population dynamic model in which inhabitants can change their residential behavior depending on the properties of their neighborhood, neighbors and the whole city has been built (Benenson 1998). A micro-simulation model for residential location choice has been developed, in which the Monte Carlo method was employed to model individual decision rules and an Artificial Neural Network (ANN) theory has been utilized to determine individual location choice (Raju et al. 1998). It is apparent that the principle of ABM has brought numerous researchers into the field of residential mobility simulation. Since residential location has been abundantly simulated, some researchers have begun to consider the environmental influences and landscape changes caused by household residential location choice. A framework called HI-LIFE, to be used for simulating and modeling residential demand for new housing by considering household interactions taking life cycle stages into account was argued for in 2009 (Fontaine and Rounsevell 2009). Within this work household residential location choices have been simulated to predict regional landscape pressure in the future. Furthermore, an ABM framework integrating spatial economic and policy decisions, energy and fuel use, air pollution emissions and assimilation has been developed for urban sustainability assessments (Zellner et al. 2008).

Land and Housing Market Simulations

For researchers to use ABM to simulate land and housing markets is quite usual now. In research by D. C. Parker, a local land market was portrayed as a special conceptual residential market, and the stakeholders and households were agents within it. This research combined traditional deductive optimization models of behavior at the agent level with inductive models of price expectation formation (Parker and Filatova 2008; Filatova et al. 2008). As implemented in this research, households make decisions on their housing behavior by evaluating the house utility and finally determining their willingness to pay for it. Another researcher simulated relocation processes and price setting in an urban housing market through modeling households' decision-making on relocation based on perceptions of housing market probabilities (Ettema 2011). In this study, utility was also an important factor for households' preference evaluations.

Such simulations can be quite helpful for local governments making policies affecting urban housing and land markets or making decisions about related policies. However, there is still little exploration into planning behaviors using ABM approaches.

Simulation of Planning Behaviors in the Planning Process

Limitations of Current Urban Models

We reviewed some typical simulations of urban policy or urban phenomena based on the ABM approach described above. These studies are typically aimed at helping planners or decision makers work out special planning policies. Within these models, urban policies are mostly imported as simulation factors, viewing the policies as preconditions for simulation. These simulations concentrate mainly on representing possible urban changes that could be influenced by policy implementation. However, there is too little information about how a policy making process could be implemented and how would it influence organizations. Thus, our focus has come to be how to support policy makers with practical planning behavior models, through which planning behaviors can be automatically introduced into complex organizational systems evolving in space and time.

The Garbage Can Model

To solve the problem discussed above, we introduce the garbage can model. The garbage can model of organizational planning behavior allows structuring of planning issues so that control is to be investigated, rather than merely imposed externally. It is thus particularly appropriate for investigating planning in organizations. Planning interventions or actions are at least partially substitutable for aspects of organizational design. Both affect the coordination of decisions. Thus to investigate planning, we must be able to manipulate aspects of organizational design and of planning interventions in that design. We first explain the original garbage can model and then introduce planning as an extension of the model.

The original formulation of the garbage can model of organizational choice considers four elements: choices, solutions, problems, and decision makers (Cohen et al. 1972). Choices are situations in which decisions can be made, that is, commitments are made to take certain actions. In organizations, votes to spend money or signatures on forms to hire or fire persons are examples of actions on choices. Solutions are actions that might be taken, such as tax schedules that might be levied or land developments that might be approved. Solutions are things that choices can commit to enact, things we have the capacity to do directly. Problems are issues that are likely to persist and that decision makers are concerned with resolving, such as homelessness, unfair housing practices, congested highways, or flooding. Note that choices enact solutions; they do not solve problems. We cannot merely choose not to have homelessness. We cannot "decide a problem." We can choose to spend money on shelters or to hire social workers, which may or may not affect the persistence of homelessness as a problem. Decision makers are units of capacity to take action in choice situations.

A garbage can is a choice opportunity where the elements meet in a partially unpredictable way. Solutions, problems, and decision makers are thrown into a garbage can and something happens. There is, however, no simple mapping of decision makers to problems or of solutions to problems. Further, an organization has many interacting garbage cans, many interacting choice opportunities. The original model was used to investigate universities as an example of "organized anarchy." Structure as control can be increased from this starting point, however, which makes possible the investigation of a wide range of types and degrees of organizational structure (e.g., Padgett 1980). Planning and organizational design are at least partially substitutable strategies for affecting organizational decision making. Organizational design and planning are both means for "coordinating" related decisions. Thus the garbage can model provides a useful starting point for investigating planning behaviors in organizations.

The major assumption of the models is that streams of the four elements are independent of each other. Solutions may thus occur before the problems these solutions might resolve are recognized. Choice opportunities may occur because regular meetings yield decision maker status, independent of whether solutions are available.

Cohen et al. (1972) reported their results by focusing on four statistics: decision style, problem activity, problem latency, and decision difficulty. The three decision styles were resolution, oversight, and flight. Resolution meant that a choice taken resolved all the problems that were thrown into the garbage can at that choice opportunity. If a decision was taken for a choice to which no problems were attached, it was classified as oversight. All other situations constituted flight. Cohen et al. were able to demonstrate the sensitivity of organizational behavior to various access structures and decision structures.

The decision process was quite sensitive to net energy load. Net energy load is the difference between the total energy required for a problem to be resolved and that available from decision makers. With the general formulation of a decision process considering net energy load, Cohen et al. (1972) ran a simulation addressing four variables: net energy load, access structure, decision structure, and energy distribution. Different net energy loads, roughly analogous to organizational capacity in the form of decision makers relative to organizational demand, should yield differences in organizational behavior and outcomes. Access structure is the relationship between problems and choices. A zero-one matrix defines which problems can be resolved by which choices. Different access structures vary in the number of choices that can resolve particular problems. Decision structure defines which decision makers can address which choices and thus how the total energy capacity of the organization can be brought to bear in resolving choices.

Planning Behaviors in Garbage Can Models

The original garbage can model implies that the organization does not have control over the occurrence of problems and choices. In particular, the organization is not capable of generating choice opportunities to deal with problems that have just arisen in a given time step. The arrival of choice opportunities and the arrival of problems are both random. One way of introducing planning to the model is to allow the organization to purposefully create choice opportunities for resolving problems. This choice–problem dependence is a matter of degree with one extreme being the case of the original garbage can and the other extreme a complete mapping of arriving problems to created choice opportunities. This is equivalent to being able to compare garbage cans and choose one to act in at each time period of the simulation. What effect would the ability to choose among choice opportunities over time so as to match current problems have on the simulation results?

Lai (1998, 2003) ran a prototypical simulation to illustrate this approach. He assumed that, at a given time step, the planner is able to acquire complete information about the structure of the organization, except for the arrival of problems in that time period. The planner knows the decision structure and access structure, and the relationships among the elements. Thus the planner can predict which decision makers and problems will be in which garbage cans (choice opportunities) and how much energy will be accumulated in and spent by decision makers in each one. Choice opportunities in this case are related choices that the planner can select from based on the difference between the energy required to make a decision and the energy available from decision makers. The planning criterion is thus to select the choice opportunity (the garbage can) that results in the smallest energy deficit. Planning thus defined involves choosing the entry times for choices, without considering problems, decision makers, or solutions. Simulation results were sensitive to interventions based on this definition of planning. In the pilot study, such planning resulted in increasing the efficiency with which choices were made, meaning more choices were made with less energy expended, but fewer problems were resolved. Problems, choices, and decision makers tended to remain attached to each other in the case where planning occurred more than in the case without planning.

Lai's work was only tentative because of the small size of the simulations. Also, his scheme is only one way of introducing planning into the original model. Control as structure over other elements could also be considered. A combination of partial controls in experimental design on the four elements might yield planning possibilities that would result in more useful analyses of simulation results. Regardless of these details, Lai (1998, 2003) was able to demonstrate the possibility of gleaning instructive results for understanding planning effects because he showed that it is possible to add structure to decision making without increasing the organization's ability to resolve problems. The result suggests that this simulation modeling approach incorporates sufficient degrees of freedom to discover counterintuitive results.

Incorporation of Spatial Relationships

All planning behaviors have to be conducted within a certain urban space. Thus, there is no doubt that when planners try to make a plan there will be an interaction between the urban area and planning behaviors. How to better represent the spatial process of planning behavior is now a problem. In this session, we first review how spatial processes are explored using CA and ABM approaches in current research reports and then present two examples for discussion on how spatial datasets can be integrated into ABM simulations. We present a possible solution using a Prisoner's dilemma game to simulate spatial evolution in planning behaviors, such as planning in organizations promoted by the Garbage can model.

Spatial Process of Urban Modeling

Urban Modeling and Spatial Processes

In urban modeling, spatial processes can be simulated through automata. There are automata of different types, but simply put, each automaton can be defined as a discrete processing mechanism with internal states. When the state of one automaton has been changed by its own characteristics or through input from outside conditions, such as urban policies, this change will be transmitted to other automata through a predetermined transition rule. Thus, researchers can represent a spatial process by defining the spatial features of automata and the transition rule between them. A typical application of this principle is CA simulation, in which a real urban space is modelled as a cell in the simulation. Each cell has its own spatial characteristics, and urban policies can be input as simulation conditions. Changing conditions will finally result in a cell state change. This type of model has been used in research simulating strategic spatial plans, with cells' spatial attributes including landscape, land-use zoning, slope, urban plan, and land price (Ma et al. 2010). These attributes will determine the state value of a cell, and so, along with the transition rule, determine any changes of state it may undergo.

Spatial processes within ABM can be achieved through the interaction between agents and space. Spatial information for a simulation model can be gained by coupling geographic information systems with the model. Some researchers in this field have argued that a simulation approach named the geographic automata system, in which a MAS can be combined with CA and which takes advantages of GI Science to model complex geographic systems that are comprised of infrastructure features and human objects (Torrens and Benenson 2005). Within geosimulation, the most common implementation of multi-agent models is for the agents to act as objects within a spatial framework (Albrecht 2005; Benenson and Torrens 2004). This approach also can be employed for residential mobility simulation (Torrens 2007).

As most simulations of residential location do, the spatial features of a cell are utilized to calculate the location utility for agents. Thus, the interaction between agents and space can be within the simulation model. As in the research on residential pressure on landscape change we reviewed before, regional space is represented by a regular lattice of grid cells indexed by their geographic coordinates (i, j) in the matrix space $\{I \times J\}$. Each cell is considered to be a homogenous land unit with key spatial information including three groups, *like available properties, land accessibility, and environmental amenities*. Interaction between agents and space takes place as agents choose their new location by evaluating the utility of cells. Household location change will further change the landscape of a cell. Thereby, landscape pressure can be evaluated (Fontaine and Rounsevell 2009).

GIS and Spatial Datasets

The ABM simulations can (e.g., Jiang 2000) be run on a platform combining geographic information systems and ABM or CA, the former being the ArcGIS system of ESRI whereas the latter is an agent-based or CA software such as StarLogo, created by MIT or AUGH, developed by Cecchini (1996). Two examples of such coupling are provided here. In the first example, a land-use change model was constructed with the StarLogo software programmed by MIT coupled with GIS (Lai and Han 2009). The research assessed the probability of development based on economic property right indices, and the probability of possible land uses allowed in the zoning system was embedded in the simulation rules. Finally, the research used parameters developed over 100 generations of a genetic algorithm method to calibrate the simulation model. The main results from this research are as follows: (1) Using the parameters gained from the genetic algorithm method, the model was indeed able to simulate, at least partially, the pattern of land uses for the Taipei metropolitan area. (2) The zoning system in the simulated area does influence the appearance and pattern of land uses. It limits the development of industrial land use and affects the fractal pattern of commercial land use. (3) After comparing the spatial patterns of simulation results and conducting one-way ANOVA analysis, it can be concluded that zoning affects specific locations, but not the fractal pattern of land uses.

Figure 4.1 shows the logic of how the use type of a particular parcel (cell) is determined. Note that the use types are checked against zoning regulations in which mixed uses are allowed. Figure 4.2 is a sample illustration of the simulation and Fig. 4.3 is the interface.

In the second example, research was conducted, grounded on a microscopic simulation approach to studying how decisions made locally give rise to global patterns (Lai and Chen 1999). CA provide the simplest bottom-up way to study discrete systems and complex urban spatial systems. Based on the coupling of a CA model and GIS of a small town in central Taiwan, Minjian Township, the research focused on an agent-based simulation approach to considering land-use and transportation networks as two traits of the evolution of complex urban spatial systems.

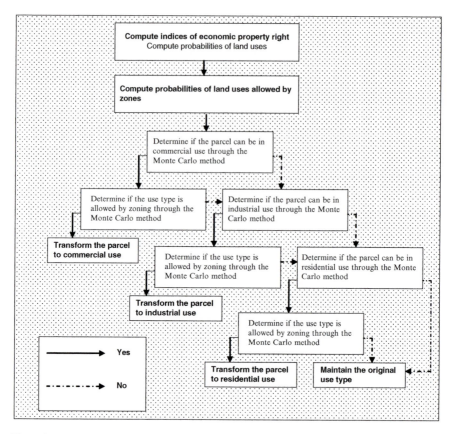

Fig. 4.1 Logic of computing and determining land-use transitions

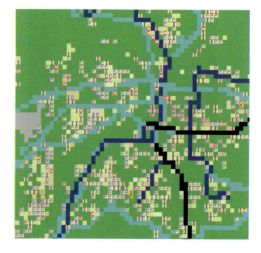

Fig. 4.2 Sample simulation plot (Black = road; blue = transit line; I = industrial; C = commercial; R = residential; E = vacant)

4 Planning in Complex Spatial and Temporal Systems: A Simulation Framework   83

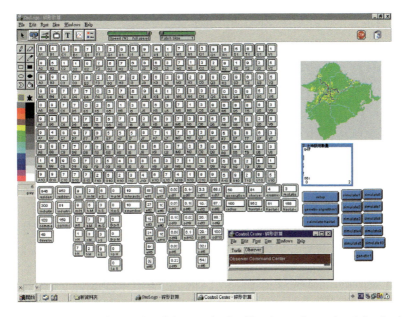

Fig. 4.3 Simulation interface. Rules of the genetic algorithm (*upper buttons*) and the simulated maps and necessary data (*lower buttons*) are displayed

The simulation views land development decisions as simple rules characterized by degrees of complexity and diversity. Three factors, the numbers of transition rules (N), the diversity of transition rules (D), and the numbers classifications of the transition rules (n), are derived from theories of measures of complexity and evolution in general system theory (GST). Simulating systems behavior based on the transition rules classified and sorting out the results by the three factors shows that when the diversity of rules increases, the urban structure will grow in a complex, fractal way. Figure 4.4 shows the simulation framework and two illustrations of the simulation are given in Fig. 4.5. Note that the simulation was run on the AUGH platform (Cecchini 1996). Both rules of the simulation in Fig. 4.5 are for high diversity, measured by and derived from different levels of complexity.

Space Evolution in Planning Behavior

Prisoner's Dilemma Game in Space

Planning in the context of urban development, both physical and social, must acknowledge the significance of spatial effects of association and competition. Recent work on evolution of behavior, characterized as games in space, provides one starting point for incorporating space in simulations (Nowak 2006). Here, we first present the model of Nowak and May (1993) and then introduce planning to the

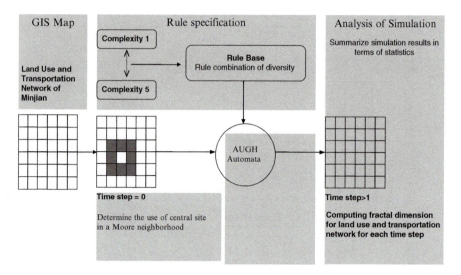

Fig. 4.4 Simulation framework (complexity 1 through 5 represent degrees of complexity as measured by fractals)

Fig. 4.5 Coupling of a CA model with GIS to explore land use and transportation interaction (yellow = residential; red = commercial; gray = road; blue = river)

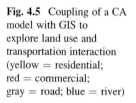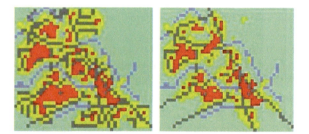

model. Allen and Sanglier (1981) considered some aspects of urban spatial evolution in a similar framework.

Nowak and May (1993) investigated the spatial evolution of a set of actors in a square lattice as actors in a sequence of prisoner's dilemma games. The prisoner's dilemma game presents each player with two options: cooperate or defect. The payoffs are determined by the combination of plays such that the values of the payoffs for player one decrease in the following order $DC > CC > DD > CD$ where C signifies cooperate and D signifies defect. The first strategy is the action of player one, and the second is the action of player two. Player two faces a symmetric situation. The dilemma is that it is in each player's individual interest to defect regardless of the action of the other, but in doing so they both end up worse off, since DD has a lower payoff for each player than CC. Interaction among players (agents) in a spatial configuration based on this simple decision rule generates

4 Planning in Complex Spatial and Temporal Systems: A Simulation Framework

complex spatial patterns given different relative payoff levels. The effects of spatial configurations can be investigated by comparing the results of non-spatial sequences of prisoner dilemma games.

Arthur (1989, 1994) interpreted a similar sequence in terms of increasing returns to market share. He showed that if the payoff for adopting a strategy increases with the number of agents adopting the same strategy, it is impossible to predict the eventual evolutionary outcome of the resulting trajectory of market share. One typical illustration of this phenomenon is the adoption of particular computer software packages. The payoff increases with the number of other adopters because of the greater likelihood of additional compatible packages, knowledgeable users, and continuing upgrades. Similarly, consumers seem to have made choices about Betamax versus VHS in video formats at least partially on the basis of likelihood of available videotapes to play rather than on picture quality.

Planning in a Spatial Evolution Game

The payoff in the prisoner's dilemma game will then vary depending on the number of agents in the "neighborhood" choosing a particular option: cooperate or defect. This combination can be described using the fractal concept of space (see Mandelbrot 1983; Batty and Longley 1994; Batty 2005). This approach allows simulations to characterize spatial relations across continuous dimensions, which can represent a richer variety of urban geographic relationships or organizational structures more effectively than Euclidean space.

Consider a continuum of space-filling agents residing in a space of fractal dimension who act based on the payoff in the prisoner's dilemma game and the principles of garbage can simulation. Each agent makes one of two choices, defect or cooperate, as in the usual prisoner's dilemma game. The payoffs, however, are not fixed, but depend on the numbers of these choices adopted in the system. The payoffs for player one are depicted below, where (p,q) denotes the initial payoff or preference for choice p made by an agent interacting with another agent making choice q; r is the rate of change of return relative to the number of agents making a particular choice; and $n(p)$ is the total number of agents choosing choice p. The rate r can be positive, negative, or zero, representing increasing, decreasing, or constant rates of return, respectively. Note that $a(d,c) > a(c,c) > a(d,d) > a(c,d)$ (Table 4.1).

The agent chooses for the next time step so that it yields the maximum payoff based on the choices among its neighbors at the current time step. The neighbors are the agents, including the agent under consideration, located within a radius R from

Table 4.1 Payoff matrix

Player 1/Player 2	Cooperate	Defect
Cooperate	$a(c,c) + rn(c)$	$a(c,d) + rn(c)$
Defect	$a(d,c) + rn(d)$	$a(d,d) + rn(d)$

the site where the agent is located in fractal space. The agent can also make a choice by selecting the maximum payoff among the agents located within its "neighborhood" or the square lattice over time period T. R and T are thus indicative of planning investment in space and in time according to our definition. That is, they denote the scope of related choices considered in space and time, respectively. Let M (R,k) be agents standing for the mass of the fractal sub-space with radius R and center k. We have M (R,k) = uR(D) where u is a uniform density and D is a fractal dimension (Mandelbrot 1983). The total payoff for an agent j located at the center of M(R,j) is the sum of the payoffs for that agent interacting with all agents i in M(R,j), including j itself. That is,

$$P(j, t, R, T) = \sum \sum [(a(c(j, t), c(i, t)) + rn(c(j, t))]$$

where P(j,t,R,T) = the cumulative payoff function for j over time period T at time t within M(R,j), and c(j,t) = the choice made by j at time t.

The first summation is over t of elements of the set T and the second is over i elements of the set M(R,j). The decision rule for any agent k at time t + 1 is to adopt the choice made by the agents in k's neighborhood that yields the maximum payoff. That is,

$$c(k, t + 1) = c(j \text{ I Max } P(j, t, R, T)), \text{ for } j \text{ is an agent of } M(R, k).$$

This form of simulation of spatially structured behavior provides a basis for incorporating space into a simulation model similar to the garbage can model discussed above.

Idea for Integrating the Garbage Can Model with a Spatial Evolution Game for Planning Simulation

To incorporate the garbage can model into a spatial model, consider a continuum of decision makers (agents) in a fractal space. There are finite numbers of problems and choices. Define a decision structure of relationships among decision makers and choices, an access structure of relationships between problems and choices, and a solution structure between solutions and problems. These zero-one matrices of relationships have the same meaning and range of forms as in the original garbage can model and are givens external to each simulation run. These structures can be varied as described earlier to discover their effects on choice-making behavior.

The initial payoffs for all decision makers are the same, but these payoffs vary with respect to two variables in the simulation. The first variable is the number of agents that adopted that choice in the particular time step of the simulation. The second variable is the problems associated with that choice at that time. Because the problems arrive in a random sequence, the payoffs are subject to random

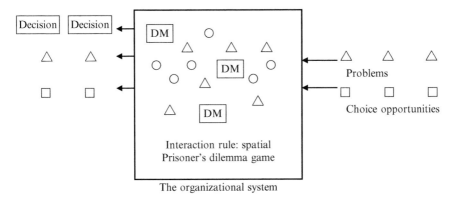

Fig. 4.6 Relationship between the garbage can model and the prisoner's dilemma game model (*DM* decision maker)

fluctuations and the evolution is not deterministic. The payoff table is therefore different from that of the prisoner's dilemma game because in the spatial version there is interaction among the agents involved. The decision rule for an agent adopting a particular choice is the same as that in our spatial model: the best choice is the one that yields the maximum payoff considering the choices among the agent's neighbors in space and time.

This spatial version of the prisoner's dilemma model can then be used to consider the effects of space on a garbage can model that also incorporates planning behaviors as suggested in Fig. 4.6. We have not yet run such simulations in a structured way so as to test the sensitivity of this model to spatial scope or temporal consideration of related choices. We can identify, however, the types of questions that might be addressed. Different types of organizational structures, from strictly hierarchical to "matrix" structures could be considered as partial substitutes for planning intervention. For example, does a hierarchical organizational structure benefit less from planning than a "matrix" organization? Does planning that is focused on considering more related choices (that is more garbage cans) yield more problem resolutions than planning focused on generating solutions for fewer choices (garbage cans)? What differences arise from increasing the size of the neighborhood in space relative to increasing the size of the neighborhood in time? A spatial version of the Garbage can model has been provided (Lai 2006), and the simulation framework suggested here can be considered as a sequel to that model.

Conclusions

We have proposed modified versions of two previously proposed simulation models to allow consideration of the effects of planning in complex, spatial, temporal organizational systems. We have extended the garbage can model of Cohen,

March, and Olsen so as to consider a particular definition of planning behavior. Recent simulation runs suggest that the revised model is sensitive to these planning interventions. We have also proposed a revised version of the prisoner's dilemma spatial game taking into account space and planning. In particular we have considered increasing returns, planning investments, and fractal space. Such simulations can be coupled with GIS to yield policy implications for real world situations. The major work of running structured sets of simulations so as to discover and elucidate systemic principles remains.

Simulations of this type are of interest because of the abstract form of questions that can be considered. The intent is not to simulate concrete, specific cases, but to understand the functioning of systems. The simulation result is encouraging in that it implies planning interventions might increase the efficiency of choice making without increasing the number of problems resolved. This suggests that useful, counterintuitive properties might be discovered. Such systemic understanding must then be interpreted in concrete terms for organizational behavior.

References

Albrecht, J. (2005). "A new age for geosimulation," Transactions in GIS. 9(4): 451–454

Allen, P. M. and M. Sanglier (1981). "Urban evolution, self-organization, and decisionmaking," Environment and Planning A. 13: 167–183.

Arthur, W. B. (1994). Increasing returns and path dependence in the economy. Ann Arbor: The University of Michigan Press.

Arthur, W. B. (1989). "Competing technologies, increasing returns, and lock-in by historical events," The Economic Journal. 99: 116–131.

Batty, M. (1976). Urban modelling: Algorithms, calibrations, predictions. Cambridge, UK: Cambridge University Press.

Batty, M. (2005). Cities and complexity: understanding cities with cellular automata, agent-based models, and fractals. Cambridge, Massachusetts: The MIT Press.

Batty, M. (2008). Fifty years of urban modelling: macro statics to micro dynamics. In S. Albeverio, D. Andrey, P. Giordano and A. Vancheri, Editors, The dynamics of complex urban systems: an interdisciplinary approach. Physica-Verlag, Heidelberg, DE, 1–20.

Batty, M., H. Couclelis and M. Eichen (1997). "Urban Systems as Cellular Automata," Environment and Planning B: Planning and Design. 24(2): 159–164.

Batty, M. and P. Longley (1994). Fractal Cities. New York: Academic Press.

Batty, M., Y. Xie and Z. Sun (1999). "Modeling urban dynamics through GIS-based cellular automata," Computers, Environment and Urban Systems, 23(3): 205–233.

Benenson, I. (1998). "Multi-agent simulations of residential dynamics in the city," Computers, Environment and Urban Systems. 22(1): 25–42.

Benenson, I and P. M. Torrens (2004). "Geosimulation: object-based modelling of urban phenomena," Computers Environment & Urban Systems. 28: 1–8.

Cecchini, A. (1996). "Urban modeling by means of cellular automata: generalized urban automata with the help on-line (AUGH) model," Environment and Planning B. 23:721–732.

Chin, N. (2002). "Unearthing the roots of urban sprawl: A critical analysis of form function and methodology," Centre for Advanced Spatial Analysis (CASA) working paper series. London: University College London.

Claramunt, C. and B. Jiang (2001) "A qualitative model for the simulation of traffic behaviours in a multi-lane environment," Geographical Sciences. 11 (1): 29–42.

Clarke, K. C. and L. J. Gaydos (1998). "Loose-coupling of a cellular automaton model and GIS: long-term growth prediction for the San Francisco and Washington/Baltimore," International Journal of Geographical Information Science, 12(7): 699–714.

Cohen, M. D., J. G. March and J. P. Olsen (1972). "A garbage can model of organizational choice," Administrative Science Quarterly. 17(1):1–25.

Crooks, A., C. Castle and M. Batty (2008). "Key challenges in agent-based modelling for geospatial simulation," Computers, Environment and Urban Systems. 32(6): 417–430.

Ettema, D. (2011). "A multi-agent model of urban processes: modelling relocation processes and price setting in housing markets," Computers, Environment and Urban Systems. 35(1): 1–11.

Ewing, R. H., (1994). "Characteristics, causes and effects of sprawl: a literature review," Environmental and Urban Issues. 21(2): 1–15.

Ewing, R. H. (1997). Is Los Angeles-style sprawl desirable? Journal of American Planning Association. 63(1): 107–126.

Fang, S., G. Gertner, Z. Sun and A. Anderson (2005). "The impact of interactions in spatial simulation of the dynamics of urban sprawl," Landscape and Urban Planning. 73(4): 294–306.

Feitosa, F. F., Q. B. Le and P. L. G. Vlek (2011). "Multi-agent simulator for urban segregation (MASUS): A tool to explore alternatives for promoting inclusive cities," Computers, Environment and Urban Systems. 35(2): 104–115.

Ferdinando Semboloni et al., (2004) "CityDev, an interactive multi-agents urban model on the web," Computers, Environment and Urban Systems, 28(1–2): 45–64.

Filatova, T., D. C. Parker and V. D. A. Veen (2008). "Agent-based urban land markets: agent's pricing behavior, land prices and urban land use change," Journal of Artificial Societies and Social Simulation. 12(13), <http://jasss.soc.surrey.ac.uk/12/1/3.html>

Fontaine, C. M. and M. D. A.Rounsevell (2009). "An agent-based approach to model future residential pressure on a regional landscape," Landscape Ecology. 24: 1237–1254.

Friend, J. and A. Hickling (2005). Planning under pressure: the strategic choice approach. New York: Elsevier Butterworth Heinemann.

Hopkins, L. D. (2001). Urban development: the logic of making plans. London: Island Press.

Jager, W. and H. J. Mosler, (2007). "Simulating human behavior for understanding and managing environmental resource use," Social Issues. 63 (1): 97–116.

Jagera, W. (2007) "The four P's in social simulation, a perspective on how marketing could benefit from the use of social simulation," Journal of Business Research, 60(8): 868–875.

Jiang, B. (2000). "Agent-based approach to modeling environmental and urban systems within GIS," Proceedings of the 9th International Symposium on Spatial Data Handling. 1–12.

K A. Raju, P.K Sikdar and S.L Dhingra, (1998) "Micro-simulation of residential location choice and its variation," Environment and Urban Systems, 22(3): 203–218

Lai, S.-K. (1998). "From organized anarchy to controlled structure: effects of planning on the garbage-can processes," Environment and Planning B: Planning and Design. 25: 85–102.

Lai, S.-K. (2003). "Effects of planning on the garbage-can decision processes: a reformulation and extension," Environment and Planning B: Planning and Design. 30: 379–389.

Lai, Shih-Kung (2006). "A spatial garbage-can model," Environment and Planning B: Planning and Design, 33 (1): 141–156.

Lai, Shih-Kung and Chen, Yo-Jen (1999). "A dynamic exploration into land uses and transportation networks interaction based on cellular automata simulations," paper presented at the 1999 International Symposium of City Planning, Tainan, Taiwan.

Lai, Shih-Kung and Haoying Han (2009). Complexity: New Perspectives on Urban Planning. Beijing: China Architecture and Building Press.

Li,X.,Yeh,A.G.O., (2000) "Modeling sustainable urban development by the integration of constrained cellular automata and GIS," International Journal of Geographical Information Science, 14(2): 131–152.

Ligtenberg A., et al., (2010) "Validation of an agent-based model for spatial planning: A role-playing approach," Computers, Environment and Urban Systems, 34(5):424–434.

Mandelbrot, B. B. (1983). The Fractal Geometry of Nature. New York: W.H. Freeman.

Ma Y., et al., (2010) "Urban Growth Simulation for Spatial Strategic Plan of Chuangdong Area, China," 18th International Conference on GeoInformatics, pp 1–6.

Neuman, M., 2005, The compact city fallacy, Planning Education and Research 25(1), pp. 11–26.

Nowak, M.A. and R. M. May (1993) "The spatial dilemmas of evolution," International Journal of Bifurcation and Chaos. 3(1): 35–78.

Nowak, M. A. (2006). Evolutionary Dynamics: Exploring the Equations of Life. Cambridge, Massachusetts: The Belknap Press of Harvard University Press

Padgett, J. F. (1980). "Managing garbage can hierarchies," Administrative Science Quarterly, 25(4): 583–604.

Parker D. C. and Filatova T., (2008) "A conceptual design for a bilateral agent-based and market with heterogeneous economic agents", Computers, Environment and Urban Systems, 32(6):454–463.

Saarloos D. J.M., Arentze T. A., Borgers A. W.J. and Timmermans H. J.P., (2008) "A multi-agent paradigm as structuring principle for planning support systems," Computers, Environment and Urban Systems, 32(1): 29–40.

Schaeffer, P. V. and L. D. Hopkins (1987). "Planning behavior: The economics of information and land development," *Environment and Planning A*, 19: 1211–1232.

Torrens P. M., (2007) "A Geographic Automata Model of Residential Mobility," Environment and Planning B: Planning and Design, 34: 200–222

Torrens P. M. and Beneson I., (2005) "Geographic Automata Systems," International Journal of Geographical Information Science, 19(4): 385–412

Wu, F., (2002) "Calibration of stochastic cellular automata: the application to rural–urban land conversions," International Journal of Geograghic Information Science, 16(8): 795–818.

Zellner M. L. et al., (2008) "A new framework for urban sustainability assessments: Linking complexity, information and policy," Computers, Environment and Urban Systems, 32(6): 474–488

Zhang H., Zeng Y.N., Ling B., Yu X.J., (2010) "Modelling urban expansion using a multi agent-based model in the city of Changsha," Geographical Sciences, 20(4): 540–556.

Chapter 5
Reaching Consensus Among Stakeholders on Planned Urban Form Using Constrained CA

Ying Long and Zhenjiang Shen

Introduction

In this chapter, we propose an approach for negotiation between planners and developers by identifying the required spatial policies for predefined alternative plans. Recently, the constrained cellular automata (CA) method has been extensively applied to simulate urban growth for predicting future urban forms (Wu 1998; Ward et al. 2000; Yeh and Li 2001; He et al. 2008). However, an alternative plan, which is prepared by planners during the plan compilation process, has been used in the decision-making process of the urban master plan worldwide. In China, practical urban growth often deviates from the planned urban form, resulting in the planned urban form being broken. Local planning authorities who lack policy guidance have little knowledge of the exact policies related to the planned form and their differences with current policies. Given this situation, urban policies required for the planned urban form are a governmental concern. If the simulated urban form using constrained CA can be accepted as a reasonable form by planners, then the simulated results can be possibly used for the deliberation in the local planning committee among planners, developers, and other stakeholders.

Urban form must be given ample attention because it guarantees sustainable development. Inappropriate urban forms such as sprawled urban forms may have negative effects on society, including overconsumption of land resources (Kahn 2000; Landis 1995; Johnson 2001); increased commuting distances and traffic jams (Newman 1989; Kahn 2000; Ewing et al. 2002); decreased availability of affordable housing (Danielsen et al. 1999); increased urban infrastructure construction costs (Speir and Stephenson 2002); reduced water supply (Otto et al. 2002);

Y. Long (✉)
Beijing Institute of City Planning, School of Architecture, Tsinghua University, Beijing, China
e-mail: longying1980@gmail.com

Z. Shen
School of Environmental Design, Kanazawa University, Kanazawa, Japan

Z. Shen, *Geospatial Techniques in Urban Planning*, Advances in Geographic
Information Science, DOI 10.1007/978-3-642-13559-0_5,
© Springer-Verlag Berlin Heidelberg 2012

poor neighborhood interaction (Ewing 1997; Freeman 2001); poor public health (Ewing et al. 2003); and increased social inequity (Bullard et al. 1999). The policy parameter regulates and controls the impact of the corresponding spatial constraint on urban growth. Long et al. (2009) proposed the term "form scenario analysis" to identify optimal policy parameters through a logistic regression-based approach for four planned forms in the Beijing Metropolitan Area using constrained CA. However, the optimal solution of policy parameters for a predefined alternative plan is not unique. Therefore, in this chapter, we focus on all the policy parameters required to satisfy an alternative plan.

The simulated urban form using constrained CA is only reasonable from the viewpoint of history. All parameters in the transition rule of CA simulation are estimated using a historical dataset, which can be recognized as an optimal solution for simulating the future urban form. However, if the historical trend cannot be replicated in the future, then the estimated parameters cannot be used to simulate the future urban form. In this sense, if the future development is considered to be different from the historical trend, then the estimated parameters will not be useful. Thus, knowing how to define parameter sets becomes a precondition for simulation.

In planning practice, planners should consider all possible factors for drawing an urban form. These factors are actually proposed and prepared in different sections of the local government. If the planners have different opinions on the future urban form, then there will be numerically possible parameters for those factors. Defining the parameters of a form is possible via negotiation among planners, developers, and other stakeholders. Thus, this chapter focuses on how to retrieve all parameters for a future form through negotiation, supposing that planners opt to use CA simulation. This remains unexplored by existing research.

A number of studies have investigated the impact of policy parameters on the urban form using a scenario analysis approach in traditional land-use models, such as the California Urban Future model (CUF) (Landis and Zhang 1998a, b), *What If?* (Klosterman 1999), and CA-based models (Li and Yeh 2000; Yeh and Li 2001; Long et al. 2008). In previous studies, less attention has been given to identify the types of spatial policies that are available for the planned alternative. In essence, the impact of constraints on the possible urban form has yet to be investigated, and research into the impact of constraint parameters on urban form will be useful for investigating whether an alternative plan can be achieved. Therefore, identifying appropriate policy parameters for the alternative plan is necessary, which can be done via negotiation between planners and stakeholders.

We describe a methodology for the policy parameter identification (PPI) process in Sect. "Approach". The resulting urban forms from various constraints are illustrated in Sect. "How to Present the Negotiation Process Among Stakeholders Using CA", in which the impact of constraints on the resulting urban form is evaluated. The PPI method will be tested in a virtual city in Sect. "Presenting the Negotiation in the Virtual City", and we discuss the PPI method and present some concluding remarks in Sect. "Conclusion".

Approach

We assume that planners are making an urban master plan in a city, for which they have prepared an urban form based on their preferences. On the other hand, planners in the environmental department have established some environmental conservation areas, and some developers are considering the locations of their development projects. Thus, they need to cooperate with each other when making an urban form that meets their distinct requirements. The CA method is employed to combine their requirements by simulating the spatiotemporal dynamic urban growth process.

First, we retrieve all the parameters that match an urban form as suggested by an urban planner. A CA approach is then used as follows:

$$\{X^*|F = f(X,A), X^* = X_{\max(\text{GOF}(F_h,F))}\}, \tag{5.1}$$

where f is the transition rule of the constrained CA; A is spatial constraints; X is the intensity of spatial policy on spatial constraints; F_h is the planned urban form; F is the simulated urban form using A and X as the parameters of constrained CA; and X^* is the estimated policy parameter after a calibration process with the maximum goodness-of-fit (GOF) between F_h and F. Each simulated urban form generated by constrained CA is compared to the predefined urban form using the *KAPPA* as a GOF indicator. A benchmark is required to evaluate the degree of matching between the simulated and predefined urban forms. We set an 80% benchmark in this chapter based on our experiences, that is, two urban forms can be regarded as being quite similar when their *KAPPA* value approaches 80%. Consequently, all the sampled parameter sets can be classified into two clusters, namely, the valid cluster and invalid cluster for a predefined alternative plan.

In this chapter, we can generate a valid cluster that includes all possible solutions for the policy parameters through the following steps.

1. Define the initial distribution of each policy parameter x in X. To reflect the proportional relationship among policy parameters, each x is assumed to be uniformly log-distributed from 10^{-3} to 10^3.
2. A sample policy parameter x is randomly drawn from its parent probability distribution.
3. The constrained CA model is run with x to obtain a simulated urban form F.
4. The *KAPPA* of the simulated urban form F and its predefined form F_d is calculated, and F is evaluated as being valid or invalid.
5. Repeat steps 3 through 5 until *KAPPA* reaches convergence.
6. Classify F and x into F_B and x_B and F_{NB} and x_{NB} based on the validation results, where x_B is the identified policy parameter for the predefined urban form F_d.
7. Assess the accepted parameter vector x_B to quantify the parameter uncertainty.
8. Check the significance s to determine whether the two distributions of x_B and x_{NB} are distinctly separated by the K-S (Kolmogorov–Smirnov) test using its Z value.
9. Rank the parameters according to significance s to assess the global sensitivity of each parameter.

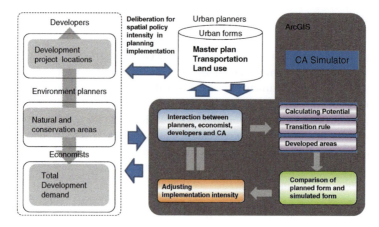

Fig. 5.1 Framework of our approach using the CA simulator

Second, we need to let the planners and stakeholders select their parameters from the identified solution space in the previous process (steps 1–9). In the present work, we define the negotiation process among stakeholders as the adjustment of parameters in CA simulation (Fig. 5.1), which is a negotiation or cooperation process used to reach a consensus on the planned urban form among stakeholders. The planned urban form is assumed to be a goal expected to be realized in the planning period. For implementing the planned form, various stakeholders, including urban planners, developers, economists, and environment planners in the local government, can ask for their respective policy intensities to be strengthened or weakened by comparing the planned and simulated urban forms.

We only investigate the possibility of solutions of the different parameters of spatial constraints for negotiation among stakeholders; the practical negotiation process among them is not discussed in this chapter. In the next section, we describe a virtual city and the corresponding spatial constraints for retrieving all solutions of the policy parameter sets that match the planned urban form.

How to Present the Negotiation Process among Stakeholders Using CA

Planned Urban Form and Spatial Constraints in a Virtual City

The virtual city has 400 square cells, where the size of each cell is 100 m × 100 m (Fig. 5.2). The virtual city is 2 km wide and 2 km long. Three cells are developed in the initial urban form of this virtual city, and we expect to develop 150 cells in the future. For the constraints in the virtual city, we consider two point constraints: PP and NP; two line constraints: PL and NL; and two polygon constraints: PG and NG. Table 5.1 lists these constraints in the virtual city together with their detailed descriptions and

5 Reaching Consensus Among Stakeholders on Planned Urban Form

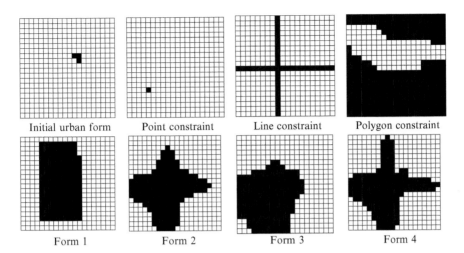

Fig. 5.2 Predefined urban forms in the virtual city

spatial distributions. Each constraint has a corresponding spatial distribution, which largely originates from natural land use, planning, or policy files (e.g., eco-protected zones) or the results of spatial analysis (e.g., proximity of road networks). Table 5.1 shows that these constraints have positive or negative effects (promoting or restricting urban growth) on the future urban form, thus shaping the future urban form due to dynamic interactions with the neighbor effect in the simulation process.

We set four predefined urban forms (Forms 1–4, Fig. 5.2), which are intended as alternative land-use plans in the virtual city. The four urban forms, which vary greatly from each other, will be used as various future images in the virtual city for urban growth simulation with single and multiple constraints.

Policy Intensity for Spatial Constraints Reflecting Negotiation Among Various Stakeholders

Urban form is defined as the geometric shape of the built-up urban space. The urban form at time $t + 1$ can be generated based on the urban form at time t, as follows:

$$F^{t+1} = F^t \bigcup \Omega(P^t | P^t \geq P_{Threshold}), \qquad (5.2)$$

where F^{t+1} is the urban form at time $t + 1$; F^t is the urban form at time t; Ω is the entire region; P^t is the probability of development in the region at time t; and $P_{Threshold}$ is the development threshold (0–1) as the lowest allowed development suitability above which urban development will occur. $\Omega(P^t | P^t \geq P_{Threshold})$ stands for the region in which the urban development probability is greater than or equal to the development threshold.

Table 5.1 Constraints in the virtual city

	Type	Name	Description	Stakeholder(s)
Global	Total number of developed cells	TNDC	Total urban development demand	Economist
Point	Positive	PP	Promoting urban growth, such as town centers, subway stations, and public service centers	Private or public developers
	Negative	NP	Restricting urban growth, such as geological disaster sites, pollution sources, solid waste/wastewater treatment facilities, and other not-in-my-backyard facilities	Environment planner
Line	Positive	PL	Promoting urban growth, such as road networks	Transportation planner
	Negative	NL	Restricting urban growth, such as high-voltage power lines, seismological fault lines, and administrative division boundaries	
Polygon	Positive	PG	Promoting urban growth, such as officially planned areas and special policy zones	Land use zoning planner
	Negative	NG	Restricting urban growth, such as agricultural land, ecologically protected areas, and steep areas	Environment planner

In CA simulation, the urban form generally results from multiple constraints rather than a single constraint (Wu and Webster 1998; Li and Yeh 2002; Liu et al. 2008). A multi-criteria evaluation (MCE) approach can be applied for identifying the influence of each constraint. In the MCE, the weight of a constraint is used to estimate the effect of constraint on the urban form. The status transition rule for the proposed constrained CA for urban growth simulation is expressed as follows:

$$s^t = \sum_{k=1}^{m+n} x_k{}^* a_k + x_N{}^* a_N^t$$

$$P_g{}^t = \frac{1}{1 + e^{-s^t}}$$

$$P^t = \exp\left[\alpha\left(\frac{P_g{}^t}{P^t{}_{g\max}} - 1\right)\right]$$

$$F^{t+1} = F^t \bigcup \Omega(P^t | P^t \geq P_{Threshold}), \tag{5.3}$$

where x_k is the policy parameter of constraint a_k (x_k can be regarded as the policy impacts on constraint a_k); $a_N{}^t$, which ranges from 0 to 1, is the number of developed cells within the neighborhood of the cell, excluding the cell itself, divided by 8 (we used the Moore neighborhood with a 3×3 cell configuration for the constrained CA); x_N is the parameter for the neighbor constraint that is set to 1 and remains static across repeated simulations (because x_k is assumed to be log-uniformly distributed to reflect the proportional relationship between the constraint parameters in the parameter identification process); s^t is the development suitability of the cells at time t; $P_g{}^t$ is the development potential of the cells at time t; $P_{g\max}^t$ is the maximum $P_g{}^t$ across the entire lattice space at iteration t; and α is the dispersion parameter, which ranges from 1 to 10, indicating a rigid level of development that regulates urban growth speed. The value of α is herein set to 2. In addition, the positive and negative point and line constraints are assigned effect values using $a_k = e^{-\lambda*dist^a_k{}^+}$ and $a_k = 1 - e^{-\lambda*dist^a_k{}^-}$ for the positive and negative constraints, respectively, to normalize the constraint values to range from 0 to 1. For the positive polygon constraint, the value of the cell is set to 1 within the constraint and 0 outside the constraint. For the negative polygon constraint, the value of the cell is set to 0 within the constraint and 1 outside the constraint.

Presenting the Negotiation in the Virtual City

Concept of Negotiation While Considering the Neighbor Effect

The negotiation process in the planned urban form among various stakeholders (Fig. 5.3) includes the following: (1) the economist sets the total development demand as the total number of developed cells (TNDC); (2) planners argue for spatial policy intensity in planning implementation, such as the planning policy on road networks and environmental conservation areas that are defined as certain spatial constraints; (3) developers argue for the development policy on their selected locations, which can be defined as a point constraint. Therefore, we consider that changing policy parameters on spatial constraints in simulation can reflect negotiation among stakeholders who are concerned with the different kinds of spatial constraints.

In the following context, we focus on the relationship between spatial constraints and their parameters to gain some insights on the negotiation among stakeholders. Given that the neighbor effect is an important factor in CA simulation, we cannot check the simulation results without considering the neighbor effects. In the virtual city, we conducted two types of experiments: one with no constraints and the other that considers single constraints. In the no-constraint scenario, the development suitability in (5.3) is expressed as $s^t = x_N{}^*a_N{}^t = a_N{}^t$ (as described in (5.3), where x_N is set to 1). See Fig. 5.4 for the results of the no-constraint scenario.

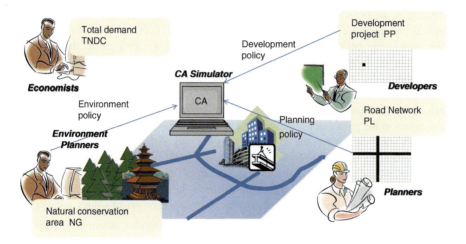

Fig. 5.3 Negotiation process among stakeholders in CA

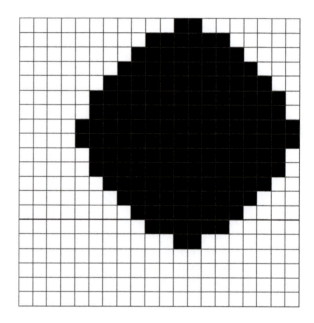

Fig. 5.4 Simulated urban form with no constraints in the virtual city

For a single constraint, the suitability obtained is calculated using $s^t = \beta^* a_1 + a_N{}^t$, where a_1 is the single constraint. We herein consider the constraint PP as the single constraint. The proportion between the constraint and neighbor parameter is a key factor that influences the simulated form. Therefore, the proportional relation between the single constraint parameter and the neighbor parameter is employed using $\beta = x/x_N = x$, where x is the single constraint parameter, x_N is the parameter of the neighbor effect (set to 1), and β is the reflection of the effect of the single constraint compared with the neighbor effect. To identify the effect of the

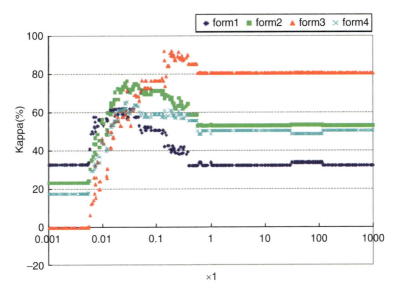

Fig. 5.5 Scatter plots of *KAPPA* and β for constraint PP in the virtual city

constraint on the simulated urban form, we set β to 10^{-3}, 10^{-2}, 10^{-1}, 10^0, 10^1, 10^2, and 10^3, respectively, thereby obeying the log-uniform distribution as described in the approach section.

Figure 5.5 shows whether the urban forms simulated using the single constraint PP can be fitted to the four predefined forms in terms of the *KAPPA* indicator. Form 3 is the only form that can be achieved by the single constraint PP when its β value ranges from 0.1 to 0.6, whereas the other three forms cannot be achieved with the single constraint PP, regardless of the β value. For β values less than 0.01 or greater than 1, a relatively steady *KAPPA* value will be achieved for all predefined forms. Thus, *KAPPA* is sensitive when β is between 0.01 and 1, where the effect of the constraint becomes larger and closer to the neighbor effect. When β is greater than 1, then the constraint effect becomes the dominant factor during simulations.

One Planning Policy for Single Spatial Constraint

We ran six single-constraint tests for the six constraints using constrained CA, respectively. In the constrained CA with a single constraint, the simulation process runs while considering the dynamic interactions between the neighbor effect and single constraint. Table 5.2 shows the simulated results of various experiments using constrained CA ($P_{Threshold}$ was set at 0.999) with different β values. Based on the simulated urban forms, we find that single constraints can change the simulated urban form and that β has a strong effect on the simulated urban form.

We can conclude that different spatial constraints have different significant impacts on urban growth. Table 5.2 shows that we only checked the one-policy parameter on one spatial constraint, which highlights the importance of checking the interaction among multi-policy parameters on different spatial constraints. We then assume that the one planning policy represents one planner. Thereby, if more than two policy parameters are taken into account, then there should be a negotiation among the planners who will argue for policy intensity in planning practice. When a consensus is reached on an alternative plan during the plan compilation process, decision makers would find it helpful once they understood the potential effects of the policies on the planned urban form during the planning implementation process.

Two Planning Policies for Two Spatial Constraints

In this subsection, we test the effects of two combined spatial constraints by adjusting the policy parameters of each constraint simultaneously to demonstrate how the parameters of the combined constraints are retrieved. We use two constraints, namely, the positive point constraint (PP) and the positive line constraint (PL), in computer simulations. PP can be a location of a development project (as a starting project in a region) and PL can be a road planned by one urban planner.

In the present test, the development suitability is calculated using $s^t = x_1{}^*a_1 + x_2{}^*a_2 + a_N{}^t$, where x_1 and x_2 are the policy parameters for PP (a_1) and PL (a_2), respectively, which are simultaneously adjusted 500 times randomly. The PPI results for the four predefined forms, which vary from each other, are plotted in Fig. 5.6. For Form 1, no parameter pairs are valid, indicating that this urban form cannot be achieved with the combined PP and PL constraints. If Form 1 is set as the planned form, then practical urban spatial development will deviate from the planned form within the context of the current spatial development polices. For the other three urban forms, valid parameter pairs can be applied to realize the corresponding urban form. Considering that Form 1 cannot be archived, there is no negotiation between the planner and the developer. For Forms 2 and 4, strong policy on PP leads to strong policy on PL, which makes the balance and cooperation between the planner and developer necessary. Form 3 cannot be archived without a strong policy on PP, whereas the planner of PL has more alternatives for cooperation with the developer of PP.

Table 5.3 shows that the difference between valid and invalid parameter sets is significant as verified by the K-S test. The Z value for each policy parameter of various planned urban form stands for the sensitivity of the policy parameter. That is, a small change of a policy parameter with a high Z value will lead to a great change of the simulated urban form in terms of the *Kappa* indicator between the simulated and predefined urban forms.

Table 5.2 Simulation results of various single constraint tests in the virtual city

	Point constraints		Line constraints		Polygon constraints	
Map (*A*)					(black)	
Constraint type	PP	NP	PL	NL	PG	NG
Practical meanings	Development project	Pollutant area	Transportation plan	River	Develop promote area	Conservation area
$\beta = 0.1$						
$\beta = 1$						
$\beta = 10$						

Note: In some forms with constraint PG, the number of developed cells exceeded 150 because the number of promoted cells for PG exceeded 200

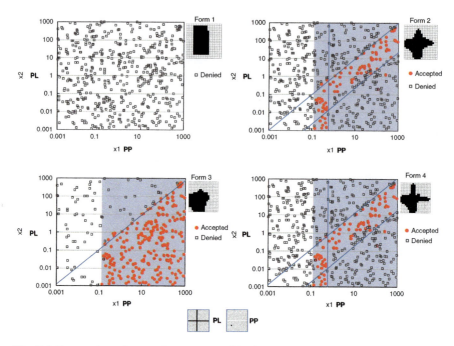

Fig. 5.6 Scatter plots of constraint parameters identified for two combined constraints in the virtual city

Table 5.3 The K-S test for four forms in the virtual city

Z value	Form 2	Form 3	Form 4
x_1	3.169[a]	7.762[a]	2.872[a]
x_2	4.161[a]	4.066[a]	5.080[a]

[a]Significant at 0.001 level (two-tailed)

Economist for Controlling the Total Number of Developed Cells (TNDC)

In this sub-section, we introduce the TNDC as the requirement of an economist in the local government, which is used for controlling the total development demand in the negotiation process among stakeholders. We argue that total development demand will have a significant impact on urban form; thus, the TNDC claimed by the economist may change the extent and shape of the future urban form. Thereby, the TNDC will change the *KAPPA* value between the simulated and planned urban forms. That is, urban forms with a TNDC of less than or greater than 150 may also be similar to the planned urban form in terms of the 80% *KAPPA* restriction.

In the test with the TNDC, we set x_1 as 0.1 for the PP constraint, x_2 as 0.58 for the PL constraint, and the desired TNDC as 400, which represents that the entire virtual city is developed as urban. For each TNDC value from 0 to 400, we run the model

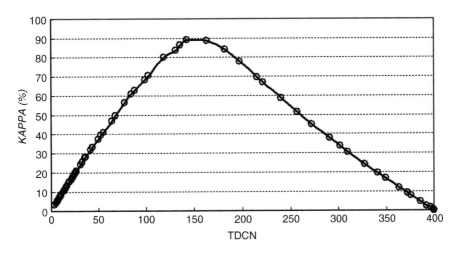

Fig. 5.7 Variations in *KAPPA* with the TNDC in the virtual city

multiple times until the accumulative average Kappa value reaches convergence. Figure 5.7 shows the dynamic variation of *KAPPA* with the TNDC during simulation, where the simulated forms with the TNDC ranging from 120 to 180 all satisfy the *KAPPA* 80% restriction. A TNDC of 150 does not represent the peak of the *KAPPA*-TNDC curve, but rather, a TNDC of 143 has the highest *KAPPA* value. This result indicates that to realize a predefined form, a TNDC value that is the same as the planned form may not be the best condition. Therefore, the TNDC is also an important parameter and should also be argued for realizing the planned urban form.

Figure 5.8 shows the identified parameter pairs for each planned urban form. Circles represent the identified constraint parameters of x_1, x_2, and the TNDC. There are also no constraint parameters available for Form 1, and we have identified numerous parameter pairs for other predefined forms. Table 5.4 presents the descriptive statistics for the three valid predefined forms, which show the overall characteristics of the identified parameters. Only 34 out of 2,000 pairs are valid for Form 2 (we ran the model for 2,000 times, which is enough for model output convergence), whereas there are 107 and 47 valid pairs for Forms 3 and 4, respectively. As a result, if an economist argues for the total demand of development in the planning process, the solution space of policy intensity on spatial constraints will be limited to smaller extent than those without the TNDC. Therefore, more meaningful and available planning policies can be reached with consensus through the cooperation among planners, developers, and economists.

As a result of the K-S test (see Table 5.5), the TNDC, as a new indicator, has a significant impact on the valid and invalid parameter groups.

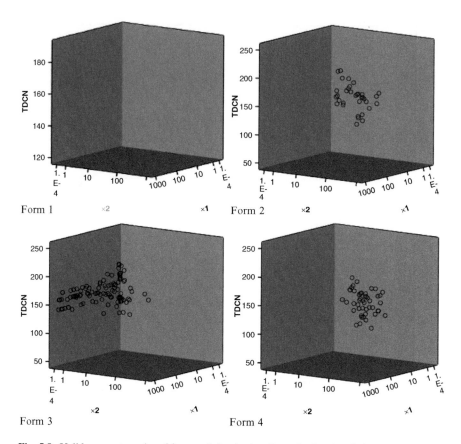

Fig. 5.8 Valid parameter pairs of four predefined urban forms in the virtual city

Table 5.4 Variables for three valid predefined urban forms in the virtual city

Form	Name	Parameter sets	Minimum	Maximum	Mean	Standard deviation
Form 2	x_1	34	0.089	716.995	9.083	177.061
	x_2		0.501	855.844	13.916	233.920
	TNDC		127	188	156.210	16.756
	KAPPA		80.088	89.173	8.357	2.344
Form 3	x_1	107	0.228	971.997	16.456	246.434
	x_2		0.001	722.178	1.491	77.460
	TNDC		125	196	159.620	17.797
	KAPPA		80.416	95.816	8.321	3.192
Form 4	x_1	47	0.022	716.995	8.310	159.772
	x_2		0.911	969.423	170.283	263.144
	TNDC		117	191	149.040	17.757
	KAPPA		80.246	92.666	8.561	3.886

5 Reaching Consensus Among Stakeholders on Planned Urban Form

Table 5.5 The K-S test including the TNDC for four forms in the virtual city

Z value	Form 2	Form 3	Form 4
x_1	1.875[a]	4.127[b]	2.439[b]
x_2	2.658[b]	2.880[b]	3.640[b]
TNDC	2.207[b]	3.729[b]	2.891[b]

[a]Significant at 0.002 level (two-tailed)
[b]Significant at 0.001 level (two-tailed)

Conclusion

In this chapter, we proposed an approach for building consensus between planners and developers for identifying the required spatial policies for planned urban form using a CA simulator (constrained CA). We focused on the relationship between spatial constraints and their parameters to gain insights on the negotiation among stakeholders using the CA simulator. We assumed that one planning policy represents one planner. Thereby, if more than two policy parameters are taken into account, then there should be a negotiation between the planners who will argue for policy intensity for their policy in planning practice. The simulated urban form using CA can reflect the policy intensity on spatial constraints, which allows the simulated results to be possibly used for deliberation in the local planning committee among planners, developers, and other stakeholders. Point, line, and polygon constraints with positive and negative effects on urban development are considered in tests in a virtual city. In addition, we also introduce the total number of developed cells (TNDC) as a requirement of economists in the local government, which is used for controlling the total urban development demand in the negotiation process among stakeholders.

In summary, our proposed approach applies to the desired urban form in the foreseeable future by combining various stakeholders in an integrated CA simulator. This tool, as a planning support system (PSS) in planning practice, enables these stakeholders involved in urban development negotiating with each other for realizing a planned urban form.

Acknowledgments We would like to thank the National Natural Science Foundation of China (No. 51078213) for the financial support.

References

Bullard, R. D., Johnson, G. S., Torres, A. O., 1999 "Sprawl Atlanta: Social equity dimensions of uneven growth and development" Atlanta, GA, Clark Atlanta University, The Environmental Justice Resource Center

Danielsen, K., Lang, R., Fulton, W., 1999 "Retracting suburbia: Smart growth and the future of housing" **Housing Policy Debate** 10(3) 512–40

Ewing, R., 1997 "Is Los Angeles style sprawl desirable?" **Journal of the American Planning Association** 63(1) 107–126

Ewing, R., Pendall, R., Chen, D., 2002 "Measuring sprawl and its impact" **The Smart Growth America**, Washington, DC.

Ewing, R., Schmid, T., Killingsworth, R., Zlot, A., Raudenbush, S., 2003 "Relationship between urban sprawl and physical activity, obesity, and morbidity" **American Journal of Health Promotion** 18(1) 47–57

Freeman, L., 2001 "The effects of sprawl on neighborhood social ties: An exploratory analysis" **Journal of the American Planning Association** 67(1) 69–77

He, C., Okada, N., Zhang, Q., Shi, P., Zhang, J., 2008 "Modelling dynamic urban expansion processes incorporating a potential model with cellular automata" **Landscape and Urban Planning** 86 79–91

Johnson, M. P., 2001, "Environmental impacts of urban sprawl: A survey of the literature and proposed research agenda" **Environment and Planning A** 33(4) 717–735

Kahn, M., 2000 "The environmental impact of suburbanization" **Journal of Policy Analysis and Management** 19(4) 569–586

Klosterman, R. E., 1999 "The *What if?* Collaborative planning support system" **Environment and Planning B: Planning and Design** 26(3) 393–408

Landis, J. D., 1995, "Improving land use futures: Applying the California urban model" **Journal of the American Planning Association** 61(4) 438–457

Landis, J. D., Zhang, M., 1998a "The second generation of the California urban future model, Part1: Model logic and theory" **Environment and Planning B: Planning and Design** 25(5) 657–666

Landis, J. D., Zhang, M., 1998b "The second generation of the California Urban Future model, Part2: Specification and calibration results of the land-use change submodel" **Environment and Planning B: Planning and Design** 25(6) 795–824

Li, X., Yeh, A. G. O., 2000 "Modeling sustainable urban development by the integration of constrained cellular automata and GIS" **International Journal of Geographical Information Science** 14(2) 131–152

Li, X., Yeh, A. G. O., 2002 "Neural-network-based cellular automata for simulating multiple land use changes using GIS" **International Journal of Geographical Information Science** 16(4) 323–343

Liu, X., Li, X., Liu, L., He, J., 2008 "A bottom-up approach to discover transition rules of cellular automata using ant intelligence" **International Journal of Geographical Information Science** 22(11-12) 1247–1269

Long, Y., Shen, Z., Du, L., Mao, Q., Gao, Z., 2008 "BUDEM: An urban growth simulation model using CA for Beijing Metropolitan Area" **Proceedings of the SPIE** 71431D-1-15

Long, Y., Shen, Z., Mao, Q., Du, L., 2009, "Form scenario analysis using constrained CA" **Proceedings of the Conference of Computers in Urban Planning and Urban Management, Hong Kong**

Newman, P. a J. K., 1989 "Gasoline consumption and cities: A comparison of U.S. cities with a global survey" **Journal of the American Planning Association** 55(1) 24–37

Speir, C., Stephenson, K., 2002 "Does sprawl cost us all? Isolating the effects of housing patterns on public water and sewer costs" **Journal of the American Planning Association** 68(1) 56–70

Otto, B., Lovaas, D., Stutzman, H., Bailey, J., 2002 "Paving our way to water shortages: how sprawl aggravates drought" **The American Rivers, The Natural Resources Defense Council, and The Smart Growth America** (Washington D C)

Ward, D. P., Murray, A. T., Phinn, S. R., 2000 "A stochastically constrained cellular model of urban growth" **Computers, Environment and Urban Systems** 24(6) 539–558

Wu, F., 1998 "Simland: a prototype to simulate land conversion through the integrated GIS and CA with AHP-derived transition rules" **International Journal of Geographical Information Science** 12(1) 63–82

Wu, F., Webster, C. J., 1998 "Simulation of land development through the integration of cellular automata and multicriteria evaluation" **Environment and Planning B: Planning and Design**, 25 103–126

Yeh, A G. O., Li, X., 2001, "A constrained CA model for the simulation and planning of sustainable urban forms by using GIS" **Environment and Planning B: Planning and Design**, 28(5) 733–753

Chapter 6
An Agent-Based Approach to Support Decision-Making of Total Amount Control for Household Water Consumption

Yan Ma, Zhenjiang Shen, Mitsuhiko Kawakami, Katsunori Suzuki, and Ying Long

Introduction

In this chapter, we present an agent-based model of household water consumption simulation (HWCSim) for the visualization of policy effectiveness of total amount control for household water consumption and as a guide for sustainable water resource management. Within this model the volume of household water consumption is regulated through a negotiation process regarding water price adjustment between household and government. Water consumption in an urban area is examined as a closed local water market, and a water price negotiation process between the supply side and the demand side is simulated. This process reflects how the supply side (government) and the demand side (households) reach a consensus on water price.

The world urban population is predicted to reach 5.0 billion in 2030 (United Nations 2007). As a result of massive immigration to urban areas, in the past decade, household-level population activities and environmental influences have become major fields in population-environment research (Sherbinin et al. 2008). People's increasing interest in environmental protection has caused sustainable development to become a popular topic. Sustainable development in this context means that energy supplies are readily available at reasonable cost, but are also sustainable over the long term and can be used for all required tasks without negative social effects (Ibrahim 2010). Water is the most basic resource. No matter how much urban development advances and technology improves, water remains a basic necessity for all life on Earth (Hildering 2004). Despite its importance, the amount of clean water is limited. As reported by the U.S. Geological Survey

---

Y. Ma (✉) • Z. Shen • M. Kawakami • K. Suzuki
School of Environmental Design, Kanazawa University, Kanazawa, Japan
e-mail: shenzhe@t.kanazawa-u.ac.jp

Y. Long
School of Architecture, Tsinghua University, Beijing, China

Z. Shen, *Geospatial Techniques in Urban Planning*, Advances in Geographic
Information Science, DOI 10.1007/978-3-642-13559-0_6,
© Springer-Verlag Berlin Heidelberg 2012

(Gleick 1996), only 0.8% of Earth's water can be considered to be fresh water. Therefore, a serious problem emerges in that, although there is a limited supply of fresh water, the population is increasing sharply. This problem results in a conflict between the water supply side and the water demand side and also results in water shortages.

The quantity of water resources per capita in Japan is half the world average. Tokyo restricted water usage for 42 months, from October 1961 to March 1965 (World Bank Analytical and Advisory Assistance (AAA) Program 2006). The administrative structure in Japan is such that the overall framework for water resource development is set by the national government, whereas actual implementation and management is accomplished at the local level. The national government in Japan implements the Comprehensive National Water Resources Plan (hereinafter, the Water Plan) to manage water resources. This plan is generally implemented every year in order to determine the supply of water and the budget for water supply development. The goal of the administrative structure of Japan is to have the water supply approach the water demand. The national government actually does not play a very large role in local water management; rather, in order to avoid price increases as a result of competition between water users, the government supplies local utilities with significant subsidies (World Bank AAA Program 2006). This leads to relatively low water prices, which may have a negative effect on water conservation. The relatively low price of water determined by local utilities usually does not reflect the exact cost of water. Furthermore, the government incurs a huge cost associated with the water subsidies.

In order to reduce the budgetary burden on the government associated with water supply development, Japan usually subsidizes local water supplies with an appropriate sliding scale, where the subsidies for domestic water use and industrial water use are relatively lower than those for agriculture and sewage treatment. Thus, if the budget for domestic water supply is limited, a balance between water supply and water demand must first be reached. Basically, if the demand can be controlled, then the subsidy for domestic water supply can be better controlled or reduced, for example, through a policy to control the total amount of household water consumption. In this chapter, we suppose that this policy would help the government to coordinate water supply capacity and water demand by affecting personal water consumption.

In order to better evaluate the effectiveness of this policy, we reviewed a number of related studies. These studies can basically be classified into two types. In the first type, researchers analyze the relationships between policies, the location and intensity of urban activities, and urban environmental problems (Alberti 1999; Chin 2002; Ewing 1994, 1997; Neuman 2005). These studies were mainly qualitative and quantitative analyses, which cannot flexibly reflect household decisions. In the second type, the agent-based model (ABM) has proven to be a useful simulation method for reproducing the activity of those who can make their own decisions, and this method has been widely used to reflect the flexible actions of human beings (Fontaine and Rounsevell 2009; Brown et al. 2008; Torrens 2007). Thus, a number of studies have concentrated on simulating the dynamic interactions between household behavior, policy making, and environmental influences (Vlek 2000; Jager and Janssen 2003). These simulations have significantly contributed to the study of behavior–environment interactions and have provided a valuable tool for exploring the effectiveness of policy measures in complex environments (Jager and Mosler 2007).

The goal of this research is to supply governments with the simulation model HWCSim to allow them to visualize the effectiveness of government policies and in this way help governments make policy decisions. In this chapter, we concentrate on domestic water consumption only.

An Agent-Based Approach for Supporting Decision-Making of Local Governments

Local governments are assumed to control the total amount of household water consumption so as not to exceed the threshold determined by local water supply capacity. In order to achieve this target, local governments adjust the price of water in order to regulate household water consumption. In the present research, we construct an HWCSim model that considers households and government to be different agents and simulates their activities as following.

First, initial values will be set for water price and household water consumption. Households will consume water according to these initial values. Then, the government agent will calculate the total amount of household water consumption in each simulation step. If this value exceeds the threshold (as determined by local water supply capacity), then the government will increases the water price in order to reduce the volume of household water consumption. This new price is referred to as the Ask Price. A special process between the government and households is required if the government wants to publish the Ask Price in the term of new policy regarding water price. This process, which in reality is referred to as a public hearing, is referred to herein as the negotiation process. During negotiation, the household side proposes a Bid Price for water according to their willingness to pay (WTP), and the household side will compare the Ask Price with their Bid Price. The comparison results yield differences in household policy attitude. Concretely speaking, if the Ask Price is less than the Bid Price or is greater than the Bid Price but less than the *WTP*, then the policy attitude of the households will be positive. Otherwise, the policy attitude of the households will be negative. After this process, the government will count the number of households that have a positive policy attitude. If more than 50% of the households have a positive policy attitude, then the Ask Price will be implemented as the new water price. Otherwise, the average value of the Bid Price by middle-income households will be set as a new water price. Finally, households will adjust their water consumption according to the new price.

The simulation loop described above will not stop until the government determines a water price that will reduce the total household water consumption to below the threshold. In this chapter, we refer to this water price as the willingness to accept (WTA).

Description of the HWCSim Model

Agent-Based Water Market

In this chapter, like land, water is considered to be a public good. In the HWCSim model, as the entity that controls the local supply of water, the local government is considered to be the supply side of the local water market. Correspondingly, the entity that consumes water is considered to be the demand side of the local water market. However, in the present study, we consider only household water consumption. Therefore, the water market in the HWCSim model is formed by the government (supply side), households (demand side), and the local water resources. In order to better reflect the water market scheme, we referred to research on the land market (Parker and Filatova 2008) and constructed the conceptual water market shown in Fig. 6.1.

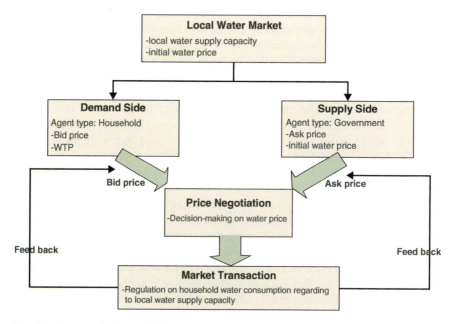

Fig. 6.1 Conceptual scheme of the water market

Demand Side of the Water Market

Budget Constraints for Household Water Consumption

Each household is assumed to know what percentage of income can be used for water consumption each month. The 5% rule reported by the World Bank assumes an elastic demand for water when the cost is less than 5% of household income and an inelastic demand when the cost exceeds 5% of household income (World Bank 1992). We hereby use 5% of household income as a maximum budget constraint for household water consumption.

In this chapter, we divide households into three groups according to income: poor households, middle-income households, and rich households. We assume that different groups have different budget constraints, and that the maximum household water consumption in each group will not exceed 5% of household income. Therefore, we obtain the budget constraints for household water consumption for poor households, middle-income households, and rich households, Y_p, Y_m, and Y_r, respectively, as follows:

$$Y = \{Y_x | Y_p, Y_m, Y_r\}, (x = p.m, r)$$
$$Y_p = 5\% * income_{poor}$$
$$Y_m = 3\% * income_{middle}$$
$$Y_r = 1\% * income_{rich} \qquad \qquad (6.1)$$

Evaluating Household Satisfaction with Water Consumption

We assume that the difference in water consumption produces different satisfaction levels for households and that there is a positive correlation between water consumption and satisfaction. As a result of the cost of water, the household water consumption will not exceed a volume whereby the total charge for water consumption approaches the *WTP* but is less than budget constraint Y. We consider the satisfaction produced by this volume of water consumption to be threshold K. Finally, the growth rate of household satisfaction S with water consumption V is given by the following logistic functional form:

$$dS/dV = rS(1 - S/K). \qquad \qquad (6.2)$$

Based on this equation, we transform the integral into function S in terms of V, as follows:

$$S(V) = K/(1 + Ae^{(-rV)})$$
$$where, A = K/S(0) - 1 \qquad \qquad (6.3)$$

where $A = K/S(0) - 1$ and K and A are constants.

Household WTP for Water

As shown in (6.3), household satisfaction is achieved if water consumption does not exceed 5% of household income. Here, we consider the WTP to be the total cash that households would like to pay for water. Therefore, we assume the WTP to be a function of individual income and the satisfaction with a certain volume of water consumption. In addition, the total cost of water consumption will be less than the budget constraint Y. We describe the dependencies of these variables as follows:

$$WTP = \frac{Y * S^2}{b^2 + S^2}, \text{ (b is a constant),} \tag{6.4}$$

where b is a constant. This equation has the same form as that used to model land prices (Filatova et al. 2008).

In (6.4), Y and S are calculated according to (6.1) and (6.3), respectively. Based on the idea proposed by Filatova et al. (2008), the individual WTP increases as satisfaction with water consumption increases but never exceed one's budget Y. Thus, according to (6.4), WTP increases monotonically to Y when $S \rightarrow \infty$. The value of b controls the slope of the function, which can be considered as a proxy for households' affordability of all other goods in their daily life, reflecting their relative influences on the WTP for water consumption. As $b \rightarrow \infty$, (6.4) becomes flatter and at the point where $S = b$, $WTP = Y/2$.

Unlike in Filatova's work, we assume that the household WTP will be influenced by the following three considerations:

- Individual WTP, which is determined by (6.1–6.4), reflects the WTP of each household without considering other factors.
- Neighborhood influences reflect the influence of neighbors on household WTP. The basic principle is that for each household, if more than 50% of the neighbors have a higher WTP than that household, then the WTP of that household will increase; otherwise, the WTP of that household will decrease. In the HWCSim model, the Moore 4 neighborhood is used.
- Global influences reflect the average WTP of all households in a society on the WTP of a single household.

Accordingly, the final WTP of each household is defined as the average value of the household WTP, the neighborhood WTP, and the global WTP.

Final Bid Price of Households

As in the study by Parker and coworkers (Filatova et al. 2008), we differentiate between the WTP and the final Bid Price of households, and between the WTA and the Ask Price of the government during the price negotiation process. The purpose of this simulation is to determine the WTA of the government such that price the total household water consumption will not exceed the local water supply capacity (the threshold for total amount control).

As in daily life, when buyer and seller try to reach a consensus on the price of a product, both the buyer and seller attempt to maximize their gains from the trade. Therefore, the Bid Price of the buyer tends to be less than the individual *WTP*, and the Ask Price of the buyer tends to be more than the *WTA*. As one of the necessities of daily life, there is no substitute for water. As such, households must pay for water consumption. However, if the government negotiates with households in determining the new water price, it is assumed that the Bid Price of households will be below their WTPs, as follows:

$$P_{bid} = WTP * (1 - \varepsilon), \text{Where we seem } \varepsilon \text{ as a constant,} \tag{6.5}$$

where ε is a constant.

Supply Side of the Water Market

WTA of the Government

Water is a natural resource because of its limited supply. As such, water is a public good in Japan. However, unlike with other public goods, such as land, the government acts as the supplier and regulator of local water market according to local water supply capacity. Accordingly, we herein consider the local government to be the single supplier of water.

We assume that the local government tries to control the total amount of household water consumption so as not to exceed the threshold of the local water supply capacity. This is accomplished by the regulation of water price. In order to set the new water price, the government has to negotiate with households and make sure that at least 50% of households can easily afford their water. As a computer experiment, the government agent will set the water price for negotiation with households in order to realize its policy on the regulation of total amount control of water consumption. In the HWCSim model, the water price set for the computer experiment is defined as the *WTA* of government.

Ask Price of the Government

In a market, the Ask Price of the seller is expected to be different than the *WTA* of the seller. In the present research, it is assumed that a water price exists, which the government decides based on the water supply capacity in the initial step. Thus, the Ask Price of the government differs in each simulation loop to reach a balance between water consumption and water supply capacity, as follows:

$$P_{t,ask} = P_{t-1}(1 + a), (0 < a < 1). \tag{6.6}$$

In (6.6), the Ask Price is denoted as $P_{t,ask}$, which is different in each simulation loop. At the beginning of the simulation, there will be an initial water price, and the government will calculate the total amount of household water consumption based on this price, if the total amount of household water consumption is larger than the threshold of the total amount control, the government will update the Ask Price using (6.6). In this equation, a is a constant that represents the government's attempt to regulate household water consumption through increasing or decreasing the water price, and a is the markup or markdown according to the water price at Step $t-1$.

Total Amount Control of Household Water Consumption

Government Total Amount Control

In the HWCSim model, the government agent will perform total amount control of household water consumption. In the first step of simulation, the initial water price is set by the government agent, and the distribution of household water consumption is assumed to be a normal distribution. Thus, the average volume of household water consumption is prepared to reflect the water consumption. The household agents determine their own consumption volume according to their attributes, i.e., the number of family members and the average volume of household water consumption within different income groups. Then, the government will calculate the total volume of household water consumption and decide whether to request a new price to regulate household water consumption activity. If the government agent decides to request a new water price, which is defined herein as the Ask Price, the simulation will enter the price negotiation stage. After the negotiation process, a new water price will be published by the government agent, and the households will adjust their water consumption according to this new price and their income at the next simulation loop. The details of this simulation flow are shown in Fig. 6.2.

Negotiation Process for Water Price

We consider the negotiation of water price between the government and households to be a bilateral trade negotiation process, in which the household side puts forward a Bid Price based on the individual *WTP* and the government side puts forward an Ask Price based on the initial water price, and the two sides negotiate the final water price as a market transaction.

The water market proposed herein is differs from traditional public goods markets in that there is a single supplier and numerous buyers. Therefore, the price negotiation between the government and each household cannot be carried out. As such, the price negotiation will take place between the government and a number of households through a policy attitude investigation. When the

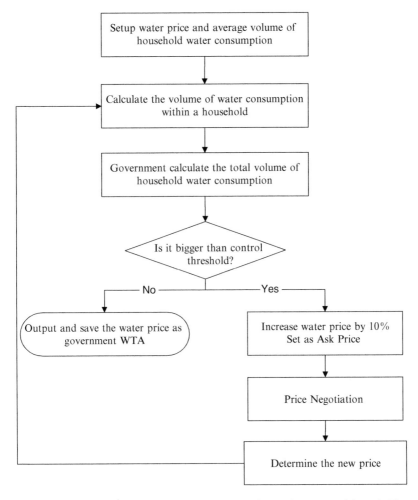

Fig. 6.2 Decision tree for the government to control the total amount of household water consumption

government requests a new water price, we assume that the government will hold public hearings to estimate the feasibility of the new price. During the public hearing, local households from different income groups will be asked to participant in the policy attitude investigation, which is carried out through the distribution of a questionnaire on household policy attitude regarding the new water price. Households fill out the questionnaire according to their WTP and expectation price, which is defined herein as the Bid Price.

The policy attitudes reported by households are defined as individual policy attitudes, which are divided into primarily two ranges representing positive and negative feedback, respectively. Table 6.1 shows the decision table for determining the individual policy attitude.

Table 6.1 Household policy attitude toward government Ask Price

	Poor households			Middle-income Households			Rich households		
Bid Price $\geq$ Ask Price	Y		N	Y		N	Y		N
WTP $\geq$ Ask Price	/	Y	N	/	Y	N	/	Y	N
Policy attitude range	[0,1]		[−1,0)	[0,1]		[−1,0)	[0,1]		[−1,0)

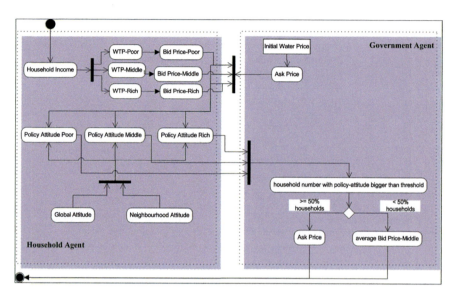

Fig. 6.3 UML state diagram of the negotiation process

The state transition in the process of water price negotiation is shown in detail in Fig. 6.3.

In the HWCSim model, the government estimates household policy attitude based on the following three considerations:

- Individual policy attitude is directly determined by households by comparing their Bid Price with the Ask Price of the government.
- Neighborhood influences reflect the influence of neighbors on household policy attitude. The Moore 4 neighborhood is used in the HWCSim model, and the average policy attitude of neighbors is defined as the neighborhood policy attitude.
- Global influences reflect the influence of society on household policy attitude. One example of such an influence is propaganda on the reduction of water consumption. The average policy attitude of all households is defined as the global policy attitude.

The final household policy attitude is the average of the individual policy attitude, the neighborhood policy attitude, and the global policy attitude. In the

6 An Agent-Based Approach to Support Decision-Making of Total Amount Control   117

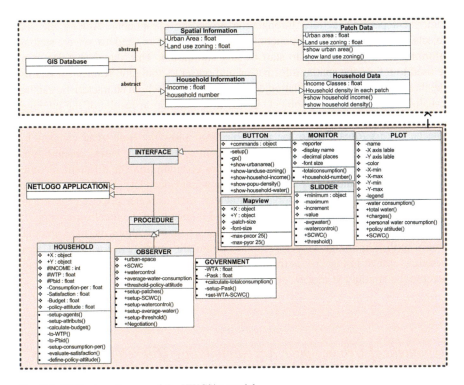

Fig. 6.4 UML class diagram of the HWCSim model

HWCSim model, if the final value of the policy attitude is larger than the threshold, which means that households agree to the Ask Price, then a threshold is used for household policy attitude evaluation.

Finally, the government will decide the Ask Price. As shown in Fig. 6.4, the basic principle is that if the threshold policy attitude is exceeded by at least 50% of households, then all of the households must accept the Ask Price. This price will be set as the new water price in the next simulation loop; otherwise, the Ask Price will be set as the average Bid Price by middle-income households.

In the following session, we discuss how to conduct the simulation using the HWCSim model.

Development and Implementation of the HWCSim Model

System Development

The HWCSim model is developed using the Netlogo platform. As shown in Fig. 6.4, there are basically two components in the Netlogo application, namely, the interface component and the procedure component. The objects in the interface component perform actions through the procedure component. Among the six objects in the interface component, the button object is used to set up spatial information on patches and to start or stop simulation. The Mapview object can be used to reflect the spatial information and present simulation results, for example, the urban space, the population density, and the household characteristic can be visualized by the Mapview object. In addition, simulation results of the average values of agents' attributes or global parameters can be presented by the plot objects. In this research, we use plot objects to show the time-trend changes in the simulation process, such as the total amount of household water consumption, the average total charges for household water consumption for different income groups, the household policy attitude, and the water price. The slider objects allow the user to adjust or present global parameters, such as water price, e.g., average household water consumption (averwater) and threshold of the local water supply. The switch object allows the user to set options with or without the total amount control in the simulation process. Another important component for the HWCSim model, as shown in Fig. 6.4, is the procedure component, which allows the user to develop the simulation model.

Hypothetical Urban Space Accommodating Household Attributes

In this simulation, the Japanese city of Kanazawa was chosen as a case study area. In order to accommodate household attributes, we prepared a hypothetical urban area and household attributes for the simulation. The spatial dataset is created based on the real dataset for Kanazawa City, including urban planning information, household attributes, and their spatial distribution for this simulation (Fig. 6.5). We assumed that the households are stable in number and location. In order to reduce the computation time, we created 1,500 household agents in the virtual urban space to represent the 450,000 households in Kanazawa City. Therefore, one household in the virtual space corresponds to 300 households in the real society, and the households are divided into the three income groups.

Parameters Setting and Model Behavior

The system interface of the HWCSim model is shown in Fig. 6.6, which is developed using Netlogo. We assumed that households are stable in number and location. In order to validate the model, we analyzed the simulation results based on the parameters in the model.

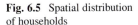

Fig. 6.5 Spatial distribution of households

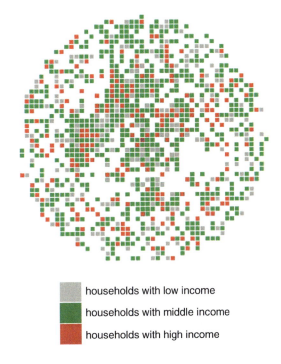

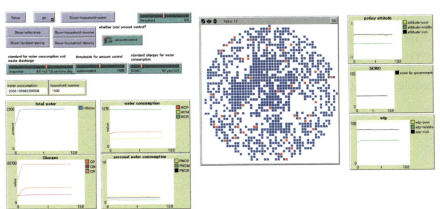

Fig. 6.6 Model interface of the Netlogo platform

Threshold of Household Satisfaction with Water Consumption

As described in Sect. "Evaluating Household Satisfaction with Water Consumption", we argued that household satisfaction with water consumption would follow a logistic functional form. Basically, households will not leave water running for the entire day in order to increase satisfaction with water consumption if the desired amount of water has already been consumed. Moreover, although the WTP or household satisfaction for water consumption is limited, there is a basic requirement for daily water consumption. These two considerations yield two thresholds for household satisfaction. As proven previously, consumer satisfaction with a product/service commonly has thresholds at both lower and upper levels (Hom 2000).

Generally speaking, people require at least 0.2 m³ of water for daily life, the satisfaction with the minimum demand for water consumption is considered to be the lower threshold. Thus, the upper threshold of household satisfaction must be defined. We assume that the household satisfaction gained by water consumption will obey the law of diminishing marginal utility. We therefore consider the relationship between the volume of household water consumption and the corresponding household satisfaction based on (6.3) in which the constants are assumed to be k = 9, A = 9, and r = 5.

Based on (6.3), we calculated household satisfaction according to water consumption from 0.2 m³ to 10 m³ of water per day. The increase in household satisfaction with increasing water consumption in increments of 0.1 m³ has also been calculated and is shown in Fig. 6.7.

As shown in this figure, the value of dS/dV (increase in household satisfaction with the increase in water consumption) is initially rising. However, dS/dV begins to decrease when V = 0.4, and the household satisfaction S continues to increase until V = 1.6. We consider a household water consumption of less than 1.6 (m³/day)/person to be the range of increasing household satisfaction with respect to water consumption. We assume that households in different income groups will seek satisfaction in this range. As such, we choose the satisfaction produced by consuming 0.6 m³, 1.0 m³, and 1.7 m³ of water as the upper levels of household satisfaction of poor, middle-income, and rich households, respectively. Thus, satisfaction values of 6.2, 8.4, and 8.9 are defined as the upper levels of household satisfaction in the following simulation.

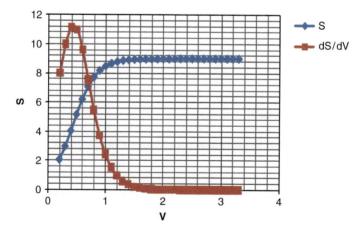

Fig. 6.7 Household satisfaction per 0.1 m³ of water consumption (K = 9, A = 9, r = 5)

Threshold of Total Water Consumption and Other Parameters Regarding Household Water Consumption

The required initial values for parameters, including "avgwater", which represents the average household water consumption, "threshold for amount control", which is the local water supply capacity, and "SCWC", which represents the standard charges for water consumption, are shown in Fig. 6.8.

We started the simulation with the initial parameter settings. As shown in Fig. 6.9, the total amount of household water consumption, i.e., "total water", increased sharply at the beginning of the simulation because households initially require high satisfaction with water consumption. However, the total amount of water consumption decreased because of the water price negotiation process and

Fig. 6.8 Initial parameter values

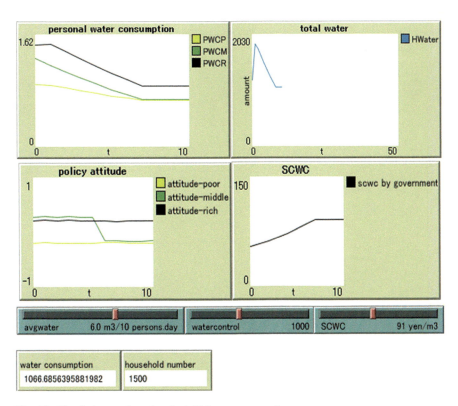

Fig. 6.9 Simulation results using the initial parameter settings

government regulations implemented in the model. Finally, the total water consumption became stable at a value below the threshold.

In order to facilitate observation of the model behaviors in the following analysis, in the following, we start the simulation without total amount control and then turn on "amountcontrol" (button shown in Fig. 6.6).

The realization of the above-described process resulted in an increase in water price, as shown by the plot of *SCWC* in Fig. 6.9, and *SCWC* grew from 50 to 91. As the water price increases, the household policy attitudes of middle-income households fluctuate significantly, which means that these households initially agree with the increased water price, but as the price approaches the *WTP*, the positive policy attitude of each household changes. Although the policy attitudes of rich and poor households are much more stable, the difference between these two types of household is that the former is always positive whereas the latter is always negative. These results appear to be reasonable because rich households have a large *WTP*, whereas poor households have a smaller *WTP*. In the following, we describe how the simulation results change with changes in the parameters.

- Model behavior: total amount control of water consumption, "watercontrol"

We continued to run the simulation and adjusted the value of "watercontrol" from 1000 to 800. As shown in Fig. 6.10, the standard charge of water consumption

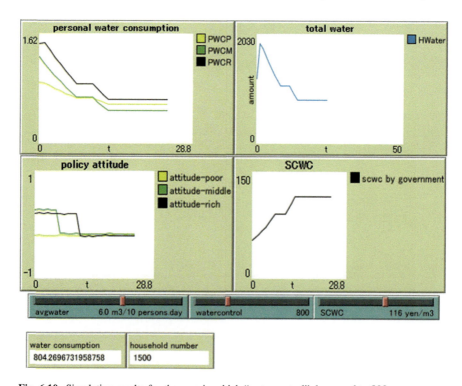

Fig. 6.10 Simulation results for the case in which "watercontrol" decreased to 800

(SCWC) increased because of the decrease in the threshold of total amount control. The policy attitude of rich households also decreased to a negative value, which means that some rich households began to disagree with the increasing water price. Thus, after the negotiation process, the new water price tends to become stable at 116 yen/m^3, where the total water volume is 804 m^3. Although this value it is not strictly below the threshold, it is also reasonable considering the randomness of personal activity.

Accordingly, when the price charged for water is 116 yen/m^3, the total amount of household water consumption will be very near the control threshold of 800 m^3. This price is considered to be the government *WTA* that should be achieved through the process of simulation.

After the simulation introduced in the last paragraph, we continued to run the simulation and increased the value of "watercontrol" to 1,200. As shown in Fig. 6.11, the values of total water and personal water consumption increased. At the end of simulation, the value of total water fluctuated to 1200 m^3 and became relatively stable at 1216.8 m^3. These results are related to the decrease in the price of water, i.e., the *SCWC*, in the present research. As indicated by the plot of *SCWC*, *SCWC* decrease greatly from 116 yen/m^3 (simulated previously) to 88 yen/m^3. Correspondingly, the policy attitude of households becomes more positive,

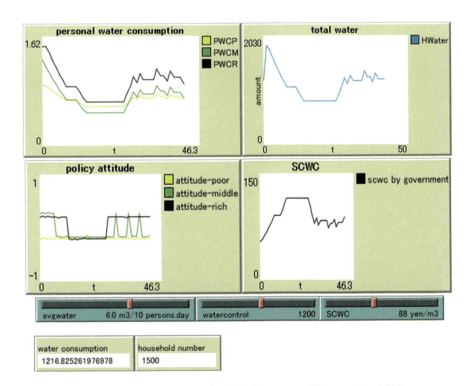

Fig. 6.11 Simulation results for the case in which "watercontrol" increased to 1,200

especially with respect to rich households. Moreover, the policy attitude of middle-income households changed in a waveform.

- Model behavior: "avgwater"

Here, we restart the simulation using the initial parameter settings in order to test the behavior of the model by adjusting the value of "avgwater", i.e., the average volume of household water consumption. Here, "avgwater" will be used to configure the different average water consumption volumes for the different income groups of households. As shown in Fig. 6.12, the simulation results are basically stable when "amountcontrol" is off. Then, "amountcontrol" was turned on as the simulation was running. As mentioned above, the function of government total amount control began to operate so that the total amount of water consumption increases and the decreases so as to obey the model function.

When "avgwater" is set to be 3.0 lower than the initial value, the total household water consumption was below the threshold, which was set to be 1,000. Then, SCWC decreased slightly when "amountcontrol" was first turned on, but the subsequent sharp increase in household water consumption associated with the upper level of satisfaction causes SCWC to increase. As shown in Fig. 6.12, after

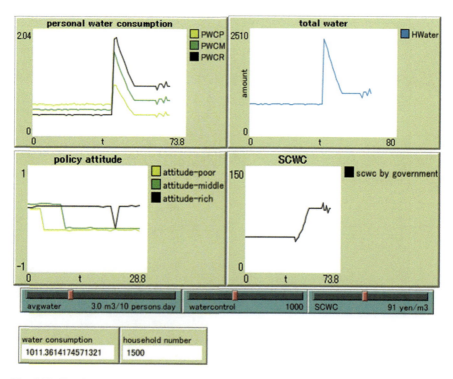

Fig. 6.12 Simulation results for the case in which "avgwater" decreases to 3

decreasing slightly, *SCWC* increased and became relatively stable at 91, where the total water consumption was 1011.36 m³, which is near the "watercontrol" threshold. During this process, different households had negative policy attitudes with regard to the increase in *SCWC*. Among these households, the policy attitudes of only the rich households return to being positive after decreasing slightly.

The simulation is restarted after adjusting "avgwater" to 9 m³/10 persons·day. The simulation results are shown in Fig. 6.13, in which the plot of *SCWC* increased noticeably because of the increase in "avgwater", i.e., average water consumption. As a result of the increase in water price, households must decrease their water consumption. Thus, as shown in Fig. 6.13, the total water consumption and the personal water consumption decreased sharply. In contrast, the volume of personal water consumption of rich households did not change noticeably during the simulation process even though the water price became higher. This phenomenon can be attributed to the large *WTPs* of rich households, which indicate relatively low sensitivity of water consumption activities to the fluctuation of water price. In contrast to rich households, poor households have a greater sensitivity to changes in water price. These findings are established by the simulation and are supported anecdotally, thus proving the effectiveness and reasonableness of the HWCSim model.

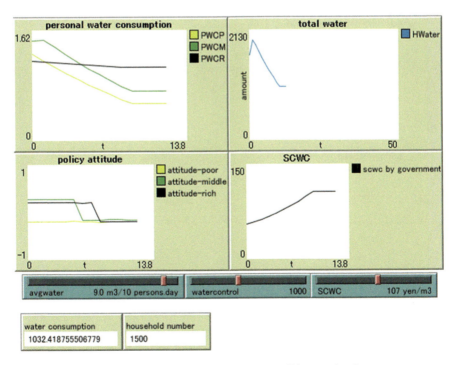

Fig. 6.13 Simulation results for the case in which "avgwater" increased to 9

Model Validation Using Real Data for Kanazawa

Validation of the HWCSim model is conducted using the real dataset of Kanazawa City, Japan. We intend to use the real data for household water consumption for Kanazawa City as the threshold of "water control" in the model in order to simulate the average volume of household water consumption and the charge in the *SCWC*. Then, the simulated results and the real data will be compared in order to validate the HWCSim model.

Based on the census survey of Kanazawa in 2009, there are 454,607 households in Kanazawa City, and the total water consumption that year was 39,295,827 m^3. Thus, each household consumed approximately 0.24 m^3 of water per day, and the total water consumption for 1,500 households was 355 m^3/day in the virtual urban area. Based on the *SCWC* published by the Kanazawa Water and Energy Center, assuming that the household water consumption was 10–20 m^3/month, the actual *SCWC* is obtained as 113 yen/m^3. The HWCSim model is validated using the parameter settings shown in Fig. 6.8, with the exception of the water control value. According to the basic principle of model validation, as described above, we set the water control to be 355 m^3. The simulation results are shown in Fig. 6.14.

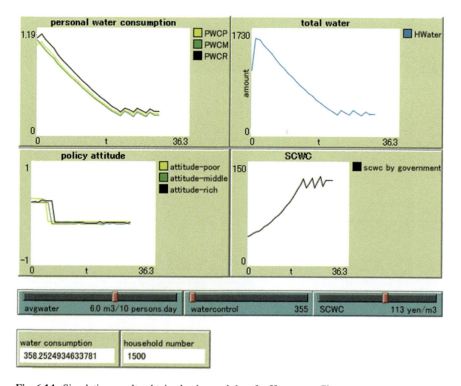

Fig. 6.14 Simulation results obtained using real data for Kanazawa City

In Fig. 6.14, the model behaves in a manner similar to that described in Sect. "Parameters Setting and Model Behavior". The household water consumption decreased because of the increasing $SCWC$. As shown by the personal water consumption chart, the difference in average household personal water consumption according to income group decreases accordingly. Since $SCWC$ is relatively lower and is acceptable to different households, we find that despite small fluctuations, initially, the policy attitudes of households were relatively stable and were all positive. Furthermore, these values are relatively high. Since $SCWC$ exceeded 90, the policy attitudes of households exhibited significant decreases, until eventually fluctuating very slightly around zero. The final results for the average volume of household water consumption, referred to as personal water consumption, for different households has been reduced to the range of from 0.22 to 0.27 m^3/household·day. This result is very similar to the real data for Kanazawa City. In contrast, the simulation results for $SCWC$ indicate that the actual $SCWC$ is just 113 yen/m^3.

Conclusion

In this chapter, the HWCSim model is designed to reflect the process of total amount control of household water consumption. This model can simulate the process of water price formation through a water price negotiation process between households and government, where the household policy attitude impacted by the water price requested by local government under the constraint of the local water supply capacity is considered through total amount control. Finally, as validated by the real data for Kanazawa City, the HWCSim model is found to be applicable for simulating the process of household water consumption regulated by government total amount control in planning practice. The simulation results, including water price and household water consumption, are very similar to the actual data obtained for Kanazawa City.

In the HWCSim model, we propose the household WTP for water consumption through evaluating the satisfaction with household water consumption. We also introduced the interaction between government agents and household agents in deciding water price, which is herein referred to as negotiation. As a result, the HWCSim model can reflect the mechanism of total amount control of the local government, while considering local water supply capacity, water price, household policy attitude, and their attributes. In the HWCSim model, households can reduce or increase their water consumptions according to their WTP and the water price requested by the government. The policy attitudes of these households will eventually impact government decision-making regarding water price.

The present research has limitations. For example, we use random factors having ranges to reflect differences in water consumption and policy attitudes among households of different income brackets, and the probability distributions of these random factors have not been taken into account. These limitations should be rectified in the future.

Acknowledgment We would like to thank the financial support of the Grants-in-Aid for Scientific Research (Nos. 23404022B and 22560602C), Japan Society of the Promotion of Science.

References

Alberti, M. (1999). Urban form and ecosystem dynamics: Empirical evidence and practical implications. In M. Jenks, E. Burton, and K. Williams, Achieving Sustainable Urban Form. London: Routledge, 84–96.

Brown, D.G., et al. (2008), "Exurbia from the bottom-up: Confronting empirical challenges to characterizing a complex system" Geoforum, vol. 39, 805–818.

Chin, N. (2002), Unearthing the roots of urban sprawl: A critical analysis of form function and methodology, Centre for Advanced Spatial Analysis (CASA) Working Paper Series, University College London, <HYPERLINK "http://discovery.ucl.ac.uk/249/1/Paper47.pdf" http//www.casa.ucl.ac.uk/working_papers/paper47.pdf>.

Ewing, Reid H. (1994), "Characteristics, causes and effects of sprawl: A literature review" Environmental and Urban Issues, vol. 21(2), 1–15.

Ewing, Reid H. (1997), "Is Los Angeles-style sprawl desirable? Journal of American" Planning Association, vol. 63(1), 107–126.

Fontaine, C.M., Rounsevell, M.D.A. (2009), "An agent-based approach to model future residential pressure on a regional landscape" Landscape Ecological, vol. 24, 1237–1254.

Filatova, T., et al. (2008), "Agent-Based Urban Land Markets: Agent's Pricing Behavior, Land Prices and Urban Land Use Change," Journal of Artificial Societies and Social Simulation, vol. 12(13). <http://jasss.soc.surrey.ac.uk/12/1/3.html>

Gleick, P.H. (1996), "Basic water requirements for human activities: Meeting basic needs" Water International, vol. 21(2), 83–92.

Hildering, A., (2004). International Law, Sustainable Development and Water Management, Netherlands: Eburon Academic Publishers, 17–23.

Hom, W., (2000), "An Overview of Customer Satisfaction Models" RP Group Proceedings, 101–110.

Jager,W., Janssen, M. A.(2003). The need for and development of behaviourally realistic agents, In J. S. Sichman, F. Bousquet, & P. Davidson (Eds.), Multi-agent based simulation II, Berlin: Springer, 36–49.

Jager, W., Mosler, H. J. (2007), "Simulating Human Behavior for Understanding and Managing Environmental Resource Use" Social Issues, vol. 63 (1), 97–116.

Neuman, M. (2005), "The compact city fallacy" Planning Education and Research, vol. 25(1), 11–26.

Parker, D. C. and Filatova, T. (2008), "A conceptual design for a bilateral agent-based land market with heterogeneous economic agents" Computers, Environment and Urban Systems vol. 32, 454–463.

Sherbinin Alex de et al. (2008), "Rural household demographics, livelihoods and the environment" Global Environmental Change, vol. 18, 38–53.

Torrens, Paul M. (2007), "A Geographic Automata Model of Residential Mobility" Environment and Planning B: Planning and Design, vol. 34, 200–222.

United Nations, (2007), World population prospects: the 2006 revision. Retrieved January 2, 2008, from United Nations Population Division Database.

Vlek, Ch. (2000), Essential psychology for environmental policy making, Psychology, vol. 35, 153–167.

World Bank (1992), World development report 1992: development and the environment, New York: the World Bank Oxford University Press, 89–104

World Bank Analytical and Advisory Assistance (AAA) Program (2006).

Yuksel, I.(2010), "Energy production and sustainable energy policies in Turkey" Renewable Energy, vol. 35, 1469–1476.

Part II
GeoVisualization and Urban Design

Visualizing design alternatives using virtual reality
Coordinating cooperative design for design guideline and regulations
Improving deliberation, decision-making and gaining consensus

Chapter 7
Review of VR Application in Digital Urban Planning and Managing

Anrong Dang, Wei Liang, and Wei Chi

Development of VR Technology

Virtual Reality (VR) technology has been applied in many fields since 1960s, such as research, education, manufacture, entertainment, medical, and urban planning. This chapter focuses on the current situation and progress of VR application in urban planning and managing, as well as on the future perspective oriented to digital urban planning and managing.

Being developed since 1960s, Virtual Reality (VR), also known as Virtual Environment (VE), is a term that applies to computer-simulated environments that can simulate places in the real world as well as in imaginary worlds, which provides 3D vision, hearing, touching, and sometimes smelling and even tasting (Lee 1993; Burdea and Coiffet 2003). The characteristics of VR are interactive, immersive, and imagination for VR users.

The Initial Concept Development Period

Although the team Virtual Reality (VR) was initially coined by Jaron Lanier, founder of VPL Research, in 1980, the concept idea of Virtual Reality (VR) could be traced back to 1962, when Morton L. Heilig developed the first VR

A. Dang (✉)
School of Architecture, Tsinghua University, Beijing, China
e-mail: danganrong@126.com

W. Liang
School of Architecture, Tsinghua Urban Planning & Design Institute, Tsinghua University, Beijing, China

W. Chi
Tsinghua Urban Planning & Design Institute, Tsinghua University, Gvitech Technologies Company Ltd., Beijing, China

Z. Shen, *Geospatial Techniques in Urban Planning*, Advances in Geographic Information Science, DOI 10.1007/978-3-642-13559-0_7,
© Springer-Verlag Berlin Heidelberg 2012

environment named Sensorama Simulator with handlebars, binocular display, vibrating seat, stereophonic speakers, cold air blower, and a device close to the nose that generates odors which fit the action in a corresponding film (Rhodes 1997). Therefore, Morton L. Heilig was named as The Father of Virtual Reality (VR) (www.mortonheilig.com).

The other two scholars are also very critical for the VR initial period. In 1963, Ivan E. Sutherland developed a man–machine graphical communication system which named Sketchpad in MIT. In 1965, Ivan E. Sutherland presented his achievements as the Ultimate Display concept in the conference of International Federation for Information Processing (IFIP). And in 1973, the graphic scenic producer was developed by Evans and Sutherland (http://accad.osu.edu/), which is the foundation of VR environment.

The Major Equipment Development Period

From 1970s to 1980s, amount of funds were invested to develop VR equipments by America military, such as flying helmet for flying simulation training and modern military simulator. These kinds of military simulation equipments pushed the VR technology development though most of them were transferred their usage to civil later. In 1881, National Aeronautics and Space Administration (NASA) of USA developed Helmet-Mounted Display systems (HMD) with LCD, known as VIVED (Virtual Visual Environment Display). And in 1988, it was developed to Virtual Interface Environment Workstation (VIEW) with very powerful function (http://accad.osu.edu/).

Meanwhile, leaded by Thomes and Jaron, the VPL company developed the first DataGloves and became the first company to manufacture and to sale VR equipments in 1987. Later on, the Nintendo Company produced much cheaper PoweGloves. In 1988, Fisher and Elizabeth developed 3D virtual sounds multi-control system. And in later 1980s, the VPL Corporation developed the first stereo-HMD with LCD named EyePhone (http://www.fbe.unsw.edu.au/). All the above VR equipments caused the great development of VR technology in this period.

The System Research Development Period

In 1992, a report from National Science Foundation (NSF) project expounded VR technology systematically and completely, and proposed new research topics of Virtual Reality Environment (VRE). At the same time, World Wide Web and HTML were born. All this together, lead to the substantial development of VR technology in terms of VR hardware environment and VR software system, as well as application.

7 Review of VR Application in Digital Urban Planning and Managing

Regarding to the VR hardware environment, Carolina Cruz-Neira, Daniel Sandin, Tom DeFanti, and other colleagues and students in Electronic Visualization Laboratory (EVL) of University of Illinois at Chicago (UIC), developed the new types of VR environment, named CAVE (Cave Automatic Virtual Environment) in 1992 (Cruz-Neira et al. 1993). The CAVE is a surround-screen, surround-sound, projection-based VR environment. The illusion of immersion is created by projecting 3D computer graphics into a $10'$ x $10'$ x $10'$ cube composed of display screens that completely surround the viewer. It is coupled with head and hand tracking systems to produce the correct stereo perspective and to isolate the position and orientation of a 3D input device. A sound system provides audio feedback. The viewer explores the virtual world by moving around inside the cube and grabbing objects with a three-button, wand-like device (http://accad.osu.edu/). Since then, CAVE has gained a strong following. The full-Immersion VR System, six sided CAVE was first constructed by Center for Parallel Computers at the Royal Institute of Technology in Sweden (http://www.artmuseum.net/).

Regarding to the VR software systems and application, Virtual Reality Modeling Language (VRML) was proposed in the first international conference on World Wide Web by Tim Berners-Lee and Dave Raggett in the spring of 1994. The first version of VRML 1.0 was come to the world in 1994 fall. And later VRML2.0 in 1997, named as VRML97 by ISO in 1998. GeoVRML and X3D (External 3D) developed by Web3D Union based on VRML97 became the new Web-VR standard which enlarged the application of VR technology. Meanwhile, the VR Juggler, developed by Virtual Reality Application Center, Iowa State University, is an open source Virtual Platform for Virtual Reality application development (Allen et al. 2001). Other related VR Modeling Language and platform, like Maya, Open GL, Open SG, and Vega Prime, are also developed very quickly. As well as VR application software, like Web Max, CityMaker, and Skyline are used in many fields. Meanwhile, many kinds of integrated VR system or environment based on Internet in C/S or B/S structure were developed and serviced worldwide, such as Google Earth, World Wind, Virtual Earth, Secondlife, Activeworlds, Multiusers, and GeoGlobe.

In China, CityMaker system played very important role in urban planning and managing in recent years. Designed and developed by Tsinghua Urban Planning & Design Institute and Gvitech Technologies Company Ltd., CityMaker System integrated the functions of urban planning and urban managing together. For the urban planners, CityMaker can be used to work out the planning scheme in both 3D model and 2D map with the attribute database of planning control index, such as identity of landuse, land area (LA), built-up area (BUA), greening rate (GR), building density (BD), and floor area ratio (FAR) (refer to Fig. 7.1). For the urban managers, CityMaker can be used to demonstrate the current and future urban landscape in both Macro and Micro scales, as well as to evaluate the urban development pattern, urban design proposal, and architecture construction style (refer to Fig. 7.2).

Urban Design of Jinan   District Planning of Nanjing

Fig. 7.1 CityMaker for planner (Created by Gvitech Technologies Company Ltd.)

City 3D model of Wuhan   Building 3D Model of Shanghai

Fig. 7.2 CityMaker for manager (Created by Gvitech Technologies Company Ltd.)

The High-tech Research Development Period

In recent years, the VR related research focus on new method and new technology, as well as integrating other high technology, such as high precision position technique, 3D digital multimedia display technique, 3D information inputting technique, high precision tracking technique, multi-channel interactive operation and control technique, high performance micro-projection technique, panoramic dynamic light field acquisition and processing technique, and so on. In China, the National High Technology Research and Development Program (named "863" Program) supported the following research topics (http://www.863.org.cn/):

1. Study on 3D object information input techniques with high efficiency, high precision, and very realistic characters, such as the technique for obtaining information and rebuilding the surface optical of 3D object, the technique for obtaining information and modeling the large scale and complicated scene as well as moving 3D object, and the technique for obtaining and processing the moving scene with lifelike representation.

2. Study on newly interactive stereo display techniques and equipment, such as the real 3D display and palpable technique and equipment, portable and large view field perspective headset display technique, multi-direction and large scale suspension display technique and equipment, multi-parallax stereo image acquisition and display technique and equipment, and so on.
3. Study on interactive force tactual techniques and equipment, such as force tactual technique and equipment based on physical characteristics and biological mechanism, interactive tactual technique and equipment on shapes, textures, flexibility, frictions, and temperature, man–computer interaction force tactual technique and equipment of degree of freedom and common types, and other sensory information fusion and interactive technique and equipment.
4. Study on man–computer interaction multi-channel technique and equipment, such as multiple optical touch technique and equipment, multi-channel, multi-touch, and large-scale surface interactive technique and equipment, multi-user and wide area 3D interactive technique and equipment based on vision, multi-channel interactive technique and equipment for integration of virtual and reality, man–computer interaction equipment with portable and low-power characteristics.
5. Study on new methods and new techniques of high-precision tracking, such as large-scale optical tracking technique, newly portable tracking technique, multi-feature fusion technique for video target tracking, and high precision tracking technique based on multi-source information.

Classes of VR Modeling Method

Building VR model is extremely important in the application process of VR technology. Along with the development of VR and its related technologies, the method of VR modeling is also enriched. All the method can be summed up in two classes. One is the "program modeling method" which means to construct the VR model by means of Virtual Reality Modeling Language (such as VRML) and the corresponding professional algorithm. A number of virtual simulation system which is often applied "program modeling method". The other is the "software modeling method" which means to build the VR model by means of some modeling software (such as 3Ds Max and Multi-Gen Creator). The "software modeling method" is often adopted by the VR application engineers through interactive modeling approach.

Because of the wide interests of VR application engineers, the "software modeling method" has been developed in many ways. All the "software modeling method" can be divided into the following three classes according to the type of software technique and the general technical process, such as visualization modeling method, attribute modeling method, and integration modeling method. The three classes of "software modeling method" have to be discussed in detail in the following paragraph.

Visualization Modeling Method

Visualization modeling method is characterized by application of graphics-based modeling technique and software, such as AutoCAD, 3Ds Max, Multi-Gen Creator, Sketch Up, and so on. Basically, visualization modeling method builds VR model by means of geometry elements, such as point, line, polygon surface, and 3D object.

It is very convenient to generate and edit the 3D model, as well as 3D scene rendering with suitable texture or material. However, this method is more difficult to generate large scale and complex digital terrain model (DTM), and the model always lack of attribute information. Developed by the City Simulation Group of University of California at Los Angeles (Urban Simulation Team @ UCLA), the VR model of Los Angeles downtown and El Pueblo of Los Angeles (http://www.ust.ucla.edu/) belongs to this category. In China, the earlier 3D city models, such as downtown model of Ningbo and downtown model of Nanning are all visualization models (Fig. 7.3) which were developed by Gvitech Technologies Company Ltd.

Attribute Modeling Method

Analogously, attribute modeling method is characterized by application of attribute-based modeling technique and software, such as geographic information system (GIS) like ArcGIS and MapGIS (www.mapgis.com.cn), remote sensing (RS) image processing system like ERDAS, and aerial photogrammetry system like VirtuoZo.

Generally, all the attribute-based modeling technique and software extract the three-dimensional information of terrain, buildings or constructions, and other 3D objects to generate integrated spatial database at first, and then to build 3D VR models based on the attributes in database. Clearly, the construction of the VR model is closely related to the spatial data acquisition, and it is easier to generate digital terrain model (DEM) and RS image texture. The higher spatial resolution of

Downtown Model of Ningbo   Downtown Model of Nanning

Fig. 7.3 VR model by means of visualization modeling method (Created by Gvitech Technologies Company Ltd.)

Fig. 7.4 VR model by means of attribute modeling method (Model of Xiamen City) (Created by Gvitech Technologies Company Ltd.)

RS images, the better quality of the VR model, and 3D VR model can represent a variety of properties. Usually, RS image is used to be the texture of DEM and building, therefore, this method could not suitable for creating small scale and fine model. Show as Fig. 7.4, the Model of Xiamen City, integrated DEM and Building together by Gvitech Technologies Company Ltd., was developed by means of attribute modeling method.

Integration Modeling Method

Integration modeling method is characterized by integrated application of graphics-based modeling technique and software, and attribute-based modeling technique and software at the same time, such as AutoCAD, 3Ds Max, Sketch Up, ArcGIS, MapGIS, ERDAS, and VirtuoZo. By means of the above software, spatial information of terrain and buildings can be obtained in various ways, and 3D VR model can be generated with good visualization properties. Obviously, integration modeling method overcomes the deficiencies of the former two methods, and combines the advantages of two of them. Therefore, integration modeling method is applied by many kinds of application users because it can build an ideal VR model (Doyle et al. 1998; Zhu et al. 2006). However, this method involves more kinds of techniques and requires more technical skill to support.

Developed by the U.S. Institute for Defense Analyses (IDS), the models of the quick response VR training system, such as the model of New York Downtown and the model of Washington DC, are constructed by this integration modeling method (http://virtualcities.ida.org/). In China, Gvitech Technologies and Digital City Research Center of Tsinghua Urban Planning & Design Institute developed Digital Beijing VR system, the VR models are also constructed by integration modeling

Model of the Forbidden City   Model of the Olympic Sports Center

Fig. 7.5 VR model by means of integration modeling method (Created by Gvitech Technologies Company Ltd.)

method (http://www.diciti.com/), such as the VR model of the Forbidden City which is the world cultural heritage and traditional Chinese architecture, and the VR model of the Gymnasium of National Olympic Sports Center in Beijing which is the world famous attraction and modern architecture (Fig. 7.5).

From the year 2004 to now, many kinds of web-based integrated VR application systems were worked out to service public users for both professional application and daily life requirement. Among all the systems, Google Earth is the most popular one with its integration of DEM, high-resolution RS image, multi-scale GIS map, and city 3D model with LOD, but also integrated 3D streets scenic video and business information together, which can satisfy most demands of system clients of all social levels. Others like Virtual Earth 3D, World Wind, ArcGIS Explorer, and Skyline Globe, are all integrated VR application systems for many usages. Similar web-based integrated VR application systems were developed by domestic companies in China, such as EV-Globe (http://www.ev-image.com/), GeoGlobe (http://www.geostar.com.cn/), and Diciti (http://www.diciti.com/), which represent the new trends of integrated VR.

Characteristics of VR Application in Urban Planning and Managing

In comparison, the application of VR technology in urban planning and managing started about 1980s in the Western countries, while China relatively late in later 1990s. The VR application process and characteristics appears difference from each other.

Characteristics of VR Application in the West

In the west countries, VR was first applied to urban building design rather than to urban planning in 1980s (Brooks 1986). However, in 1990s, the topics related to VR based urban planning were studied widely in detail, such as collaborative planning system (Shiffer 1992), interactive design in a virtual urban (Liggett et al. 1995), living with a virtual city (Day et al. 1996), virtual urban modeling (Jepson et al. 1996), VR as a tool for urban planning (Bourdakis 1997a), VRML for large urban models (Bourdakis 1997b), visualizing virtual urban environments (Batty et al. 1998), and VR for urban planning (Martin et al. 1998).

Furthermore, from 2000 to now, the achievements of VR application in urban revealed that VR technology was integrated with other information technology to meet more complex demands of urban planning and managing based on Internet, such as urban development simulation (Ben-Joseph et al. 2001), new interface to urban planning (Broll et al. 2003), VR as a multiuser interactive operation platform for public participation in urban planning and designing (Shen et al. 2003; Bourdakis et al. 2004), planning and designing alternatives evaluation for urban planning (Pettit et al. 2004; Drettakis et al. 2006), VR as a new tool for city planning (Sunesson et al. 2008), and VR as an online visualization tool for Internet-based urban design (Shen and Kawakami 2009).

Characteristics of VR Application in China

In China, the development process and characteristics of VR application in urban planning and managing were summarized according to incomplete statistics of the VR application achievements. Based on the information retrieval from China Knowledge Resource Integrated Database, more than 100 papers and thesis related

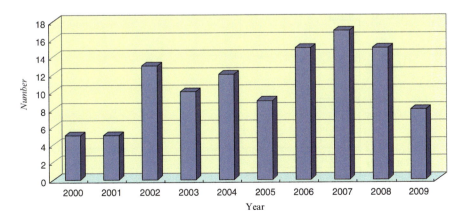

Fig. 7.6 Statistics of VR application achievements in urban planning by numbers

to VR application and urban planning and managing were published since 2000. All the achievements were classified and analyzed in following three aspects.

Firstly, by comparing the number of papers from 2000 to 2009 (Fig. 7.6), the time series characteristics of VR application in urban planning is as follows (1) the 10 years can be divided into two periods, first 5 years (from 2000 to 2004) and second 5 years (from 2005 to 2009). (2) In the first 5 years, VR application in urban planning was increasing gradually, and it became a hot topic in 2002–2004. (3) In the second 5 years, VR application in urban planning was much more popular in 2006–2008. (4) The first relatively cold year (2005) could be caused by the introspection for the development in 2002–2004, while the second relatively cold year (2009) probably because of the maturity and rational after 10 years development of VR application.

Secondly, by classifying all the papers based on the organization of the first author, such as institute and university, urban planning, urban managing, and company and business (Fig. 7.7), the characteristics were revealed that most of the achievements were published by institute and university, and much less of the achievements were published by urban planning and managing organization. The situation explains that institute and university are the backbone to push the VR application in urban planning, while the urban planning and managing organization, which could be the main subject, should be developed further more. At the same time, company and business are important partners for VR application in urban planning which should be enhanced. On the other hand, the situation shows that the VR applications in urban planning is still in the experimental study phase, and more practice needs further development, which means research opportunity for institute and university, technical challenge for urban planning and managing organization, and potential market for company and business.

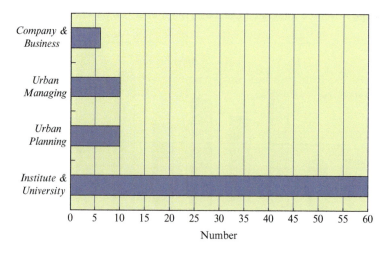

Fig. 7.7 Statistics of VR application achievements in urban planning by authors

7 Review of VR Application in Digital Urban Planning and Managing

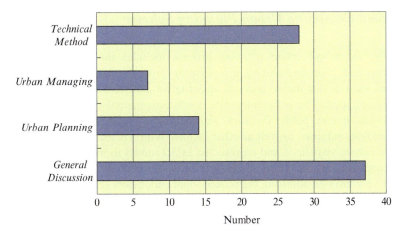

Fig. 7.8 Statistics of VR application achievements in urban planning by classes

Thirdly, by classifying all the papers based on the contents or research topics, such as general discussion, urban planning, urban managing, and technical method (Fig. 7.8), the characteristics were revealed that most of the achievements related to general discussion on the VR application in urban planning are application frameworks and technical approaches. The second large group of the achievements focused on the technical methods of VR application, especially some related key technology. Only a small amount of the achievements discussed urban planning, designing, and managing practice. The situation reveals that the current status of VR application in urban planning and managing is rather than premature, and urban planners and managers must take on the heavy responsibilities.

Aspects of VR Application in Urban Planning and Managing

The purpose of VR application in urban planning and managing can be summarized as three main aspects according to the application achievements. One is to aid urban planners to compile the urban planning or urban design scheme during the planning or design stage. Supported by VR technology, urban planners can discuss and improve the planning or design scheme in the interactive 3D visual inspection and Internet environment. The second one is to represent urban planning or urban design scheme through 3D VR model with full performance, in order to let experts to review the planning or design scheme and let public to participate planning or design process (Yang et al. 2000). The third one is to aid urban managers to implement the planning or design scheme in the urban construction process and to execute urban management duty based on the urban planning statute. VR technology has being played significant role in each of the aspects.

VR Aided Compiling Urban Planning Scheme

Basically, the whole urban planning process includes three steps: surveying, analyzing, and planning. During the planning process, many kinds of analysis methods, such as qualitative analysis, quantitative analysis, and spatial modeling analysis, are used. The population, site conditions, natural ecological resources, urban infrastructure, public facilities, and building environment, are all included in the analysis subjects, as well as other factors. With VR technology, most of the analysis can be visualized and represented in 3D model to make results easier to understand. Meanwhile, the multi-analysis and multi-alternatives can be compared interactively through adjusting the related input parameters (Zhu et al. 2003; Chen et al. 2005). Further more, the application of VR technique in urban planning allows the urban planners, planning evaluation experts, public citizens, and urban managers, to discuss within a common Virtual Environment (Shen et al. 2003; Shen and Kawakami 2009). By means of dynamic and interactive viewing the future VR buildings model, all the stakeholders mentioned above can review and evaluate the planning scheme comprehensively, which could not be achieved and could not be compared by traditional way. Therefore, along with the development of VR technology, its application in urban planning has been accumulated all over the world.

In United States, lead by Professor William Jepson, the Urban Simulation Team at the UCLA Department of Architecture and Urban Planning created an interactive city model of greater Los Angeles by integrating VR and GIS in 1995, which covered over 4,000 square miles (Liggett et al. 1995; Jepson et al. 1996). In United Kingdom, lead by Professor Martin Dodge, the Centre for Advanced Spatial Analysis (CASA) at University College London (UCL) developed Virtual London by means of VR and Internet GIS in 1998 (Martin et al. 1998). Meanwhile, some related key technology and technical methods and approaches were discussed. Some other research and practice of VR application in urban planning were done in Japan for developing Virtual Tokyo and in German for creating Virtual Berlin (Martin et al. 1998).

In the west, researchers and planners paid more attention to public participation in urban planning and design. The multimedia including text, video, and photography and spatial technology like VR and WebGIS in the Internet environment have integrated and used in urban planning and design in order to involve better public participation (Matsubara et al. 1991; Shiffer 2001). Public participation tools for planning and design were developed for participants to coordinate alternatives based on different scenarios, such as multi-user planning and design prototype for human or artificial design agents in a distributed network (Zamenopoulos et al. 2003), and public park visualization tool as an online design collaboration system (Shen et al. 2003).

In China, cities like Beijing, Shanghai, Tianjing, Guangzhou, and Shenzhen, have done related study and practice of VR application in urban planning (Wu and Pan 2002; Li and Guo 2003; Wang 2003). Among all the examples, Shenzhen

Fig. 7.9 3D VR model of Guangzhou for digital detailed urban planning (Created by Gvitech Technologies Company Ltd.)

should be the earlier one in China to practice VR application in urban planning and managing. The Urban Planning and Land Resource Bureau of Shenzhen started their VR application research and practice in urban planning in 1996. In 1998, the VR model and planning system of 8 km^2 central Shenzhen was developed with SGI workstation platform, which provided many kinds of help in urban planning and managing process (Li and Guo 2003).

In 2007, commissioned by the Guangzhou Urban Planning Bureau, the Guangzhou Urban Planning Automation Center launched a program titled Digital Detailed Urban Planning of Guangzhou, which marked VR technology applications in urban planning and design to enter a new phase (http://www.gzcc.gov.cn/). The program was planned to develop detailed VR model (Fig. 7.9) and VR system of 20 km^2 central Guangzhou based on WebGIS, RS, multi-media, simulation, and VR technology. With detailed real-time 3D VR model and planning attribute database, the urban planning process should be executed interactively, which could help urban planners to compare planning scheme and make decision much more easier (Yang et al. 2007).

VR Aided Representing Urban Planning Scheme

It was a dream for urban planners and designers to experience the urban planning and design scheme and to get the realistic feel in the old days. By using the VR technology, which characterized with interactive, immersive, and imagination, urban planning or design scheme can be represented precisely in 3D VR model

Fig. 7.10 3D VR model of Yuzhong district planning of Chongqing City (Created by Gvitech Technologies Company Ltd.)

like real 3D environment, and urban planners or designers can walk or fly through future virtual city based on the planning or design scheme (Qian and Liu, 2004; Zheng et al. 2005; Luo, 2008; Shi et al. 2008; Zhang and Chen, 2008). Not only for planners or designers, VR model of urban planning or design scheme also can help public participant, planning or design expert, and urban manager to know the real idea of planners' or designers' and understand each other (Gong et al. 2007). Because of that, from the late twentieth century, many urban planning and design institute to represent their planning or design scheme with VR technology, such as Shenzhen, Shanghai, Wuhan, Chong Qing (Fig. 7.10), and Beijing. More and more VR application achievements are accumulated.

In 2000, required by Shanghai Urban Planning Information Center, the Institute of Shanghai Urban Planning and Design represented the planning scheme of Huangpu River Sides based on Autodesk 3D MAX VIZ. The VR models of the planning scheme covered 6.7 km^2 area of both Huangpu River Sides, which allowed users to interact with various 3D virtual scenes. For example, a person, vehicle, or aircraft could be imitated, and the walking or flying route, orientation, and angle also could be selected by the user, in order to help user to do the planning analysis and evaluation. It is proved that the VR model representing of the planning scheme of Huangpu River Sides was very successful for helping most of the related aspects to observe, examine, analyze the planning scheme in realistic feel and in objective manner (Wang 2003).

In 2003, organized by China Urban Planning Society, the "national competition of VR application and 3D visual simulation in urban planning" attracted 21 urban planning related institutes submitted 40 achievements, such as urban VR model, VR system for urban planning, urban visualization system, urban simulation system, and so on. Carefully reviewed by the Competition Committee, seven achievements got the first prize (one), second prize (two), and third prize (four)

7 Review of VR Application in Digital Urban Planning and Managing 145

accordingly (Dai et al. 2003). The characteristics of the winning achievements were summarized as follows (1) VR application in urban planning is very extensive because of the winning achievements are related to different kinds of urban planning. (2) VR application in urban planning has been gradually transfer from the research to the practice phase because most of the winning achievements were participated or organized by urban planning and managing organization. (3) VR application is not just in the traditional urban planning, but has been linked with the digital urban planning.

In 2008, before the Olympic Games, digital Beijing VR system was developed and uploaded on the website by Gvitech Technologies and Digital City Research Center of Tsinghua Urban Planning and Design Institute (http://www.diciti.com/). Digital Beijing VR system integrated RS, GIS, and VR technology and covered most of Beijing construction area with 3D VR models of both reality buildings and Virtual planning or design schemes. The most important components of the VR system are Tian'anmen Square, the Forbidden City (Fig. 7.5), The Hutong historical conservation region, and the Olympic Sports Center of Beijing (Fig. 7.5). Most of the buildings or constructions are represented by vivid 3D VR models with brief attributes, which provide users full information, 3D real feels, and unforgettable experience and impression.

In 2010, one of the most important international activities in China is Expo 2010 Shanghai. In order to let people all over the world to experience Expo 2010 Shanghai by Internet, the online Expo 2010 Shanghai China platform has been developed by the Digital Exhibition Service Provider (http://en.expo.cn/index.html), named Crystal CG (http://www.crystalcg.com/). Integrated VR and multimedia technology, the online Expo 2010 Shanghai has been designed to include

Fig. 7.11 3D VR model of Yuzhong district planning of Chongqing City (Created by Gvitech Technologies Company Ltd.)

Browsing Pavilions and Experiencing Pavilions, which covering many countries like Italy, USA, Saudi Arabia, and Singapore, the municipalities of Beijing and Hong Kong SAR, the autonomous regions, and the provinces of China, and so on (http://en.expo.cn/index.html). At the same time, a VR exhibition hall was constructed with the integration of detailed 3D VR models of Expo 2010 Shanghai (Fig. 7.11) and particular attribute database (http://www.gvitech.com/), which let the Expo 2010 visitors to experience Expo 2010 Shanghai in both Virtual Environment and reality situation, as well as let the officials to manage Expo 2010 Shanghai in digital way.

VR Aided Managing Urban Development Process

Evaluating and approving urban planning, construction, and design scheme is one of the core tasks for urban management bureau. By evaluating and approving the schemes, the planning and design will become the real building or construction to form the real urban environment. In other words, the urban planning, construction, and design scheme affects the urban development and urban environment directly.

Facing the rapid urbanization and lots of approval mission of urban planning and design schemes, it is very urgent to adopt VR and related spatial information technology to aided urban managers to perform their urban management and construction duty (Guan, 2004). Only if the VR and related technology applied, the evaluation and approval can be done with more scientific, objective, and effective way. And the urban environment will be built better and better for human life. Therefore, the urban planning and managing bureau practiced a lot for VR application in urban managing.

In 2002, the VR system of Tianjin Economic and Technological Development Zone was developed by the Institute of Guangzhou Urban Information Studies. The VR system played an important role by being used to represent the planning and designing scheme, as well as aided to evaluate and approval planning and designing scheme for urban management during the urban development process (Zhu et al. 2003).

From 2006 to now, the Information Center of Zibo Urban Planning Bureau has strengthened their efforts to study VR application in urban planning and managing. A VR based 3D Spatial Information System for urban planning and managing was developed, and 3D VR model of the central zone and other five key regions of Zibo city were built. The system and the model have been used in the evaluation and approval of urban planning and designing scheme, and were certified to play very well and crucial role during the urban planning and managing process (Wang et al. 2008).

In May 2009, the Institute of Beijing Surveying and Mapping undertaken a project titled "VR application study in the detailed urban planning and management" which financed by Beijing Science and Technology Bureau. The project plans to develop detailed urban planning oriented 3D VR system, which will integrate RS, GIS, VR technology to aid urban management. The system will includes 3D VR models of both current buildings based on surveying and future buildings based on planning. All the 3D VR models will integrate with detailed urban planning attributes, such as floor area ratio (FAR), green space ratio (GSR), building density (BD), and so on. It is obvious that through the implementation of the project, the evaluation and approval process of planning and design scheme will based on 3D VR model and the VR system, which will make the urban management process more scientifically (http://www.smibj.com).

Since 1999, along with the idea and concept of digital earth and digital city, virtual world and virtual city were focused and developed very quickly. Besides the popular system like Google Earth, Virtual Earth, and World Wind, more and more virtual city or digital city were developed in China in recent years, such as Digital

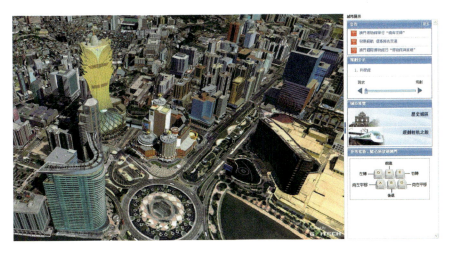

Fig. 7.12 3D VR model and interface of Digital Macao (Created by Gvitech Technologies Company Ltd. and Tsinghua University)

Fig. 7.13 3D VR model of Central Park region of Changchun City (Created by Gvitech Technologies Company Ltd.)

Fig. 7.14 3D VR model of Virtual Forbidden City (Created by Gvitech Technologies Company Ltd.)

Fig. 7.15 3D VR model of Digital Yuanmingyuan (Created by Gvitech Technologies Company Ltd. and Tsinghua University)

Beijing, Digital Macao (Fig. 7.12), and Virtual Changchun (Fig. 7.13) (http://www.gvitech.com/). The main purpose of developing digital city or virtual city is to aid urban government to manage the overall urban development by means of virtual city, which representing the current urban construction and future urban planning blue-map by 3D VR models. The virtual city also can to aid citizens to understand and to participate the city development process. Meanwhile, 3D VR model and system were also used to manage and conserve the historical region of city, such as virtual Forbidden City (the Palace Museum, Fig. 7.14), virtual Temple of Heaven (http://www.vrac.iastate.edu/), and digital Yuanmingyuan Park (Old Summer Palace, Fig. 7.15) of Beijing (http://www.gvitech.com/). Maybe the 3D VR model and system is a new way to conserve the historical region of city.

Perspective of VR Application in Urban Planning and Managing

There are two aspects to analyze the perspective of VR application in urban planning and managing: one is the demands of development and application of VR technology itself, and the other is the demands of development and update of urban planning and managing technology. In this article, the second aspect was taken into account in two phases, such as demands of current traditional urban planning, and the perspective of future digital urban planning.

Current Traditional Urban Planning and Managing Needs VR to Aid

On the one hand, with the rapid urbanization and urban development, more and more attentions are gradually focused on urban planning and managing. The application of VR technology can aid urban planners, the public, planning experts, and urban manager work together to improve the process of urban planning and managing (Wu and Pan 2002). Application of VR technology can help urban planners to work out scientific planning and designing scheme by means of quantitative analysis and scientific forecasting. It can help the public to participate the urban planning process by means of the visualization representation of the scheme. It can help planning experts to evaluate the scheme objectively by means of VR models and VR system. And it can help urban managers to approval the scheme rationally by means of ideal comparison. The final result is that an ideal urban environment will be constructed for all human beings to live, which is absolutely the ideal aim of us.

On the other hand, with the advanced development of VR technology itself, its application in urban planning and designing will be further expanded. According to the new characteristics of VR system, what it can provide for urban planning is not only 3D VR model, but VR system integrated GIS spatial analysis methods and quantitative urban planning models. Therefore, application of VR can help urban planning and managing in the many aspects, such as calculation of floor area ratio (FAR), green space ratio (GSR), and building density (BD), analysis of the building height and volume, the building sunshine, and the building style, and comparison of trees along the street, pavement of pedestrian, and sculpture of inside the park (http://www.gvitech.com/). In a word, it is believable that with the integration application of VR, GIS, and other spatial information, urban planning and managing process will be improved in full aspects.

Future Digital Urban Planning and Managing Needs VR to Support

Digital Urban Planning is a new kind of urban planning based on information infrastructure, spatial data infrastructure, and planning and managing system of digital city. It is a new developing research field which came up with the development and combination of traditional urban planning theory and modern spatial information technology (Wu et al. 2001; Liu et al. 2003; Wang and Liu 2003; Yu and Wu 2004; Cui 2005; Dang et al. 2007; Liu et al. 2009). Therefore, digital urban planning is a kinds of newly urban planning theory, method, technology, and system, which integrated many types of spatial information technology with urban planning, which will be an important component of digital city (Xu et al. 2002).

During the process of digital urban planning, all the basic materials of the city will be digital information. The purpose of digital urban planning is to determine the development goals, urban land use, urban spatial pattern, information infrastructure, spatial data infrastructure, and other integrated construction project of both realistic city and digital city. The contents of digital urban planning can be deduced as two aspects. One is Physical and Social Planning which is pay more

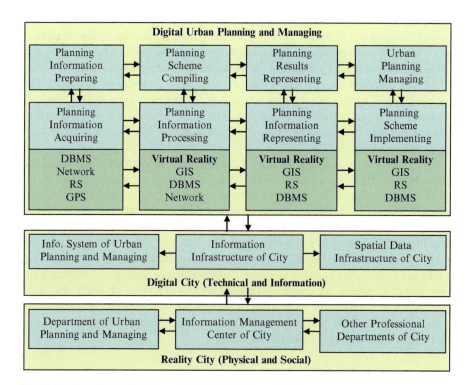

Fig. 7.16 General technical process of digital urban planning and managing

attention to the realistic city, while the other is Technical and Information Planning which is mostly concern to digital city (Fig. 7.8) (Dang et al. 2005).

Determined by the characteristics of spatial information technology (such as VR, GIS, and RS) and contents of digital urban planning and managing, the whole process of digital urban planning and managing can be divided into four periods, such as planning information preparing, planning scheme compiling, planning result representing, and urban planning managing. Many kinds of technologies need to be applied during the whole planning and managing process, and each period involves different technology (Fig. 7.16). Digital information and spatial database will be very important planning medium. And all the planning processes are supported by digital city technical platform, which is the main difference between traditional urban planning and digital urban planning (Dang et al. 2005).

Obviously, digital urban planning and managing will change the current single objective and static planning model to future multi-objective and dynamic planning model, in order to achieve integrated optimization planning objectives of physical, social, environmental, and technical (Dang et al. 2005). It is not the simple digitalization of the traditional urban planning, but to develop new theory, method, and technology to transfer the urban planning and managing into a new era (Xu et al. 2002). Among all the technologies, VR is the one which will play key role to aid digital urban planning and managing, namely, VR application will be a bright future.

Conclusion

To sum up this research, two main conclusions were worked out. One is that VR technology has been played very important role in urban planning and managing process both in the West and in China in the past two decades, especially in the recent years for the development of digital city or virtual city. However, according to the characteristics analysis of the achievements of VR application in urban planning and managing, more VR application practice needs further development to transfer the experimental study to real implementation. The other one is that not only the current traditional urban planning and managing process needs VR technology to aide, but also the future digital urban planning and managing calls VR and other spatial information technology to support the whole process scientifically. Therefore, the VR application in urban planning and managing will be a bright future.

Acknowledgements This study was supported by National Natural Science Foundation of China (51078213 and 50678088), and No. 115 National Projects of Scientific and Technical Supporting Programs Funded by Ministry of Science & Technology of China (2006BAJ14B08 and 2006BAJ07B05). Thanks for all scholars involved in the project.

References

Allen Bierbaum, Christopher Just, Patrick Hartling, et al. 2001. VR Juggler: A Virtual Platform for Virtual Reality Application Development, Proceedings of Virtual Reality 2001 Conference (VR'01), Yokohama, Japan, March 13–17, 2001.

Batty M., Dodge M., Doyle S., et al. 1998, Visualizing Virtual Urban Environments, CASA Working Paper 1. Centre for Advanced Spatial Analysis, University College London.

Beijing Institute of Surveying and Mapping, 2009. Study on Virtual Reality Application in Detailed Urban Planning and managing. http://www.smibj.com

Ben-Joseph E., Ishii H., Underkoffler J., et al. 2001. Urban simulation and the luminous planning table bridging the gap between the digital and the tangible. Journal of Planning Education and Research (21): 195–202.

Bourdakis, V. 1997a. Virtual Reality: A Communications Tool for Urban Planning, Proceedings of CAAD: Towards New Design Conventions, edited by A. Asanowicz and A. Jakimowitz, Technical University of Bialystok, 45–59.

Bourdakis, V. 1997b. The Future of VRML on Large Urban Models, Proceedings of the Fourth UK VRSIG Conference, edited by Richard Bowden, 1st November 1997, Brunel University.

Broll W., Stoerring M., & Mottram C. 2003. The Augmented Round Table: a new Interface to Urban Planning and Architectural Design. In M. Rauterberg, M. Menozzi, & J. Wesson (Eds.), Ninth IFIP TC13 International Conference on HCI-INTERACT, 1103–1104. Zurich, Switzerland: IOS Publications.

Brooks, F. P. J. 1986. Walkthrough: A dynamic graphics system for simulating virtual buildings. In F. Crow & S. M. Pizer (Eds.), Workshop on Interactive 3D Graphics. p. 9–21.

Carolina Cruz-Neira, Daniel J. Sandin, Thomas A. DeFanti. 1993. Surround-Screen Projection-Based Virtual Reality: The Design and Implementation of the CAVE. Proceedings of SIGGRAPH'93, Anaheim, California, August 1-6, 1993.

Chen Ming, Gu Chaolin, Duan Xuejun. 2005. Virtual Reality Technique Application Research in Comprehensive Plan[J]. Computer Engineering. 31(21):225–227.

Cui S., 2005. Study on City Construction Information Archives Management System based on the Digital Urban Planning Theory [J]. Journal of Construction Technology, 2005(3), 64.

Dai Feng, Mao Qizhi, ZHong Jiahui. 2003. Application of Virtual Reality and 3D Visualization simulation in Urban Planning [J]. City Planning Review. 27(8): 39–41.

Dang A., Mao Q, Shi H. 2007. Integration of Spatial Information Technology for Digital Urban Planning [J]. Proceedings of SPIE, 6754, 67540A-1–67540A-11.

Dang Anrong, Shi Huizhen, Han Haoying. 2005. Study on the System of Technical Methods for Digital Urban Planning. Proceedings of ISPRS Workshop on Service and Application of Spatial Data Infrastructure, XXXVI(4/W6):167–170. Oct.14-16, 2005. Hangzhou, China

Day A., Bourdakis V. & Robson J. 1996. Living with a Virtual City [J]. Architectural Research Quarterly, Vol.2, 84–91.

Doyle S., Dodge M., Smith A. 1998. The Potential of Web-based Mapping and VR Technology for Modeling Urban Environment [J]. Environment and Urban Systems, 22 (2): 137–155.

George Drettakis, Maria Roussou, Alex Reche, et al. 2006. Design and Evaluation of a Real-World Virtual Environment for Architecture and Urban Planning. http://citeseerx.ist.psu.edu/viewdoc/download? 2006(5): 1–21.

Gong Yong, Pu Xiaoqiong, Zhang Xiang. 2007. Improve the Degree of the Public Participation in the City Planning—Explore the Application of the Virtual Reality Technology in the City Planning [J]. Chinese and Overseas Architecture. 2007(1): 27–30.

Grigore C. Burdea & Philippe Coiffet. 2003. Virtual Reality Technology. Wiley-Interscience, A John Wiley & Sons, Inc. Publication.

Guan Ruihua. 2004. The Application of Virtue Reality Technique in Urban Planning Management [J]. Modern Computer. 2004(9): 54–57.

Jepson W., Liggett R., Friedman S. 1996. Virtual Modeling of Urban Environments [J]. Presence, 5 (1): 72–76.

7 Review of VR Application in Digital Urban Planning and Managing

Kaj Sunesson, Carl Martin Allwood, Dan Paulin, et al. 2008. Virtual Reality as a New Tool in the City Planning Process [J]. Tsinghua Science and Technology. Volume 13, Number (S1), October 2008, 255–260

Lee Adams. 1993. Visualization and Virtual Reality [M]. McGraw-Hill Publishing, NY, 1993.

Li Chunyang, Guo Yongming. 2003. The Application of Virtual Reality in Urban Planning [J]. Science of Surveying and Mapping. 2003(1): 38–44.

Liggett, R., Friedman, S. & Jepson, W. 1995. Interactive Design/Decision-Making in a Virtual Urban World: Visual Simulation and GIS, Proceedings of the Fifteenth Annual ESRI User Conference, Palm Springs, USA.

Liu Xinquan, Zhao Lin, Niu Xumiao, 2003. Digital Planning—Opportunities and Challenges Modern Urban Planning Facing [J]. Economic Geography, 2003(2), pp. 161–164.

Liu Ju, Wang Kai, Chi Yingbo. Application of VR in Urban Planning and Managing [J]. China Science and Technology Information. 2009(02): 276–278

Luo Ming. 2008. The study and develop of city planning simulation system based on 3D Virtual Reality [J]. Fujian Architecture & Construction. 2008(12): 18–19

Martin Dodge, Simon Doyle, Andy Smith & Stephen Fleetwood. 1998. Towards the Virtual City: VR & Internet GIS for Urban Planning. Virtual Reality and Geographical Information Systems, Birkbeck College, 22nd May 1998.

Matsubara M., Matsumoto N., 1991. A study on the validity of the landscape simulation methods – By investigation of bygone research and experimentation of vision and perception. Journal of the planning institute of Japan (26), 385–390, October.

Pettit C., Nelson A., Cartwright W., 2004. Recent advances in design and decision support systems in architecture and urban planning. In H. J. P. Timmermans & J.P. van Leeuwen Eds. pp.53–68. Dordrecht/Boston/London: Kluwer Academic Publishers.

Proposal Guide of National No. 863 High-tech Project, http://www.863.org.cn/863_105/index. html.

Qian Lexiang, Liu Runda. 2004. Practices of Technology of Virtual Reality in City Planning [J]. Computer Engineering. 30(11): 153–158.

Rhodes, Bradley (1997). A Brief History of Wearable Computing. MIT Wearable Computing Project. http://www.media.mit.edu/wearables/lizzy/timeline.html.

Shiffer, M.J., (1992) Towards a collaborative planning system, Environment and Planning B: Planning and Design, Vol.19, 709–722.

Shiffer, Michael J., 2011. Spatial multimedia for planning support, planning support systems. In R. K. Brail & R. E. Klosterman Eds. pp. 309–385, California: ESRI Press.

Shi Huizhen, Dang Anrong, Chi Wei. 2008. Study on Urban Planning and Design Aided by VR in Real-time [J]. Geomatics World. 2008(05): 61–66

Vassilis Bourdakis, B Rudiger, B Tournay, H Orbaek. 2004. Developing VR Tools for an Urban Planning Public Participation ICT Curriculum; The PICT Approach. Available from fos.prd. uth.gr, ISBN: 0954118324, Pages: 601–607

Wang Kanghong, Liu Li, 2003. Analysis of Digital Urban Planning System Technology [J]. Journal of Geomatics, 2003(4), pp. 42–45.

Wang Lei. 2003. Application of 3D Visualization for aiding urban planning and managing [J]. Computer Aided Design and Intelligent Building, 2003(5):66–71.

Wang Ruiguo, Jiang Xue, Wang Hui. 2008. Application of VR in Urban Planning and Planning Scheme Evaluation. Proceedings of China National Urban Planning Annul Conference. Dalian, China. 2008

Web Info: The Story so far about VR, http://www.fbe.unsw.edu.au/research/VRArch/timeline.htm.

Web Information, http://www.gzcc.gov.cn/zhzxinfo.aspx?id=27467.

Web Information: Daniel Sandin – CAVE, http://www.artmuseum.net/w2vr/timeline/Sandin.html.

Web Information: Section 17 - Virtual Reality, http://accad.osu.edu/~waynec/history/lesson17. html.

Web Information: The Father of Virtual Reality, www.mortonheilig.com.

Wu Fangsheng, Pan Zhigeng. 2002. Application of virtual reality technology in urban planning and design [J]. Journal of China Institute of Metrology. 2002(4): 264–273.

Wu Shuxia, Ai Jixi, Guo Ping, 2001. Digital Urban Planning—Concept and Support Technology. In: Theory and Practice of Digital City, edited by Lai Ming and Wang Menhui, Guangzhou: World Book Press, 2001(9), pp. 234–241.

Xu Hong, Yang Lihang, Fang Zhixiang. 2002. Study on the Support Technology System of Digital Urban Planning [J]. Journal of Wuhan University (Engineer Edition). 2002(5):43–46.

Yang Jianguo, Huang Ling, Gao Jianfeng. 2007. Application of 3D Simulation in Urban Planning [J]. Shanghai Urban Planning. 2007(6): 44–47.

Yang Kejian, Liu Shuyan, Chen Dingfang. 2000. Virtual Reality and City Planning [J]. ACTA Simulata Systematica Sinica. 2000(03):207–209, 225.

Yu Zhuo, Wu Zhihua, 2004. Foundation of Spatial Data for Digital Urban Planning [J]. Journal of Geomatics, 2004(1), pp. 28–29.

Zamenopoulos, T., Alexiou, K., 2003. Structuring the plan-design process as a coordination problem: The paradigm of distributed learning control coordination. Advanced spatial analysis. In Paul Longley & Michael Batty Eds. pp.407–426, California: ESRI Press.

Zhang Lvsong, Chen Zhigang. 2008. VR Technology Aided the Informatization in Urban Planning [J]. City Residence. 2008(7): 70–71

Zheng Weimin, Cao Wen, Cao Yongqing. 2005. Application of Virtual Reality technology in Urban Design [J]. Chinese and Overseas Architecture. 2005(1): 21–23.

Zhenjiang Shen, Mitsuhiko Kawakami. 2009. An online visualization tool for Internet-based local townscape design. Computers, Environment and Urban Systems 34 (2010) 104–116

Zhenjiang Shen, Kawakami, M., Kishimoto, K. 2003. Study on development of on-line cooperative planning and design system using VRML and JAVA – A case study on public park planning and design. The 8th international conference on computers in urban planning and urban management (CUPUM), in CD-ROM, May 2003. Sendai, Japan.

Zhu Lu, Wu Suzhi, Lin Huimin. 2003. Application and Perspective of Virtual Reality in Urban Planning: Taking Tianjin as an example [J]. City Planning Review. 2003(08): 47–48.

Zhu Weijie, Zou Zhenrong, Zou Jie. 2006. Study on Large-scale City VR Simulation System and Its Realization [J]. Acta Simulata Systematica Sinica. 29(2): 125–127.

Chapter 8
Virtual Fort San Domingo in Taiwan: A Study on Accurate and High Level of Detail 3D Modelling

Shih-Yuan Lin and Sheng-Chih Chen

Introduction

Cultural heritage is evidence of past human activity. Appropriate and correct documentation of cultural heritage is critical for the purposes of conservation, management, appraisal, assessment of the structural condition, archiving, publication and research. As the importance of cultural heritage documentation is well recognized, the necessity to document our heritage has been increasing globally (Patias 2006).

Regarding methods of documentation, monument restoration projects were traditionally performed based on archive literature review and on-site investigations. Together with interpretations of style and materials of construction, the historic buildings and monuments were repaired and preserved. Due to the rapid development of hardware and software for 3D object displaying and manipulation, digital archiving has become a trend nowadays for preserving and demonstrating heritages instead of the traditional method. Hence, this study focuses on the digital preservation of historic buildings and scenes. Digitalization techniques from the fields of geomatics engineering and virtual reality (VR) are introduced in this paper.

Models produced by techniques of geomatics engineering and virtual reality offer various advantages. The geomatics models emphasize geometric accuracy, while the models produced in virtual reality focus on the level of detail (LOD) of the buildings and scenes. To maximize the benefits contributed by the two models, this chapter aims to integrate the two techniques to provide models with both high accuracy and high level of detail.

S.-Y. Lin (✉)
Department of Land Economics, National Chengchi University, Taipei, Taiwan
e-mail: syl@nccu.edu.tw

S.-C. Chen
Master's Program of Digital Content and Technologies, College of Communication, National Chengchi University, Taipei, Taiwan

Z. Shen, *Geospatial Techniques in Urban Planning*, Advances in Geographic Information Science, DOI 10.1007/978-3-642-13559-0_8,
© Springer-Verlag Berlin Heidelberg 2012

Fort San Domingo is one of the oldest buildings designated as a first-grade historical site in Taiwan. Having been affected by the colonial domination of different countries, the outlook and structure of Fort San Domingo was consequently modified. In order to preserve history, document and record the heritage, digital archiving of Fort San Domingo was performed using the proposed integrated 3D modelling techniques. Firstly, geomatics techniques were applied to model the terrain and historical buildings across the site of Fort San Domingo. Then the geomatics models were updated using VR models, thereby providing a virtual scene with high geometric accuracy and high level of details. The 3D models of the two buildings on the test site were constructed based on integrated strategy. The overall workflow, assessment and discussion are described below.

Literature Review Regarding Geomatics Techniques for 3D Modelling

In order to investigate solutions for modelling the Fort San Domingo, geomatics techniques currently used for measurement and modelling were reviewed. The review included terrestrial surveying methods adopting total station instruments and terrestrial laser scanners, and image-based method of photogrammetry. Through the introduction of the techniques and discussion of the related applications, appropriate methods to be applied and noteworthy principles were determined.

Terrestrial Surveying with Total Station Instruments

The essential task of building modelling is to determine the 3D coordinates of feature points on the buildings of interest. This is achieved by the measurement of angles and distances to the specific points (Bannister et al. 1998; Uren and Price 2006). In order to investigate the feasibility of automated measurement of terrestrial surveying, total station instruments are selected for considering the possibility in the modelling task.

The total station first appeared in the late 1980s as a result of the development of electronics and computerisation of surveying instrumentations (Kavanagh and Bird 1992; Sakimura and Maruyama 2007). With the rapid development of electronics and computer technology in the 1990s, the total station underwent a rapid refinement in its functionality, precision and lowering of the cost. Total stations are precision electronic instruments, which incorporate a theodolite with an Electromagnetic Distance Measurement (EDM) unit, an electronic angle measuring component, data storage and a computer or microprocessor (Wolf and Ghilani 2002). With EDM equipment, either phase shift or time pulsed methods are used to measure distance. The phase shift (or phase comparison) method utilises continuous electromagnetic waves. Here the distance is measured as a phase difference with ambiguity resolution. The pulsed method uses the pulses of laser radiation to derive the distance from computing transit times. This technique is also known as the time-of-flight method. It is noticeable that the accuracy obtained with a phase shift method is typically higher than with time pulsed instruments, but the instruments using pulsed laser technology can measure longer distance than the ones using phase shift method (Uren and Price 2006).

Total station instruments have been broadly applied in various 3D modelling projects. Considering the characteristics of low cost and high effectiveness, Matori and Hidzir (2010) adopted a total station to collect terrain data for producing digital terrain model (DTM). Further applications employed resultant DTM, such as engineering simulations of cut-and-fill volume calculation and hydrology flow direction and flow accumulation, were demonstrated. Based on the same method, Bevan and Conolly (2004) used this instrument to generate a terrain model of Kythera Island in Greece for an archaeological research purpose.

In a structural documentation and analysis project, a modelling system was designed to survey the scissors arches of Wells Cathedral in the U.K. (D'Ayala and Smars 2003). Considering the on-site accessibility problems, non-contact geomatics techniques were favorable in the project. A solution was therefore identified in the form of a hybrid approach, combining elements of terrestrial survey using a total station and digital photogrammetry. The total station was used to measure points on the specific structural components. Similar applications applied total station instruments were found in the modelling of the Oudenoord building at the University of Applied Science of Utrecht in Netherlands (Goedhart and Wolters 2008) and a construction of an augmented reality environment (Newman et al. 2005).

Total stations also introduced in modelling 3D objects of interest. The Maritime Archaeology Programme carried out in University of Southern Denmark aims to protect and record the underwater cultural heritage (Maarleveld 2009). In this programme, Hyttel (2010) applied a total station to model an iron 12 pounder lifted from the wreck of British ship "St. George". Points at important angle and curve which would later be used to create 3D models were observed. As a result the gun was reserved in a 3D format and could be visualized in any digital environments.

Terrestrial Surveying with Terrestrial Laser Scanning

As defined by Boehler and Marbs (2002), terrestrial laser scanning (TLS) is a technique which allows the collection of high-density point clouds distributed over a given region of an object's surface by measuring their three-dimensional coordinates automatically in near real-time. Since the early 1990s, terrestrial laser scanning has been increasingly adopted in the surveying market as an efficient alternative 3D measurement system with respect to photogrammetry or geodetic methods (Guarnieri et al. 2005). Due to the capability of measuring an enormous amount of points at a high rate in near real-time, complete and detailed 3D models of objects can be efficiently and easily created from acquired point clouds. These features have allowed laser scanning technologies to be widely applied in industrial metrology, deformation analysis, civil engineering, city modelling, cultural heritage, topographic surveying, and engineering geodesy (Tucker 2002; Barber 2003; Schulz and Ingensand 2004).

The general laser scanning workflow involves placing the laser scanner at locations about the survey site and measuring to a number of control points (e.g., retroreflective targets or identifiable features) and the actual object of interest. The scanner is then moved to the second location and at least three common control points from the first scanner location are measured. The common measurements to the control points are then used to relate the scans together by registration (Lemmon and Biddiscombe 2006). At the last stage, the modelling of the whole combined point cloud or specific components is performed depending on the requirements of the application. In addition to the general workflow, specific details noteworthy during the scanning procedure are introduced below.

Prior to the implementation of scanning, laser scanning network design is essential if multiple scans are required. To achieve this, Lahoz et al. (2006) emphasize that the optimal station number is decided by considering the efficiency of the multi-station orientation procedures and the assurance of a global coverage of all desired surfaces. In addition, the positions of the scanner are arranged considering the incidence ray angle at the façade and the maximum scanning distance. Once the network of the scanner has been resolved, the network of the control points is designed and arranged accordingly. At the data acquisition stage, scanning density is one of the parameters required in the laser scanning settings. Hoffman (2005) suggests that an ideal spatial resolution should be equal to half of the measurement accuracy, ensuring the project receives the amount of data required.

There have been many research projects involving the scanning of historical structures for cultural heritage documentations. For example Levoy et al. (2000) used a custom laser triangulation scanner built by Cyberware to scan ten statues by Michelangelo. The images of the statues shot by a still video camera were mapped onto the resultant meshed model, rendering full-resolution and full-colour 3D models of the statues. A project sponsored by English Heritage Archaeology Commissions Team employed various terrestrial laser scanners, including a Leica HDS2500, a Zoller and Froelich Imager 5003 system and a Riegl LMS Z320 system, to measure Tynemouth Priory and Clifford's Tower in the UK (Barber et al. 2003; Bryan et al. 2004), for cultural heritage recording. Bonora et al. (2005) used a Leica HDS2500 scanner to create a 3D scanned model describing the dome intrados of the Colleoni Chapel in Italy. In addition to the 3D scanned models, 2D cross-sections were also extracted for architectural describing and investigating historical buildings in the Meldorf Cathedral in Germany, a measurement was performed with a Zoller and Froelich Imager 5003 system (Sternberg 2006). After long-term observations, deformations were determined by comparing the grid points, linear structures and surfaces acquired at different epochs. Another example for deformation monitoring using TLS technique could be found in Vezočnik et al. (2009).

Terrestrial laser scanning was introduced in geophysical and archaeological survey fieldworks conducted at the site of Tiwanaku in Bolivia, which was recognized as a birthplace for the people of the Andes (Goodmaster and Payne 2007). In this project, two scanning systems were used for long and short range scanning respectively. An Optech ILRIS-3D system was applied to collect 3D point clouds of large area with a resolution of approximately 1–3 cm. While a Konica-Minolta VIVID 9i system was adopted to perform high-resolution scanning (sub-centimeter resolution) on small artifacts, osteological elements, etc. The fusion of the scanned data formed the basis for scientific visualizations and virtual reconstructions of the site. For a detailed scanning purpose, the Maritime Archaeology Programme used FARO Arms along with a 3D laser scanner to create digital renderings of timbers from the early modern "Wittenbergen" wreck that sank in the Elbe (Stanek and Ranchin-Dundas 2010).

Image-Based Measurement Method

Among image-based measurement methods, photogrammetry is the most recognized technique in the field of geomatics. Fryer (2001) defines photogrammetry as "the science and art of determining the size and shape of objects through image analysis". The 3D coordinates of a specific object in space can be computed from the corresponding 2D information extracted from recorded photos through photogrammetric processing. Thompson (1962) further stated that photogrammetric methods of measurement are especially useful in conditions where the object to be measured is inaccessible, or when the object is not rigid and its instantaneous dimensions are required. Due to the features of rapid data acquisition, non-contact measurement, permanent recording, and capabilities of measuring deformation and movement, photogrammetry demonstrates its capacity to be used in many applications involving measurement (Torlegard 1980; Waldhausl 1992). Along with the developments in micro-electronics and semiconductor technology since the 1990s, digital photogrammetry has been driven by the development of new sensors (such as solid state cameras) and more powerful computers. The improving efficiency in image processing and automated measurement has resulted in an increasing number of fully automated systems and diversity of photogrammetric applications (Mills 1996; Wolf and Dewitt 2000; Fryer 2001; Gruen 2001).

Due to the characteristic of non-contact measurement and capability of offering accurate products, the image-based method of photogrammetry has remained a popular technique in 3D modelling for some time. According to the characteristics of the objects to be measured, cameras can be arranged on airborne platforms or on the ground to take images. For example, when performing topographic mapping tasks, a camera is fitted onboard an airborne platform and the so-called aerial photogrammetry is carried out to generate 3D digital terrain model and ortho-rectified images. This method has been well accepted. The applications and related discussions can be found in Baltsavias et al. (1996), Acharya et al. (2000), Sauerbier (2004), Zhang and Gruen (2004), Barnes and Cothren (2007) and Ion et al. (2008).

In addition to aerial photogrammetry, close range photogrammetry has also been introduced to a variety of non-topographic modelling projects. D'Apuzzo and Kochi (2003) developed an imaging system consisting of five IEEE-1394 video cameras to construct 3D human face model. The results were of great potential for automatic person identification from image data. Dunn (2009) adopted consumer-grade digital cameras to produce surface model of mining land. It was suggested that close range photogrammetry was a low cost technique but could yield accurate measurements even under a constrained environment.

In the project conducted by Psaltis and Ioannidis (2006), a concrete beam was monitored. By employing photogrammetry technique, its deformation occurred during a loading experiment was accomplished. Also, the results showed that the standards of an automatic, inexpensive, almost on-line calculation of displacements with accuracy better than 1 mm were fully met. In the research presented by Jauregui et al. (2003), photogrammetry was suggested as an attractive method for the purpose of structural health monitoring. A photogrammetric monitoring system was therefore developed to determine the vertical deflection of a bridge, and the damage occurred was further detected. The applications for modelling and monitoring using photogrammetry technique can be found in the projects conducted by Shortis (1986), Woodhouse et al. (1999), Li and King (2002), Fraser et al. (2003), Heath et al. (2004), McClenathan et al. (2006) and Luhmann et al. (2007). The broad applications are mainly due to the advantages of non-contact measurement, rapid data acquisition and capability to simultaneously sample all of the data points within the field of view of the camera (Blandino et al. 2003; Lin et al. 2008).

Summary

A review of the potential geomatics techniques for 3D modelling has been presented above. In order to produce a 3D model of the Fort San Domingo, the feasibility of terrestrial surveying with total station instruments was inspected firstly. Due to the feature of reflectorless measurement, total stations are employed in many 3D modelling applications. However, the measurement time is long, making it inefficient for the Fort modelling task. Alternatively, terrestrial laser scanning demonstrates the capability of rapid, intensive and non-contact data acquisition, and therefore is considered as a more suitable technique for undertaking the modelling task. Moreover, as the technique of convergent photogrammetry is also of potential for this mission, the feasibility will be explored as well.

To comprehensively simulate and produce the model of Fort San Domingo, the creation of terrain surface over the area is another important task in this modelling scheme. To this end, the stereo aerial imagery is applied and the photogrammetry

Methodology for Geomatics Modelling of Fort San Domingo

Introduction of Fort San Domingo

In order to understand the 380-year-long story, the history of the fort is following described. The early seventeenth century was about the time western empires extended their colonial force to the north of Taiwan. Having conquered Keelung in 1626, the Spaniards entered Tamsui and started the construction of Fort San Domingo in 1628. Fort San Domingo was located on a hilltop by Tamsui Town, previously a small fishing village in northern Taiwan, and at the mouth of the Tamsui River. The unique location made it was easy to defend but hard to besiege. The Spaniards accounted for this advantage and thus built the fort for the sake of governing (Tamsui Historic Sites Institute 2008). In 1642, the Dutch expelled the Spaniards and took over the fort. Later the Dutch rebuilt the main structure of the fort and named it Fort Anthonio. This is the structure embodying the present scale and still standing today. Since the local inhabitants in Tamsui at that time referred the Dutch as red hairs, the fort was nicknamed as the "Hong Mao Chen", meaning the fortress of red-haired people. The name was inherited to represent the historical site (Danshui Historic Sites 2006).

The Dutch was defeated by General Cheng-Gong Zheng of Chinese Ming Dynasty in 1661 and left Taiwan in 1668. Consequently Fort San Domingo remained deserted until 1683 when Zheng's army came to Tamsui. From 1683 to 1867, the Qing Dynasty Chinese government controlled the fort and built a stone wall with four gates around the fort during this period.

Following the Opium War in 1868, Tamsui harbor was open for commercial use. Due to its critical location, the British government leased the fort from the Qing government as their consulate. A throughout renovation was made accordingly for the consul business. To accommodate the need of consul and consul's family, the consul's residence was added on the east side of the fort in 1891. The consulate was closed during World War II and reopened after the end of the war. The consulate was closed again in 1972 as the British government broke off her diplomatic relations with the Republic of China (ROC) government. In 1980, the fort was formally returned to the ROC government.

As introduced above, it is realized that the site of current Fort San Domingo comprises two main historical buildings, including Fort Antonio (former Fort San Domingo) and former British Consular Residence (Fig. 8.1). The main structure of Fort Antonio is a two story square building. The wall of the building was laid with bricks and then covered with layers of stone outside and therefore provided the fort strong and effective fortification. The former British Consular Residence was

Fig. 8.1 Fort Antonio (*left*) and former British Consular Residence (*right*) located in the site of Fort San Domingo

distinguished by its Victoria style. The characteristics of red-brick structure, verandas, four-sided steep roof, and a high staircase were well reserved (Danshui Historic Sites 2006; Tamsui Historic Sites Institute 2008).

In order to provide an overall accurate scene over the site of Fort San Domingo, the solution of geomatics modelling was to produce the terrain and building models separately, to which the aerial photogrammetry was applied to generate terrain model while close range photogrammetry and terrestrial laser scanning were used to construct building models. The detailed workflows are introduced below.

Digital Terrain Model (DTM)

Digital terrain model of the test area was generated using aerial photogrammetry technique. To achieve this, a stereo pair of aerial images taken by an UltraCamD large format digital aerial camera (Leberl and Gruber 2005) was employed. The images were captured on 27th of April, 2009 from a flying height of 2,200 m with a nominal focal length of 100 mm. The image format was 11,500 pixels by 7,500 pixels and the pixel size is 9 μm × 9 μm. Given the photo scale of approximately 1:22,000, the image spatial resolution was about 20 cm × 20 cm per pixel.

The acquired imagery was input and processed in the SOCET SET workstation. SOCET SET is a digital photogrammetric and mapping software, which allows a full photogrammetric processing flowline to be performed – from import of digital imagery to the creation of DTMs, ortho-rectified images and export of CAD model. Due to the full capability, imagery from the UltraCamD camera was processed in this software throughout the experiment.

Along with the aerial images, the known interior and exterior orientation parameters were input to SOCET SET for producing the digital surface model (DSM). Subsequently the manual terrain editing was carried out to remove trees and buildings existed in the DSM. As a result, the digital terrain model showing the true terrain surface was produced. The 1-m resolution DTM represented by point cloud, triangulated irregular network (TIN) mesh and contour plot are illustrated in Fig. 8.2. Based on the DTM, an ortho-rectified image was generated as well.

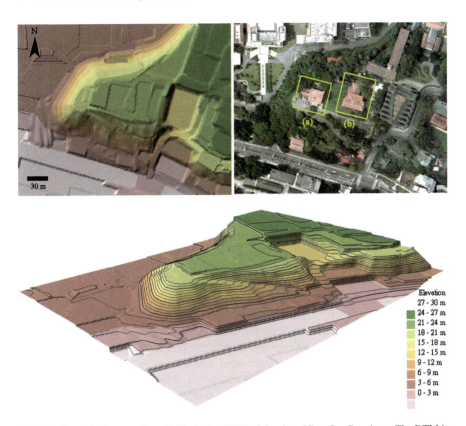

Fig. 8.2 *Top left*: The top view of hill-shaded DTM of the site of Fort San Domingo. The DTM is colorized by elevation. *Top right*: Ortho-image generated based on the DTM, in which Fort Antonio and former British Consular Residence are indicated by (**a**) and (**b**). *Bottom*: the perspective view of 1-m interval contour map superimposed on the TIN mesh DTM

Building Model

Techniques of close range photogrammetry and terrestrial laser scanning were adopted separately to generate building models located in the test site. The methods and results are following described.

Close Range Photogrammetric Building Models

A consumer-grade digital camera, Canon EOS 5D, was employed in the modelling scheme. In order to efficiently distinguish detailed features shown in images, the format of the image acquired was set as 43,68 pixels × 2,912 pixels. Additionally, a 50 mm lenses was fitted to the camera to effectively cover the buildings of interest at the test site. With this setting, the camera was calibrated using the PhotoModeler

Calibration Module before being used for image acquisition (a standard procedure independently verified in Hanke and Ebrahim (1997)). As a result, the geometric accuracy of using the non-metric digital camera could be assured.

A certain number of control points located on or near the objects of interest are required for orientating cameras and also geo-referencing the observed building models in the close range photogrammetric processing. In order to acquire absolute 3D coordinates of the control points, a two-stage method was followed in this modelling task. Firstly, a series of ground control points were selected and their absolute 3D coordinates were stereoscopic measured using the stereo aerial images in the SOCET SET workstation. To verify the accuracy of the positions of the ground control points, the distances between the ground control points were measured using a Leica TCR 803 total station (quoted angular precision of $\pm 3''$ and range precision $\pm$(2 mm + 2 ppm)) on-site. The results were then compared with the distances computed from the stereoscopic measurement. The mean value of the differences was 0.218 m, revealing that the coordinates of ground control points derived from photogrammetric stereoscopic measurement achieved about one-pixel accuracy.

At the second stage, coordinates of control points located on façades of the buildings were obtained. As it was not allowed to attach physical targets on the buildings, some reliable features on the building façades were selected as control points. To derive the absolute coordinates of these points, the Leica TCR 803 total station featuring reflectorless measurement was set up in proper positions and oriented using the ground control points acquired. Subsequently the total station was aimed at the control points on the building façades and the "Free Station" program available in the total station was executed to compute their absolute 3D coordinates. These points were mainly used for geo-referencing the observed building models.

While the control surveying was carried out, a total of 17 images of the two buildings located in the test site was taken by the Canon EOS 5D digital camera. (Refer to Fig. 8.3 for the applied camera network) The images were downloaded

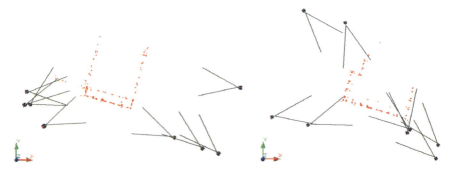

Fig. 8.3 The camera networks comprising eight and nine camera stations in Fort Antonio (*left*) and former British Consular Residence (*right*) respectively. The orientations of each camera are demonstrated. *Red points* indicate the positions of solved building features

into local computer for further processing. In order to produce geo-referenced 3D models of the buildings, 33 control points with known 3D coordinates and important feature points of the buildings with unknown 3D coordinates (e.g., end points of the rooftop, intersection points of walls, etc.) were measured in the images. The derived 2D image coordinates, along with camera's interior orientation parameters and control points' 3D global coordinates were then input into the PhotoModeler software for a bundle adjustment computation. As a result, the 3D coordinates of the feature points of the buildings measured in the images were solved. Subsequently these points were connected for constructing the building models. The resultant building models are demonstrated in Fig. 8.4.

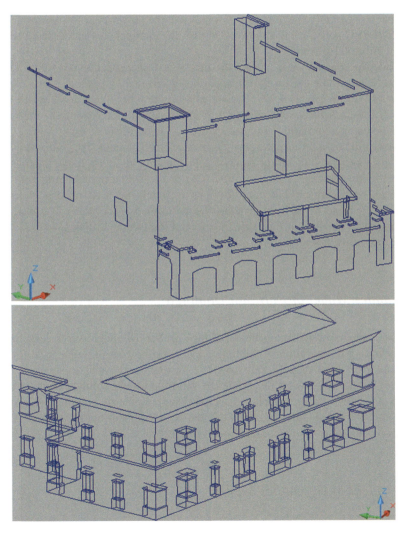

Fig. 8.4 The CAD model of Fort Antonio (*top*) and former British Consular Residence (*bottom*) produced by close range photogrammetric method

Terrestrial Laser Scanned Building Models

To acquire the as-built models on the site of Fort San Domingo, the technique of terrestrial laser scanning was also applied. In this modelling scheme, the Riegl LMS-Z420i laser scanning system was employed to observe the buildings. The scanner, with a quoted single-point position accuracy of ±10 mm, gives a 360° by 80° field of view in the horizontal and vertical directions (Riegl 2010). To completely cover the Fort Antonio and the former British Consular Residence, it was necessary to scan the two buildings from 11 scanner stations on the site. Moreover, following normal terrestrial laser scanning procedures, reflective targets were arranged and also scanned in the field of view to permit the registration of each scanned model. Once the scanning was finished, the registration was performed in RiSCAN PRO software. It was revealed that the amalgamated model has a registration error of 16.1 mm.

While the scanning was carried out on the test site, a calibrated Nikon D200 digital camera mounted on the scanner shot a series of images at each scanner station. These images were also registered with the point cloud at the processing stage in RiSCAN PRO software. As a result, the point clouds could be colorized using the color information provided by the registered images.

In order to geo-reference the amalgamated scanned model to the common global coordinate system, the control points used for producing close range photogrammetric building models (refer to Sect. "Close Range Photogrammetric Building Models") were applied. The control points appeared in the scanned point cloud were selected and the geo-referencing was performed accordingly. The resultant geo-referenced models of the Fort Antonio and the former British Consular Residence are shown in Fig. 8.5.

Geomatics Model of Fort San Domingo

The comprehensive geomatics model of the Fort San Domingo was accomplished by integrating the terrain and building models. As the control points used for the generation of these models were based in an identical spatial reference frame, the terrain and building models could be integrated without any transformation. Figure 8.6 shows the alignment of the CAD photogrammetric building models and the contour DTM. Figure 8.7 illustrates the laser scanned model along with the terrain model, which are both in the point cloud format.

Assessment of Geomatics Model

In order to evaluate the accuracy of the photogrammetric building models, independent observations from a Leica TCR 803 total station were introduced. To

8 Virtual Fort San Domingo in Taiwan

Fig. 8.5 The scanned model of Fort Antonio (*top*) and former British Consular Residence (*bottom*). The models are textured with color extracted from the imagery acquired by the digital camera mounted on the scanner

achieve this, some common identifiable points in both the photogrammetric model and physical structure are selected as check points. Then their coordinates, derived from the photogrammetric model and measured by total station, are compared to determine the accuracy. Therefore, during the control surveying for constructing close range photogrammetric building models (described in Sect. "Close Range Photogrammetric Building Models"), a total of 22 and 13 feature points on the façades of Fort Antonio and former British Consular Residence were measured

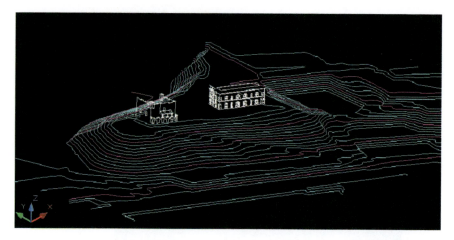

Fig. 8.6 The integration of CAD photogrammetric building models and the contour DTM

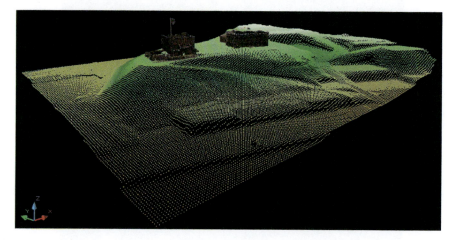

Fig. 8.7 The alignment of the laser scanned model along with the terrain model, which are both in the point cloud format

respectively. Of which 17 and 8 points were introduced correspondingly as control points in the bundle adjustment, while the remaining points were treated as check points for examining the accuracy of the resultant building models. After comparing the coordinates solved from the bundle adjustment with the coordinates measured by the Leica TCR 803 total station, it was found that the root mean square error of the check points reached 11.3 and 28.9 mm respectively.

Independent observations from the total station were also applied to determine the accuracy of the TLS model. However, as the laser scanned model was presented as a discrete point cloud, it was difficult to identify exact common points in the TLS model corresponding to those on the physical structure. To address the issue, the

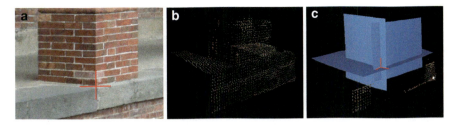

Fig. 8.8 The strategy of extracting check points from a dense point cloud

intersection of three modelled planes was used as check points. Figure 8.8 illustrates the method to locate a check point from point cloud model. Figure 8.8a is the photo showing the position of the selected check point (the red cross). In order to obtain accurate positions of the corner point, the object was scanned with a dense point cloud (Fig. 8.8b). Subsequently the plane primitive was repeated to model three faces of the point cloud (Fig. 8.8c). The corner point could be extracted by selecting the intersection of the three modelled planes and it was taken as the check point for examining the model accuracy. In this evaluation four check points in the two scanned building models were applied. The overall root mean square error of the check points compared with total station measurements was solved as 25.8 mm. From the accuracy assessment, it was realized that the resultant photogrammetric models and the laser scanned models was capable of achieving cm-level accuracy. The comparable accuracy can be observed in Fig. 8.9 by viewing the fitness of the two models.

Integration of Geomatics and VR Models

Virtual Reality Modelling

The applications of architectural design in virtual reality have increased significantly worldwide. For applications in heritage documentation, historical buildings can be preserved and shown via the images derived from the database of archival digital imagery. Furthermore, virtual reality also provides the possibility for modeling the buildings and scenes of interest. By introducing the digital image data, with further processing, 3D structures, buildings and information can be constructed and visualized in a virtual reality environment (Chen 2008). The method has been successfully used in heritage documentations conducted by Obertreiber and Stein (2006), Jabi and Potamianos (2007) and White et al. (2007). The resulting models are of benefit in regard to the high level of detail, making the interactions between browsers and virtual objects highly achievable.

The VR modeling task follows two-stage processing. The first step is the data collection of traditional media. In terms of data collection, the aim is to collect

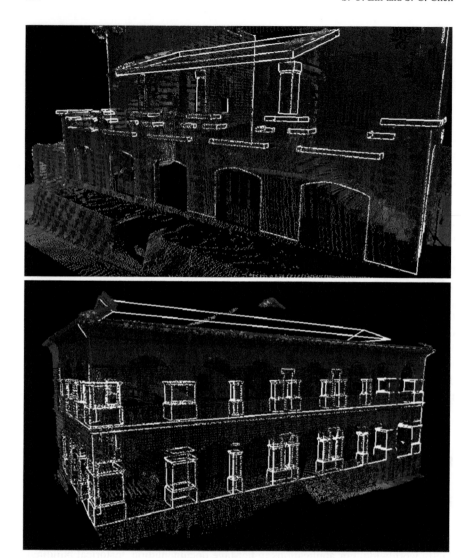

Fig. 8.9 The alignment of the photogrammetric CAD model and laser scanned model

historical descriptions of each building such as text, pictures, photos, plane, façade, architectural section and other useful information that can comprehensively illustrate the building. At the second stage, digital simulation technology is applied to construct the 3D plane, façade and architectural section of the building based on the data acquired. Images acquired on-site are applied subsequently to perform the texture mapping. The resulting models can be demonstrated by any suitable virtual reality software and environment. The workflow of VR modeling is shown in Fig. 8.10a.

8 Virtual Fort San Domingo in Taiwan   171

After the accurate geomatics models covering Fort San Domingo were produced, high LOD VR models were constructed to simulate the physical buildings. In order to integrate advantages inherited from the two models, an integration of the geomatics and VR models was performed. The integrated strategy was proposed as follows, and the 3D models with characteristics of high geometric accuracy and level of detail were produced in the end.

Integrated Strategy

For the model integration, two phenomena including pre- and post-VR modelling integration, are considered. For the former case, the integration is conducted before VR models are produced. In the original workflow the 3D modelling task is accomplished using digital image data. However, the image data is replaced by the introduction of CAD structures at the modelling stage in this integration operation. It is noted that the acquisition of digital image data is still necessary as they are required for model mapping. The flowchart of this case is illustrated in Fig. 8.10b. The post-VR modelling integration refers to the integration with existing VR models. In this case CAD structures are introduced prior to the mapping stage. The scale and orientation of the VR models are examined and adjusted accordingly if there is any disagreement between the two models. The workflow is shown in Fig. 8.10c.

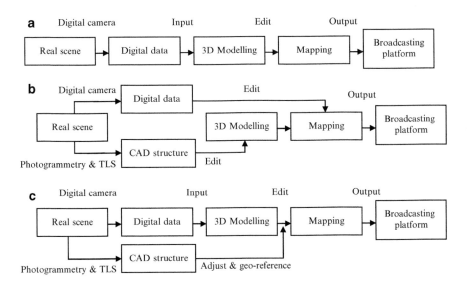

Fig. 8.10 Integrated procedure of the topographic scene making and historical building assembly

Illustration of Integration and Comparison of Geomatics Models and 3D VR Models

Case of Fort San Domingo

As the VR models of Fort San Antonio and the former British Consular Residence were constructed, the post-VR modeling integration was performed in this paper. A reference model was created using the geomatics models mentioned in Sect. "Building Model" (shown in Fig. 8.11, the one in white). The model was then imported and compared to the VR model in 3ds Max software. The misalignment of the two models is observed from various viewing angles in Fig. 8.11. As a deviation in scale of model remained unsolved after the initial adjustment (refer to Fig. 8.12), further improvement was carried out using control points derived from free-form deformation (FFD) modifier, which is similar to building scaffolding in real life. The processing of refinement of the building in the virtual simulation is shown in Fig. 8.13. Once the overall adjustment was finished, the accurate VR model was derived (Fig. 8.14).

Case of the Former British Consulate

The integration of the geomatics model and 3D VR models for the former British Consular Residence was carried out using the same method applied in the adjustment of Fort San Antonio. The model comparison, adjustment and final refinement are illustrated in Figs. 8.15, 8.16 and 8.17.

One of the goals of this study was to achieve a new spatial experience through the integration of image visualization and various 3D digital techniques. From the successful implementation of the model integration, of which the accurate

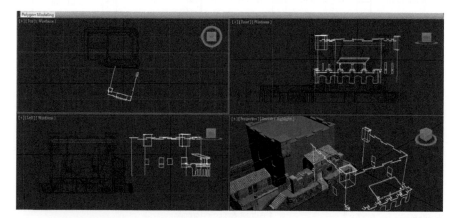

Fig. 8.11 Misalignment of the geomatics model and the original 3D VR model of Fort San Antonio (displayed in 3ds Max software)

8 Virtual Fort San Domingo in Taiwan

Fig. 8.12 Differences, in particular the model sale, are shown by visualizing the superimposition of the two models

Fig. 8.13 Refinement of the VR building model, in which the free-form deformation is performed in 3ds Max software

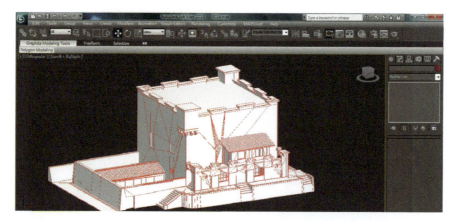

Fig. 8.14 Final 3D uncolored model after the overall adjustment

Fig. 8.15 Misalignment of the geomatics model and 3D VR model of the former British Consular Residence displayed from various angles

photogrammetric CAD models were utilized for geo-referencing, adjusting and refining the high LOD VR model, the goal was fulfilled satisfactorily. The final 3D model of Fort San Domingo and its surrounding areas are shown in Fig. 8.18.

Discussion

Non-contact Geomatics Techniques

Due to the limited accessibility to the heritage, non-contact geomatics techniques were proposed to construct 3D models of the Fort Antonio and former British

8 Virtual Fort San Domingo in Taiwan

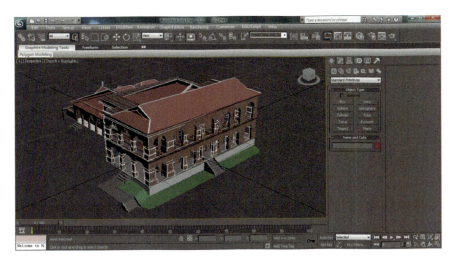

Fig. 8.16 An improved fitting achieved after the initial adjustment

Fig. 8.17 Detailed evaluation and refinement of the VR model are performed using the accurate geomatics model

Consular Residence. Based on the results demonstrated in Sects. "Close Range Photogrammetric Building Models" and "Terrestrial Laser Scanned Building Models", close range photogrammetry and terrestrial laser scanning were proved very capable of producing as-built models. However, there were also some issues during the operations that are now discussed.

Fig. 8.18 Uncolored (*top*) and texture mapped (*bottom*) VR models of the Danshui historic site

Efficiency of the Techniques

The first issue relates to the efficiency at surveying and processing stages. At surveying stage, close range photogrammetry and terrestrial laser scanning both need observations of control point. For the former technique, consideration of number and position of control points are required as these are critical factors for correctly solving orientation parameters of camera stations. Furthermore, the solved orientation will affect the quality of intersected feature points. For the TLS technique, control points are used for model co-registration. Ideally the control point should be arranged in the overlapping areas between each scan. Some other techniques (e.g., surface matching) are also applicable for co-registration even if no control points are available. Therefore the requirement of control point in TLS operation is not as serious as in close range photogrammetry task. As a result it takes less time to arrange control surveying when performing terrestrial laser scanning. Nevertheless, regarding the time required for data acquisition of the whole buildings, it is found that photogrammetric acquisition is more efficient than TLS technique.

Main task in photogrammetric processing is to select corresponding points, including control points and building features, among multiple images in the software. Once these points are verified and computed, they are connected to derive the CAD model. To achieve this, no special techniques but manpower and cost of time are required. As for processing of laser scanned data, the co-registered point cloud can be accomplished manually or automatically in the specific software. However, it is worthwhile noting that the format of point cloud is not fully accepted yet in the field of 3D modelling. Hence a heavy modelling work is necessary for extracting CAD models from point cloud. In summary, the overall effort to derive a photogrammetric CAD model or TLS point cloud model is comparable. However, considering the required format of 3D building model, the employment of TLS technique takes more time to achieve if a CAD format is the final product required.

Completeness of Geomatics Models

Secondly, the level of completeness of the resultant models is discussed. Due to various characteristics of the models derived, the discussion focuses on completeness of CAD structure of photogrammetric model and surface continuousness of laser scanned model respectively. As shown in Sect. "Close Range Photogrammetric Building Models", the façade of photogrammetric model of Fort Antonio facing north and façades of former British Consular Residence facing north and northeast were not constructed. The failure was caused by obstructions of heavy vegetations in the test site. Due to the constraint, the setup of camera stations and a clear view field of important building features became unachievable. The camera network and the positions of observable building features illustrated in Fig. 8.3 explained the influence of limited exposure stations. Since a comprehensive image acquisition was not available, the level of completeness of constructed photogrammetric model was affected accordingly.

Regarding TLS performance, as dense scanning was carried out to collect point cloud of Fort Antonio and former British Consular Residence, building models with high level of completeness were obtained (Fig. 8.5). However, some areas with broken surface were also observed. This was mainly caused by obstructions occurring between the building and the scanning system. Normally such discontinuities could be re-constructed by repeated scanning from different stations, but as the same constraints met in photogrammetric modelling, some broken surfaces still failed to be covered due to unavoidable vegetation and difficult scanning angles. As such on-site constraints are unavoidable in the applications of the two geomatics techniques, this is treated as natural limitation and should be carefully considered in the fieldwork plan.

Accuracy of Geomatics Models

The final discussion involves measurement accuracy. The accuracy of the two resultant models was examined by independent observations using total station instruments. It was indicated that the models were able to achieve a cm-level accuracy although a specific accuracy was not requested in this case (refer to Sect. "Assessment of Geomatics Model" for details). For the request of advanced accuracy in close range photogrammetric models, increasing the number of exposure stations would be beneficial. However accompanying issues, such as image storage space, processing efficiency and on-site limitations, would also need to be considered. The final solution would be a compromise between required accuracy, processing time and budget limitations.

Level of Visual Application of VR Models

As the purpose of application of VR model is to provide the highest visual quality and most realistic presentation, a sophisticated method and high resolution images are required for modelling and simulation. To achieve this, the level of detail of the models and operation of light in the VR scene are critical factors. For high LOD models, very high resolution images are required for constructing detail components and also simulating the materials of the buildings. When simulating light effects in this paper, it was found that the use of advanced "ray tracing" and "radiosity" techniques provided the best simulation results. However, it was also realized that the application of such data and techniques occupied heavy computer resources and reduced the performance of display of VR models and scenes. In order to improve the displaying efficiency, the techniques of texture mapping, culling, fog effects and polygon reduction were applied to upgrade the instant screen display. The performance of the virtual presentation of the models was improved by sacrificing the visual effect of the resultant simulations in virtual reality.

The human-computer interaction (HCI) also benefitted the modification. In this 3D digital heritage project, one of the objectives was to surpass the traditional presentation of 2D pictures/animation for building structure and offer visitors an interactive experience closest to reality. For instance, when a person is visiting an architectural space in real life, he is free to walk around instead of observing the space passively. The instant 3D virtual reality simulation allows users to participate actively and freely during visits, which is vastly different from the traditional 2D web browsing. As the efficiency of computer computing capacity was improved by applying the proposed method, smooth interactions between visitors and VR heritage models are achievable in the VR environment. In summary, it is realized that the optimum level of visual effect should be determined considering the balance between the necessary LOD and the computer capacity being employed.

Data Interoperability Issues

In this project three types of 3D building models were produced, including VR model, photogrammetric CAD model and laser scanned point cloud. As the

8 Virtual Fort San Domingo in Taiwan

Table 8.1 The techniques, software and compatible data formats involved in this project

Technique	Software	Compatible data format
VR modelling	Autodesk 3ds Max	3D Studio (*.3DS)
		Autocad DWG (*.DWG)
		Autocad DXF (*.DXF)
		Autodesk FBX (*.FBX, *.DAE)
		IGES (*.IGS)
		Wavefront (*.OBJ)
		StereoLitho (*.STL)
		Shockwave 3D (*.W3D)
		VRML (*.WRL)
		VIZ Material XML (*.XML)
Close range photogrammetry	PhotoModeler	ASCII (*.*)
		Autocad (*.DXF)
		Wavefront (*.OBJ)
		VRML (*.WRL)
		3D Studio (*.3DS, *.PRJ)
Terrestrial laser scanning	RiSCAN Pro	ASCII (*.*)
		3DD with SOP (*.3DD)
		point cloud (*.3PF)
		Autocad (*.DXF)
		Wavefront (*.OBJ)
		VRML (*.WRL)
		PLY (*.PLY)
		LAS (*.LAS)
		PointCloud for AutoCad (*.PTC)
		Google Earth (*.KMZ)

principles of modelling techniques are different, specific software are applied to handle building modelling accordingly. Furthermore, according to the integrated strategy proposed in Sect. "Integrated Strategy", the photogrammetric and TLS CAD models are treated as the primitive structure for geo-referencing and adjusting the VR model. However, as reported by Kolbe et al. (2005), 3D building modelling techniques have been continuously developed in various fields. Each of which has established its own standard. Hence the appropriate software for performing the model integration should be considered.

The techniques and software involved in this project, and their compatible data formats are listed in Table 8.1. It is noted that several data formats are exchangeable among the three pieces of software. That is, the three pieces of software are all of potential for performing model integration. However, it was found that some components of the models were missing or misaligned during the processing of data exchange. To address the issue, the three pieces of software were assessed to examine stability of model interoperability and efficiency of model processing. Based on the result of the assessment, Autodesk 3ds Max software was selected as the main working environment for model integration. Also, the Autocad DXF file was used as the common format of building models.

Hence, the photogrammetric and TLS CAD models are saved as drawing exchange format DXF and then imported into 3ds Max directly for model editing and integration. Once the VR model is adjusted, the final product with high accuracy and LOD is constructed.

Conclusions and Suggestions for Future Work

In order to accomplish digital preservation of historic buildings and scenes, digitalization techniques from the fields of geomatics engineering and virtual reality were introduced. The background, overall workflow and assessment of the techniques, as well as the resultant models have been demonstrated in this paper. Moreover, an appropriate procedure capable of integrating the advantages inherited from the geomatics and VR modelling techniques was proposed. The feasibility of the method was proved through the integration and assessment in the 3D modelling scheme carried out in the historical site of Fort San Domingo. The overall techniques and integrated method have great potential to be transferred and applied in various areas.

Although the successful modelling method was demonstrated, a number of limitations occurring during the modelling tasks were raised and discussed. To overcome the limitations and broaden the potential field of applications, two further topics are suggested.

The first work aims to improve the efficiency of TLS modelling applications. As summarized in Sect. "Efficiency of the Techniques", intense effort was required for extracting CAD models from the laser scanned point cloud and this may reduce involvement of the TLS technique in 3D modelling. To address the issue, the images shot on-site can be of helpful. Once the images and the laser scanned point cloud are orientated in an identical reference system, the orientated images can then be introduced as a visual tool supporting the extraction of CAD models from the point cloud. The rationale has been introduced in commercial software, such as Phidias, Kubit PointCloud and Pointools Edit, and the modelling efficiency can be improved accordingly. Another issue mentioned in Sect. "Completeness of Geomatics Models" was broken scanned surface caused by obstructions occurring between the building and the scanning system. To overcome the limitation, an introduction of a terrestrial laser scanner which is capable of receiving multiple return signals is suggested. As the density of points being reflected from the scanned object is greatly increased, detection of the objects obstructed by vegetation becomes possible. The completeness of the laser scanned surface can be upgraded. Overall, the solutions proposed will be examined for increasing the involvement of TLS in 3D modelling tasks.

As discussed in Sect. "Level of Visual Effect of VR Models", the optimum level of visual effect was determined by the trade-off between necessary LOD and the computer capacity employed. Hence, innovative approaches capable of increasing efficiency and saving resources should be developed to achieve enhanced visual effect and smoother HCI. To this end, potential research topics include:

1. Innovative visualization of multiple data for understanding and analysis
   In order to broaden the applications of HCI, multiple data, such as archival audio and video files, information of user's behaviors and characteristics, etc., are collected and stored in the data storage system. For turning such large and complex data into visual presentation efficiently, innovative approaches are required. The results are also expected to be helpful for analysis of multi-information analysis.
2. Suitability and practicality of the operation and analysis environment
   As described above, the collection of multiple media data becomes essential in HCI applications. Since each data set has its own specific processing software and required hardware, a combination of various technical requirements is inevitable. Hence, an ideal environment capable of handling data storage, data processing and data conversion will be studied. To perform the innovative visual presentation, this topic should be considered comprehensively.
3. Task-oriented designed and user-friendly graphic user interface
   User interface is a critical element for successful HCI applications. To avoid confusion occurring during operation and functions with low access frequency caused by ambiguous interface design, development of task-oriented design will be introduced in the HCI in the near future. Along with innovative design and upgrading of new hardware devices, the gap between the digital environment and users can be eliminated.

Acknowledgements The authors would like to acknowledge Strong Engineering Consulting Co., Ltd. (Taiwan) for providing the terrestrial laser scanned data. The authors also would like to thank Tamsui Historical Museum (Taiwan) for enabling access to the study area used in the paper.

References

Acharya, B., Fagerman, J. and Wright, C., 2000. Accuracy assessment of DTM data: a cost effective approach for a large scale digital mapping project. International Archives of Photogrammetry and Remote Sensing, 33(B2): 105–111.

Baltsavias, E.P., Li, H., Stefanidis, A., Sinning, M. and Mason, S., 1996. Comparison of two digital photogrammetric systems with emphasis on DTM generation: case study glacier measurement. International Archives of Photogrammetry and Remote Sensing, 31(B4): 104–109.

Bannister, A., Raymond, S. and Baker, R., 1998. Surveying, seventh edition. Addison Wesley Longman Limited, Essex, UK. 502 pages.

Barber, D.M., 2003. Terrestrial Laser Scanning for the Metric Survey of Cultural Heritage Structures. PhD Thesis, University of Newcastle upon Tyne, Newcastle upon Tyne, UK (unpublished): 222 pages.

Barber, D., Mills, J. and Bryan, P., 2003. Towards a standard specification for terrestrial laser scanning of cultural heritage. CIPA International Archives for Documentation of Cultural Heritage, 19: 619–624.

Barnes, A. and Cothren, J., 2007. Mapping Tiwanaku: Using modern photogrammetric methods to analyze an ancient landscape. http://cast.uark.edu/assets/files/PDF/DO_workshop_Barnes&Cothren.pdf [assessed 15th October 2010].

Bevan, A. and Conolly, J., 2004. GIS, archaeological survey and landscape archaeology on the Island of Kythera, Greece. Journal of Field Archaeology, 29: 123–138.

Blandino, J.R., Pappa, R.S. and Black, J.T., 2003. Modal identification of membrane structures with videogrammetry and laser vibrometry. AIAA Paper 2003-1745: 11 pages.

Boehler, W. and Marbs, A., 2002. 3D scanning instruments. Proceedings, International Workshop on Scanning for Cultural Heritage Recording. Corfu, Greece: 9–12.

Bonora, V., Colombo, L. and Marana, B., 2005. Laser technology for cross-section survey in ancient buildings: A study for S. M. Maggiore in Bergamo. Proceedings, CIPA 2005 XX International Symposium. Torino, Italy: 6 pages.Briese, C., 2006. Structure line modelling based on terrestrial laser scanner data. International Archives of Photogrammetry, Remote Sensing and Spatial Information Sciences, 36(5): 6 pages.

Bryan, P.G., Barber, D.M and Mills, J.P., 2004. Towards a standard specification for terrestrial laser scanning in cultural heritage – one year on. International Archives of the Photogrammetry, Remote Sensing and Spatial Information Sciences, 35(B7): 966–971.

Chen, S.-C., 2008. Virtual Fort San Domingo: A Preliminary Process of Digital and Virtual Preservation. Ph.D. Thesis, National Chiao Tung University, Hsin-Chu, Taiwan (unpublished): 144 pages.

Danshui Historic Sites, 2006. Fort San Domingo. http://www.tshs.tpc.gov.tw/fore/en1-1.asp [assessed 19th October 2010].

D'Apuzzo, N. and Kochi, N., 2003. Three-dimensional human face feature extraction from multi images. Proceedings, Optical 3-D Measurement Techniques VI (Eds. A. Gruen and H. Kahmen). Zurich, Switzerland. Volume I: 140–147.

D'Ayala, D. and Smars, P., 2003. Architectural and structural modelling for the conservation of cathedral. Journal of Architectural Conservation, 9(3): 51–72.

Dunn, M.L., 2009. Recent developments in close range photogrammetry for mining and reclamation. http://www.techtransfer.osmre.gov/ARsite/Publications/Dunn_billings_crp1abstract.pdf [assessed 17th October 2010].

Fraser, C., Brizzi, D. and Hira, A., 2003. Vision-based, multi-epoch deformation monitoring of the atrium of Federation Square. Proceedings, the 11th FIG Symposium on Deformation Measurements. Santorini, Greece: 6 pages.

Fryer, J.G., 2001. Introduction. Close Range Photogrammetry and Machine Vision (Ed. K. B. Atkinson). Whittles Publishing, Caithness, UK. 371 pages: 1–7.

Goedhart, W. and Wolters, M., 2008. 3D models, University of Applied Sciences Utrecht, Faculty of Nature and Technology. http://www.michelwolters.nl/huen/huen.html [assessed 17th October 2010].

Goodmaster, C. and Payne, A., 2007. 3D laser scanning at Tiwanaku: Potentials for documentation, visualization, & analysis. http://cast.uark.edu/assets/files/PDF/DO_workshop_Goodmaster&Payne.pdf [assessed 15th October 2010].

Gruen, A., 2001. Development of digital methodology and systems. Close Range Photogrammetry and Machine Vision (Ed. K. B. Atkinson). Whittles Publishing, Caithness, UK. 371 pages: 78–104.

Guarnieri, A., Pirotti, F., Pontin, M. and Vettore, A., 2005. Combined 3D surveying techniques for structural analysis applications. International Archives of Photogrammetry, Remote Sensing and Spatial Information Sciences, 36(5/W17): 6 pages

Hanke, K. and Ebrahim, M.A.-B., 1997. A low cost 3D-measurement tool for architectural and archaeological applications. International Archives of Photogrammetry and Remote Sensing, 32(5C1B): 8 pages.

Heath, D.J., Corvetti, J., Gad, E.F. and Wilson, J.L., 2004. Development of a photogrammetry system to monitor building movement. Proceedings, 2004 Annual Technical Conference of the Australian Earthquake Engineering Society (Eds. K. McCue, M.C. Griffith and B. Butler). Mt. Gambier, South Australia: 19 pages.

Hoffman, E., 2005. Specifying laser scanning services. Chemical Engineering Progress Magazine, 101(5): 34–38.

Hyttel, F., 2010. Guns in 3D. http://maritimearchaeologyprogramdenmark.wordpress.com/2010/01/06/guns-in-3d/[assessed 15th October 2010].

Ion, I., Dragos, B. and Margarita, D., 2008. Digital photogrammetric products from aerial images, used for identifying and delimiting flood risk areas. International Archives of Photogrammetry, Remote Sensing and Spatial Information Sciences, 37(B7): 361–364.

Jabi, W. and Potamianos, I., 2007. Geometry, Light, and Cosmology in the Church of Hagia Sophia, International Journal of Architectural Computing, 5(2): 303–320.

Jauregui, D., White, K., Woodward, C. and Leitch, K., 2003. Noncontact photogrammetric measurements of vertical bridge deflection. Journal of Bridge Engineering, 8 (4): pp 212–222.

Kavanagh, B.F. and Bird, S.J.G., 1992. Surveying Principles and Applications, third edition. Prentice-Hall, Inc., New Jersey, USA. 667 pages.

Kolbe, T.H., Grger, G. Plumer, L., 2005. CityGML – interoperability access to 3D city models. Proceedings, the 1st International Symposium on Geo-information for Disaster Management (Ed. P. van Oosterom, S. Zlatanova and E. M. Fendel). Delft, Netherlands: 883–899.

Lahoz, J.G., Aguilera, D.G., Finat, J., Martinez, J., Fernandez, J. and San Jose, J., 2006. Terrestrial laser scanning metric control: assessment of metric accuracy for cultural heritage modelling. International Archives of Photogrammetry, Remote Sensing and Spatial Information Sciences, 36(A5): 6 pages.

Leberl, F. and Gruber, M., 2005. ULTRACAM-D: Understanding some noteworthy capabilities. Photogrammetric Week 2005 (Ed. D. Fritsch): 57–68.

Lemmon, T. and Biddiscombe, P., 2006. Laser scanners improve survey workflow. Professional Surveyor Magazine, 26(3): 14–18.

Levoy, M., Ginsberg, J., Shade, J., Fulk, D., Pulli, K., Curless, B., Rusinkiewicz, S., Koller, D., Pereira, L., Ginzton, M., Anderson, S. and Davis, J., 2000. The digital Michelangelo project: 3D scanning of large statues. Proceedings, SIGGRAPH 2000. New Orleans, USA: 131–144.

Li, C. and King, B., 2002. Close range photogrammetry for the structural monitoring. Journal of Geospatial Engineering, 4(2): 135–143.

Lin, S.-Y., Mills, J.P. and Gosling, P.D., 2008. Videogrammetric Monitoring of As-built Membrane Roof Structures. Photogrammetric Record, 23(122): 128–147.

Luhmann, T., Robson, S., Kyle, S. and Harley I., 2007. Close Range Photogrammetry: Principles, Techniques and Applications. Whittles Publishing, Caithness, Scotland, UK. 528 pages.

Maarleveld, T., 2009. Boomstamboot Kadoelerveld Opgravingsrapport. University of Southern Denmark, Esbjerg, Denmark. 31 pages.

Matori, A.N. and Hidzir, H., 2010. Low cost DTM for certain engineering purposes. Proceedings, Map Asia 2010 & ISG 2010. Kuala Lumper, Malaysia.

McClenathan, R.V., Nakhla, S.S., McCoy, R.W. and Chou, C.C., 2006. Use of photogrammetry in extracting 3D structural deformation/dummy occupant movement time history during vehicle crashes. SAE 2005 Transactions Journal of Passenger Cars: Mechanical Systems. 3099 pages: 736–742.

Mills, J.P., 1996. The Implementation of a Digital Photogrammetric System and Its Application in Civil Engineering. PhD thesis, University of Newcastle upon Tyne, Newcastle upon Tyne, UK (unpublished): 252 pages.

Newman, J., Fraundorfer, F., Schall, G. and Schmalstieg, D., 2005. Construction and maintenance of augmented reality environments using a mixture of autonomous and manual surveying techniques. Proceedings, Optical 3-D Measurement Techniques. Vienna, Austria.

Obertreiber, N. and Stein, V., 2006. Dokumentation und Visualisierung einer Tempelanlage auf Basis von Laserscanning und Photogrammetrie, in: Luhmann, T. and Mueller, C., eds., Photogrammetrie Laserscanning Optische 3D-Messtechnik - Beitrage der Oldenburger 3D - Tage 2006, Wichmann Verlag, Heidelberg: 317–323.

Patias, P., 2006. Cultural Heritage Documentation. http://www.photogrammetry.ethz.ch/summerschool/pdf/15_2_Patias_CHD.pdf [assessed 16th October 2010].

Psaltis, C. and Ioannidis, C., 2006. An automatic technique for accurate non-contact structural deformation measurements. International Archives of Photogrammetry, Remote Sensing and Spatial Information Sciences, 36(5): 242–247.

Riegl, 2010. Terrestrial scanning. http://www.riegl.com/nc/products/terrestrial-scanning/produktdetail/product/scanner/27/[assessed 15th October 2010].

Sauerbier, M., 2004. Accuracy of automated aerotriangulation and DTM generation for low textured imagery. International Archives of the Photogrammetry, Remote Sensing and Spatial Information Sciences, 35(B2): 521–527.

Sakimura, R. and Maruyama, K., 2007. Development of a new generation imaging total station system. Journal of Surveying. Engineering, 133(1): 14–22.

Schulz, T. and Ingensand, H., 2004. Influencing variables, precision and accuracy of terrestrial laser scanners. Proceedings, INGEO 2004 and FIG Regional Central and Eastern European Conference on Engineering Surveying. Bratislava, Slovakia: 8 pages.

Shortis, M.R., 1986. Close range photogrammetric measurements for structural monitoring, deformation surveys and engineering surveillance. Australian Journal of Geodesy, Photogrammetry and Surveying, 45: 55–64.

Stanek, A. and Ranchin-Dundas, N., 2010. Recording in 3D. http://maritimearchaeologyprogramdenmark.wordpress.com/2010/04/30/recording-in-3d/ [assessed 15th October 2010].

Sternberg, H., 2006. Deformation measurements at historical buildings with terrestrial laser scanners. International Archives of Photogrammetry, Remote Sensing and Spatial Information Sciences, 36(5): 6 pages

Tamsui Historic Sites Institute, 2008. Fort San Domingo: General information. Taipei County, Taiwan. 2 pages.

Thompson, E.H., 1962. Photogrammetry. The Royal Engineers Journal, 76(4): 432–444.

Torlegard, A.K.I., 1980. An introduction to close range photogrammetry. Developments in close range photogrammetry, volume I. (Ed. K.B. Atkinson). Elsevier Applied Science, London, UK. 222 pages: 1–14.

Tucker, C., 2002. Testing and verification of the accuracy of 3D laser scanning data. Proceedings, ISPRS Technical Commission IV Symposium 2002. Ottawa, Canada: 6 pages.

Uren, J. and Price, W.F., 2006. Surveying for Engineers, fourth edition. Palgrave Macmillan, Hampshire, UK. 824 pages.

Vezočnik, R., Ambrožič, T., Sterle, O., Bilban, G., Pfeifer N. and Stopar, B., 2009. Use of terrestrial laser scanning technology for long term high precision deformation monitoring. Sensors, 2009(9): 9873–9895.

Waldhausl, P., 1992. Defining the future of architectural photogrammetry. International Archives of Photogrammetry and Remote Sensing, 29(B5): 767–770.

White, M., Petridis, P., Liarokapis, F. and Plecinckx, D., 2007. Multimodal Mixed Reality Interfaces for Visualizing Digital Heritage. International Journal of Architectural Computing, 5(2): 321–338.

Wolf, P.R. and Dewitt, B.A., 2000. Elements of Photogrammetry: with Applications in GIS, third edition. McGraw Hill, New York, USA. 624 pages.

Wolf, P.R. and Ghilani, C.D., 2002. Elementary Surveying: An Introduction to Geomatics, tenth edition. Prentice-Hall, Inc., New Jersey, USA. 900 pages.

Woodhouse, N.G., Robson, S. and Eyre, J.R., 1999. Vision metrology and three dimensional visualization in structural testing and monitoring. The Photogrammetric Record, 16(94): 625–641.

Zhang, L. and Gruen, A., 2004. Automatic DSM generation from linear array imagery data. International Archives of Photogrammetry, Remote Sensing and Spatial Information Sciences, 34(B3): 128–133.

Chapter 9
Web-Based Multimedia and Public Participation for Green Corridor Design of an Urban Ecological Network

Zhenjiang Shen, Mitsuhiko Kawakami, and Kazuko Kishimoto

Introduction

In this chapter, we analyze the effectiveness of multimedia in participation over the Internet through online deliberation for reviewing plans of the ecological network district in downtown Kanazawa City. In order to improve the understanding of stakeholders with respect to the planning concepts for reaching a consensus, we attempt to support planners in presenting their planning concepts during virtual meetings using web-based multimedia materials. In this case study, we focus on planning considerations for a natural environment and an urban environment in a green corridor, which incorporates a boulevard (Hirosaka Boulevard) and a traditional irrigation channel. This green corridor, which connects Kenroku Mountain and Central Park in Kanazawa City, is an important part of the ecological network in the city and contributes to the realization of the eco-city on the urban district level.

Generally speaking, an eco-city (Register 1987) is a city that is designed while considering environmental impact and inhabited by people who are dedicated to the minimization of the consumption of energy, water, and food and the production of waste. In planning, the concept of the eco-city is expressed in a local environmental planning system through sustainable development programs. The objectives of sustainable development projects include the conservation and restoration of natural environments, the creation of landscapes and parks, and the promotion of environment-friendly industries. Among these objectives, we focus on planning support regarding the creation of landscape and parks for the conservation of the natural environment in a built-up urban area, which incorporate the planning of the ecological network and urban design.

Planning of green corridors in urban areas is relative to both the ecological network and urban public spaces. The term "green corridor" has frequently been

---

Z. Shen (✉) • M. Kawakami • K. Kishimoto
School of Environmental Design, Kanazawa University, Kanazawa, Japan
e-mail: shenzhe@t.kanazawa-u.ac.jp

Z. Shen, *Geospatial Techniques in Urban Planning*, Advances in Geographic Information Science, DOI 10.1007/978-3-642-13559-0_9,
© Springer-Verlag Berlin Heidelberg 2012

used to reference open spaces in towns and in the countryside. Urban green spaces have been analyzed both on the level of the entire habitat network and the level of individual patches for sustaining small animals (Jepsen et al. 2005). Mörtberg and Wallentinus (2000) investigated whether the remnants of forest in a city and a system of green space corridors could support target species in order to determine whether target species would be able to exist in such habitats near cities. On the other hand, the results of a case study of East Cheshire in northwest England suggest that "green corridors" (Groome 1990) can provide traffic-free recreational routes and local open spaces, which could be used for developing the principles and guidelines for the design of urban green space corridors. As such, the analysis of the habitat network and the effect of the patch and corridor configuration is important for ecological network planning. In addition, patches or green corridors located in urban areas are also possible to be planned as public spaces.

In recent years, local governments have tended to improve public participation in the planning process. In citizen partnerships, it is important to involve citizens in investigating the environmental status and sharing information through education programs, in order to build the local eco-city. In Japan, in order to encourage the creation of green spaces and animal networks for sustainable local communities, since 2001, municipalities and cities have been competing for the title of Japan's top eco-city. In this chapter, as a case study carried out in downtown Kanazawa City, we discuss the effectiveness of online multimedia materials on green corridor design in planning practice from the viewpoint of citizen partnership.

With respect to participation, the formalization of "bottom-up" community involvement in environmental management projects has been driven by past failings of "top-down" approaches (Fraser et al. 2006). In the case of net participation, participants can join freely without spatial or temporal limitations during online meetings through the Internet (Shen and Kawakami 2010). Germain et al. (2001) presented case studies on public participation in areas such as informing, manipulation, consultation, collaborative decision-making, and delegated power and citizen power, which are several types of one-way flow of information to citizens for use in the decision-making process. In the context of spatial and environmental planning, Tippett et al. (2007) reported the existence of five major participatory planning processes, namely, inform, design, consult, deliver, and monitor, as well as the learning cycle from implementation. These stages of the planning process are different from strategic planning stages (Eppler and Platts 2009), such as the analysis stage, the development stage, the planning stage, and the implementation stage. We attempt to argue the issues of planning support from a five-stage conventional process scheduled during planning meetings in Japanese planning practice, such as opening up planning information, learning planning information, collecting public opinions and planning requirements, reviewing all proposals, and finding a planning solution.

Generally speaking, planners and officers should make proposals and find a final solution in order to meet the planning requirements of planning committees. In the planning process, collecting public opinions should be carried out after opening up the planning information with the objectives proposed by the local government and planners. Among the prevalent discussions in urban planning, the learning program in the participatory process can help to construct and maintain collaborative platforms (Ataöv and Kahramana 2009). We herein focus on the effectiveness of online multimedia materials for representing alternative plans and designs and improving participants' understanding of alternative plans.

An online multimedia tool that incorporates planning information is expected to help participants to find better solutions and improve the proposed plans through net participation. In the process of collecting public opinions, an investigation of the ecological environment conducted by experts requires a sophisticated plan, and the result of the investigation can be made open to the public for learning and understanding. Oura et al. (2001) collected public opinions in an urban redevelopment project area through a "card workshop". In the case of ecological network planning, planners can attempt to find ideas from information voluntarily input by residents in order to formulate a planning framework of the ecological network from the viewpoint of the human–environment relationship.

Normally, the learning process is scheduled such that stakeholders can understand the current situation of a planning site as well as the proposed planning objectives. Stakeholders can consider planning requirements and present opinions at planning meetings. Moon (2003) proposed an online multimedia tool combining Geographic information system (GIS), Virtual reality (VR), photographs, and video for public learning and collecting public opinions in South Korea. Recently, various studies on online design cooperation have demonstrated that participants can make design proposals over the Internet. For instance, Shen et al. (2002) developed an online park design game, which can be used to clarify the current situation and collect design proposals from the public. In addition, an interactive urban design (IUD) tool, which is not an online design tool, was implemented in a platform of Desk-Cave allowing for direct manipulation of masses and with immediate feedback on the urban planning in order to improve the compliance of the urban design with the planning requirements (deVries et al. 2005).

Citizens who participate in workshops or are appointed as planning committee members, are invited to take part in the process of reaching a consensus among stakeholders. Manabe et al. (2003) obtained comments from participants regarding a Webgis map submitted by residents and analyzed the advantages and disadvantages of net participation. A BBS or chat tool can be used to reach a consensus in the online deliberation of planning committee meetings (Shen 2009). However, compared with conventional planning committee meetings and workshops, Internet participants could not talk with each other face to face over the Internet. As such, the participants might have difficulty understanding facial expressions conveyed over the Internet and sometimes only type comments or questions in a chat room without talking in real time. Moreover, messages sent to the chat room may be ignored, misunderstood, or interpreted as being frivolous (Shen et al. 2002).

In this chapter, we designed a system to support the process of net participation in various stages, i.e., providing planning information over the Internet to support the decision-making of the green corridor design while considering the natural and human environments in the downtown urban ecological network. Among the above planning processes, this chapter focuses on how the online multimedia tool can improve public learning and understanding in the learning process and how the online multimedia tool can help a facilitator to organize a process by which to reach consensus for finding a reasonable solution that satisfies all of the planning requirements.

Hirosaka Green Corridor

In this case study, we focus on net participation while referring to conventional planning committee meetings scheduled at prescribed times. We conducted a social experiment and required participants to participate in these meetings through the Internet. Our goal was to verify whether net participation has advantages with respect to planning decision in the case of a green corridor if the tool we provided is used.

The Green Corridor incorporates Kenroku Mountain (core area), Central Park, and Hirosaka Boulevard, where a traditional irrigation channel was constructed when Kanazawa City was founded. The traditional irrigation channel was constructed as a regional water supply network, which is currently home to several types of fish species. We herein focus on how to support the planning committee on the reconstruction plan of an irrigation channel in which water is currently flowing through Hirosaka Boulevard and through the entire downtown area.

In an effort to support the Hirosaka Green Corridor of the Kenroku District Ecological Network in Kanazawa City, we suggested that the local government develop an online tool based on the process of conventional commission meetings, including providing planning information and reviewing alternative plans.

Unlike the planning materials used in conventional commission meetings, the materials that are available on the Internet, including virtual reality using Virtual reality modeling language (VRML) to represent the design schemes, digital photographs, and audio files explaining the design scheme, were used in this research project. Thus, the four types of multimedia, namely, text, VR, digital photographs, and audio files, were provided on the Internet, and an online bulletin board system (BBS) was set up for the purpose of discussing the green corridor design scheme. As shown in Fig. 9.1, Hirosaka Street is the green corridor in the downtown area of Kanazawa City that is a city landmark located near Kanazawa City Hall and Central Park. The planning commission was set up in 2003 by Ishikawa Prefecture in order to revitalize the downtown business center by means of the green corridor design, which is an important part of the Kenroku District Ecological Network.

The planning committee included officers of Kanazawa City, planners of local consulting companies, and citizens. During the period when planning information was first released, Kanazawa City had a statutory obligation to seek the views of its residents regarding the proposed plan for Hirosaka Green Corridor. During the planning meetings, we suggested the use of a visualization tool for representing

Fig. 9.1 Hirosaka green corridor in Kanazawa City

alternative plans in order to openly consider the various designs (Eppler and Platts 2009) and clarify the opinions of individuals. Moreover, a social experiment using the proposed system, which was developed according to planning documents and paper drawings presented in conventional planning meetings, was carried out while the real meetings proceeded. Meanwhile, the virtual planning committee meetings were conducted over 37 days from February 2 to March 12, 2004, during which the planning committee members, including all planners and officers, joined the virtual meetings conducted over the Internet. There were 48 experimental participants (Table 9.1) comprising officers of Ishikawa Prefecture, officers of Kanazawa City, and planning consultants involved in the project. In order that participants would take responsibility in the virtual meetings, we asked the participants to register using their real names. In the virtual meetings held over the Internet, the participants were organized into three groups to attend these meetings. The groups were divided based on the materials used into the document group, the audio group, and the VRML group. For the document group, only text and photographic documents were prepared. For the audio group, text, photographic, and audio files were prepared. Finally, for the VRML group, all types of media, including text, photographs, audio files, and VRML files, were used in the virtual meeting. Based on the deliberations during the virtual meetings, the effectiveness of sharing various types of media over the Internet for clarifying and evaluating the design schemes was verified.

At the same time, a student experiment was implemented in order to verify the validity of sharing planning information over the Internet using the online system. In the student experiment, we focus on the effectiveness of expressing the concepts of the green corridor plan, emphasizing the function of Hirosaka Green Corridor, which includes a traditional irrigation channel, a boulevard that connects Kenroku Mountain with Central Park in the downtown area of Kanazawa City. In the student experiment, 65 Kanazawa University students of the School of Environment

Table 9.1 Planning committee members

		Document group	Audio group	VRML group	Total
Officers	With planning background	5 (31.3%)	4 (25.0%)	4 (25.0%)	13 (27.1%)
	Without planning background	2 (12.5%)	2 (12.5%)	2 (12.5%)	6 (12.5%)
Consultant company	With planning background	1 (6.3%)	5 (31.3%)	5 (31.3%)	11 (22.9%)
	Without planning background	1 (6.3%)	1 (6.3%)	1 (6.3%)	3 (6.3%)
Other business companies		1 (6.3%)	1 (6.3%)	2 (12.5%)	4 (8.3%)
Others		6 (37.5%)	3 (18.8%)	2 (12.5%)	11 (22.9%)
Total		16 (100.0%)	16 (100.0%)	16 (100.0%)	48 (100.0%)

Fig. 9.2 Plan site

Design, took part in the virtual planning committee meetings, which were organized into four phases in the planning process from January 8 to February 13, 2004. This 34-day period is referred to as the total experimental period.

System Framework and Social Experiment

Development of the Online Multimedia Tool

Regarding the deliberation of conservation Hirosaka Green Corridor (Fig. 9.2), which is composed of a traditional irrigation channel and a boulevard, four zones for site analysis were discussed during the planning meeting (1) the irrigation channel and tree-lined space as the central separation along the boulevard, (2) the roadway, (3) the left sidewalk, and (4) the right sidewalk. Officers and planners should provide planning information, collect opinions and proposals, and draw up alternative plans. On the other hand, participants in the planning committee meetings should carry out a learning process stage to understand the design scheme and should be able to present their planning requirements and review all of the alternative plans being while consulting planners and experts. Finally, officers and planners, as well as planning committee members, should find a final solution through consensus. In this chapter, we develop an online multimedia tool and examine its effectiveness from the viewpoint of the participants, rather than officers or planners. Thus, the following arguments were organized based on the results from the learning process stage and from the reviewing of alternative plans taking place during the planning meetings.

Online Multimedia for Providing Planning Information for Learning

The Hirosaka Green Corridor was discussed considering the above four zones for site analysis. In this case study, considering the zones and the size limitation of the display screen of the personal computer, the planning information is divided into the four zones and organized as HTML files, embedded with text, audio, photographs, and VRML files, which are linked to each other. In the virtual meetings, planning information was prepared just as in the real meetings. For example, in order to present the surroundings of the four zones for site analysis, the planners integrated the planning information for the irrigation channel, the tree-lined space, the roadway, and the sidewalk (left and right sides) on paper for presentation in the planning meetings. In the web page, voice narrations were organized as audio explanations for each zone, and participants were able to open the web page for reviewing the site plan, as shown in Fig. 9.3-1(a–d), in which text descriptions for each space are provided as part of the site plan. The plan drawings represent the overall environment of the draft plan, and the VRML file represents the entire area of the project. Thus, the participants can experience the virtual world from different viewpoints, and move in different directions as shown in Fig. 9.3-1(d). Moreover, there are also URL links to audio files in some locations of the virtual world. As shown in Fig. 9.3-2, participants can compare different alternative plans using the web pages of the online system.

The file size of voice data, which is saved in .WAV format, is approximately 800–1,200 kB. The VRML file sizes for alternative plans, existing buildings for the background of all alternatives, and the combination of people and cars are approximately 5.6–7.2 MB, 391 kB, and 4 MB, respectively.

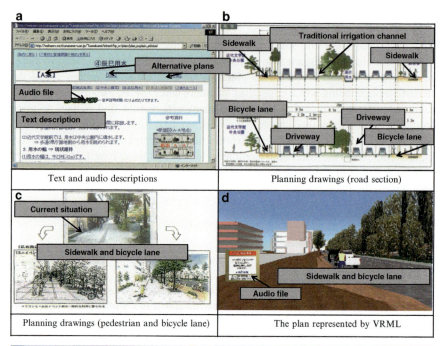

Fig. 9.3 Planning information using multimedia

BBS for Reviewing Alternative Plans

Participants can make suggestions and discuss alternative plans using a BBS tool from their PC at home or any other location after reviewing the provided planning information by planners and officers. Participants can also consult experts to learn to edit web content and gain a better understanding of the planning process. The planners and officers can answer questions and update information on the board as responses to participants. Through deliberations on alternative plans over the internet, the alternative plans can finally be reviewed over the Internet (Fig. 9.4).

Social Experiment Design

Virtual planning committee meetings based on the process of actual planning committee meetings, were scheduled in three stages (1) the learning process, which consists of understanding the current situation and problems involving the planning site, (2) reviewing alternative plans, which involves discussing the proposed plans and discussing the planning requirements, and (3) reviewing the final plan, which involves summarizing all opinions and revised proposals based on the discussions. The facilitator must collect the opinions at each stage for different design elements, summarize the overall trend of participants, and update the information over the Internet. Referring to the updated information on the web, the planner can improve the planning and design for the next stage (Table 9.2).

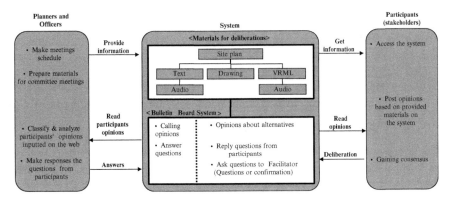

Fig. 9.4 Online multimedia tool and stakeholders

Table 9.2 Planning process

Step		Objective
Learning process (Jan. 8 to Feb. 13, 2004)		Understanding and checking planning information presented by planners and officers regarding the current state and planning requirements
Review process	Deliberation on planning alternatives (Feb. 13–30, 2004)	Review alternative plans presented by the planners
	Deliberation on final plan (Mar. 1–15, 2004)	Check the revised final plan

Learning and Reviewing

Planning Issues in the Learning Process

Eppler and Platts (2009) argued that "seeing openly", "seeing correctly" and "seeing clearly" are new challenges for visualization, which are critical for the interactive visual representations in a strategic-planning process. The design principle of the Hirosaka Green Corridor planning is to conserve the green corridor and appropriately maintain the balance between heavy traffic and natural space. Trees and shrubs in the tree-lined space are cultivated for landscape design. However, the irrigation channel is hidden by the shrubs and the branches of taller trees that overhang the roadway and scrape car windows. Therefore, the Hirosaka Green Corridor Project was started by the Planning Authority of Ishikawa Prefecture in order to solve this problem. Since the green network hampers the flow of traffic, Ishikawa prefecture is considering whether it is necessary to move the irrigation channel and tree-lined space to the left or right sidewalk. In this section, the effectiveness of multimedia materials is discussed through analysis of participants' learning about the design criteria and planning principles of the ecological network of the Hirosaka Green Corridor and focus on how the multimedia materials can help participants to understand the planning issues related to the traditional irrigation channel and the boulevard. As mentioned above, conflicts between shrubs and the irrigation channel and between the roadway and tree branches are explained through text, and participants can confirm the descriptions by means of scanned paper drawings on the web site. A VRML file is edited to represent the design scheme so that participants can confirm that the plan is improved.

A questionnaire conducted after the social experiment included questions such as "Did you understand the planning principles described by the digitized drawings on the web?" and "Were you able to deliberate with others regarding the contents of the design schemes?" As a result, as shown in Table 9.3, both text descriptions and scanned paper drawings of the site plan were helpful in clarifying the planning principles from the respondents of all groups. Regarding the audio file, 24 of the 31 respondents (77%) could "easily image the site plan based on the text description and drawings" and "did not listen to the audio explanation on the PC". Thus, the voice narration is likely unnecessary. However, some participants stated that, "In order to save time, I did not read the documents and only listened to the audio file", or that "The audio file was helpful for easily reading the text". Therefore, for individuals who do not like to read long sentences, such as the elderly or children, the audio files may be useful. "Supporting seeing", namely assisting visualization (Eppler and Platts 2009) is an advantage of audio files as using a visualization medium. Some respondents reported that they "failed to hear the audio introduction". Thus, reducing the size of audio files is helpful, and the explanation of how to control the volume settings should be carefully displayed on the web page.

9 Web-Based Multimedia and Public Participation for Green Corridor Design   195

Table 9.3 Learning about the planning information using multimedia materials (%)

	Document group		Audio group		VRML group	
Text description	17	(81.0)	12	(92.3)	12	(85.7)
Scanned site plan	10	(47.6)	8	(61.5)	9	(64.3)
Roadway cross-section	5	(23.8)	6	(46.2)	11	(78.6)
Scanned paper drawings	11	(52.4)	10	(76.9)	6	(42.9)
Audio file for guidance			0	(0.0)	1	(7.1)
VRML					4	(28.6)
Sum (number/person)	43	(2.0)	36	(2.8)	43	(3.1)
Number of respondents	21	(100)	13	(100)	14	(100)

Among the members of the VRML group, 79% experienced the VRML world in order to "understand" the plan principles, whereas 64% viewed the scanned paper drawings over the web.

Deliberation in the Review Process

There were several opinions regarding the different plans that were exchanged through the BBS over the Internet. Here, participants forged a consensus in reviewing Plans A, B, and C. Through deliberation on the planning principles regarding the irrigation channel, shrubs, and central tree-lined space, agreement on the planning principle was reached. Finally, although most of the participants chose Plan A, the participants did not want to widen the central tree-lined space. In order to solve the problem of traffic flow and obstructions caused by tree branches, most participants preferred keeping the trees in the central tree-lined space. Thus, since clear ground space is necessary in order to cut tree branches, shrubs in the central tree-lined space should be removed. For landscape design, it was suggested that part of the irrigation channel be moved to Central Park. The form of the Hirosaka Green Corridor was conserved, and part of the irrigation channel path was changed. As a result, all of the trees were left intact, although the shrubs were removed. Fish are still able to live in the irrigation channel, and nothing changed with respect to the birds in the Kenroku District Ecological Network. As shown in Table 9.4, stakeholders are able to achieve "seeing differences" and "seeing correctly", namely understand the planning principles and differences between alternative plans correctly using visualization medium (Eppler and Platts 2009) through comparing alternative plans, thereby assuring difference, consistency, and comprehensiveness between alternative plans in visualized contents.

As indicated in Table 9.5, the tool was reported by 67% of the members of the document group, 69% of the members of the audio group, and 57% of the members of the VRML group to be very useful or fairly useful in the deliberation regarding the Hirosaka Green Corridor. As such, the system can be said to be effective, and there was no significant difference among the various groups. Compared with the

Table 9.4 Discussion on alternative plans

	Plan A	Plan B	Plan C	Consensus
Irrigation channel	Remove to the central park, can be seen from the park	No change	Move to the left side beside the sidewalk	Move part of the irrigation channel to Central Park. The fish species network can be conserved
Shrubs	Remove	Remove	Remove	Remove
Sidewalk (left side, park site)	Widen from 2.6 m to 4 m (bus stop), divided into bicycle and pedestrian lanes	Pedestrian lane only	Bicycle and pedestrian lanes	Widen the sidewalk of the central park site, and separate the sidewalk into pedestrian and bicycle lanes
Sidewalk (right side, shopping)	Bicycle lane	No bicycle lane	Bicycle lane	Bicycle lane
Central tree-lined space	Widen 0.5 m, cut tree branches	Narrow Remove the pine trees, but leave the cherry trees	Narrow Remove the pine trees, but leave the cherry trees	Leave the width unchanged, and cut tree branches frequently. The green corridor can be maintained unchanged. Fish can continue to live in the irrigation channel, and nothing will change for birds

9 Web-Based Multimedia and Public Participation for Green Corridor Design

Table 9.5 Effectiveness of deliberation (%)

	Document group		Audio group		VRML group		Total	
Completely effective	1	(4.8)	2	(15.4)	0	(0.0)	3	(6.3)
Effective	13	(61.9)	7	(53.8)	8	(57.1)	28	(58.3)
Moderately effective	5	(23.8)	2	(15.4)	5	(35.7)	12	(25.0)
Less effective	2	(9.5)	2	(15.4)	1	(7.1)	5	(10.4)
No effective	0	(0.0)	0	(0.0)	0	(0.0)	0	(0.0)
Total	21	(100.0)	13	(100.0)	14	(100.0)	48	(100.0)

learning process as shown in Table 9.3, the scanned paper drawings were used by 29% of the respondents during the review process, which is 14% less than in the learning process. In particular, 86% of the respondents used the roadway cross-section during the reviewing process, which increased seven points more than in the learning process, whereas other options in Table 9.3 were kept no significant changed. The roadway cross-section was more helpful for deliberations regarding tree branches and the roadway in the Hirosaka Green Corridor.

Analysis of the Chat Log

In the previous section, we examined the effectiveness of the system by means of a questionnaire. In this section, we analyze the deliberation during the virtual planning committee in order to test the effectiveness of the online multimedia tool.

Log Analysis of Online Chat During Virtual Meetings

Chats during the virtual planning committee meeting, which were achieved through a chat room on the Internet, are divided into several 6-h time periods, as shown in Table 9.6. Chat room deliberations considered bus lanes, tree branches, and the width of the sidewalk incorporating a bicycle lane, where the focus was on conflicts between human activities and the natural space in the Hirosaka Green Corridor. For example, the discussions focused on the removal of shrubs from the central tree-lined space, cutting tree branches, and the removal or covering of the irrigation channel. A number of the participants argued that the density of trees in the central tree-lined zone was too high, so that car drivers were unable to see other vehicles when attempting to turn. These participants argued that the central tree-lined space was dangerous and that the green corridor should be completely eliminated. In addition, a number of the participants claimed that the width of the sidewalk was insufficient and that the central tree-lined space should be moved in order to incorporate a bicycle lane and more pedestrian space. There are very few comments on animal network, in which some committee members argued about if fish and

Table 9.6 Chat logs of virtual meetings

Time period	Midnight and early morning		Morning		Afternoon		Night		Total	
	0:00–6:00		6:00–12:00		12:00–18:00		18:00–24:00			
Number of speakers	7	(30.4)	14	(60.9)	19	(82.6)	16	(69.6)	23	(100.0)
Number of messages	11	(6.6)	51	(30.5)	57	(34.1)	48	(28.7)	167	(100.0)

Table 9.7 Checking planning information during virtual planning committee meetings

	Current state and problems of maintenance		Alternatives	
Read completely	4	(8.3)	6	(12.5)
Read	30	(62.5)	23	(47.9)
Read partially	6	(12.5)	8	(16.7)
Read only slightly	5	(10.4)	9	(18.8)
Did not read	3	(6.3)	2	(4.2)
Total	48	(100.0)	48	(100.0)

birds network will not be cut off even though moving the route of irrigation channel to the central park. There were conflicting opinions regarding wildlife conservation in the Hirosaka Green Corridor. As shown in Table 9.6, the arguments presented over the Internet were inputted during the different time periods.

As shown in Table 9.6, 23 of the 48 participants posted their topics. Among the 23 participants, 60.9% input their opinions during the period from 6:00 a.m. to 12:00 a.m., and 82.6% input their opinions during the period from 12:00 a.m. to 18:00 p.m. Therefore, the participants likely expressed their opinions using computers from their offices. In addition, 69.6% of the 23 participants input their opinions during the period from 18:00 p.m. to 24:00 p.m.

With respect to the numbers of messages, among the 167 messages, 35% were input during the period from 18:00 p.m. to 6:00 a.m., and 64.6% were input during the period from 6:00 a.m. to 18:00 p.m. Therefore, these participants appear to have joined the discussion from their homes during the evening after work or during working hours using office computers. Unlike the conventional meeting, which has a prescribed time and location, using the Internet-based system, the participants were not limited to a specific location or time. As such, virtual planning committee meetings expanded the participation to allow residents who would be unable to attend the conventional planning committee meetings.

In the virtual meeting, most of the participants selected Plan A with some modifications. In making their final decisions, 70% of the participants reviewed the planning information as shown in Table 9.7, which was edited via the web page as the "current status and development issues", and approximately 60% of participants reviewed the planning information once more before making their final decisions. In the conventional meetings, a large volume of paper materials was distributed to each member in the meeting room, and it was difficult to look through all of the materials in a short time. If the participants used the system over the Internet, there was sufficient time to read the planning documents that were edited in digital form. Moreover, participants were able to review the planning documents according to their time schedule. In addition, participants provided their opinions over the web by messages that usually contained approximately 321 Chinese characters, and the longest among the posted messages are 2071 Chinese characters. In the chat room deliberations, certain topics were covered in detail, and each topic received at least three comments. Thus, relatively comprehensive discussions were conducted. Accordingly, the final decisions were made at the end of the Internet meetings.

Role of the Facilitator in the Virtual Meetings for Gaining Consensus

In the conventional meeting, facilitators can increase the attendance rate by negotiating with members in making the meeting schedule, and facilitators should organize and encourage members to express their opinions during planning committee meetings. If members attempt to present a long speech, the facilitator can ask the speaker to be brief so that opinions from other participants can be heard. However, over the Internet, even though a facilitator can encourage participants to express their opinions, it is not necessary for members to respond in real time. If there is a lack of individual motivation and or a lack of interest, the members do not have to respond to the facilitator. The document group, the audio group, and the VRML group each had 16 members. However, during each stage, the maximum number of participants was ten, which decreased slightly in the following stage, as shown in Fig. 9.5. In particular, starting from the stage of reviewing the alternative plans, the number of participants decreased sharply.

In addition, thinking over the opinions offered by other participants over the Internet before responding can be advantageous. Therefore, active participants in the virtual meetings posted responses of more than 1,000 Chinese characters.

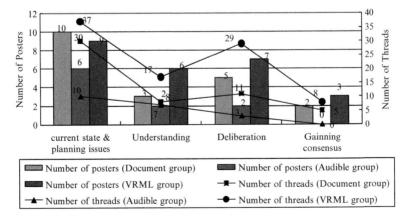

Fig. 9.5 Evaluation by participants of different processes

Table 9.8 Role of the facilitator

	Document group		Audio group		VRML group	
Comments on the messages of other participants	12	(22.2)	6	(30.0)	16	(17.6)
Call for messages	13	(24.1)	3	(15.0)	3	(3.3)
Number of messages by the facilitator	23	(42.6)	10	(50.0)	27	(29.7)
Total numbers of messages	54	(100.0)	20	(100.0)	91	(100.0)

On the other hand, other participants reported that it "became harder to make brief remarks regarding opinions posted during the meeting, and they eventually tired of thinking how to post their own opinions". Therefore, in order to make it easier for all participants to express their opinions, a rule may be necessary in order to force participant to express their opinions briefly and with a clear focus. Furthermore, by looking only at messages posted in the chat room, participants occasionally had difficulty in responding because of unclear messages.

As in the case of a conventional meeting, the character and ability of the Internet facilitator should also be considered. Thus, the contributions of the facilitators in each group were analyzed. As shown in Table 9.8, the number of posts from the facilitator of the audio group was less than half those of the other groups, whereas there were few posts from participants in the audio group. In the document group, the facilitator asked for opinions for 13 times, and commented on posted opinions 12 times. The facilitator of the VRML group commented on posted opinions 16 times, which is 1.7 times that of the document group.

Therefore, the comments in these two groups were posted after the cited sentence. For example, comments in VRML group always started from "says..." with the name of the cited participant. In interviews after the virtual meetings, participants reported that "it was difficult to determine whether others participants checked my posts, and so it was difficult to decide whether to continue the discussion". Thus, it would be helpful for the facilitator to comment on posts while citing the name of the contributor. In order to share information and deliberate over the Internet, the real name, gender, age, and occupation of each member should be provided, and the connection of each member to the planned target area should be clear to other participants.

Effectiveness of Online Multimedia

As shown in Fig. 9.6, the respondents reported that the tool was very effective or reasonably effective for improving public participation in local planning commission. As shown in Fig. 9.6, a small percentage of the members of the document group reported the system to be effective for use in the planning process. During the final stage, 57% of the members of the document group report the system to be effective, whereas 93% of the VRML group members report the system to be effective.

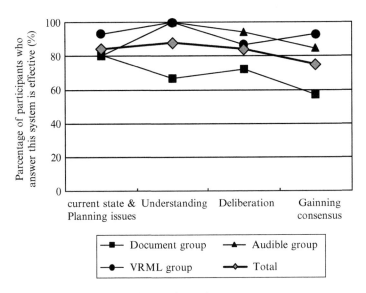

Fig. 9.6 Comprehensive evaluation of the online tool

Therefore, the mixed-multimedia materials are considered to be more effective in the planning process.

Comparison of the Virtual Meeting and the Conventional Meeting

After investigating the effectiveness of the online multimedia tool, we compared the advantages and disadvantages between the conventional meeting and the virtual meeting. The biggest difference in participating in the virtual meeting as compared to participating in the conventional meeting is whether the meeting is held at the same place and the same time, and in the case of the Internet, the meetings can be held at different times and different places. As such, members can simply post comments over the Internet and enter into discussions of the BBS (Table 9.9).

The planning committee members, who are experts living in other cities and members who are out of Kanazawa City on business, have difficulty in attending conventional meetings scheduled at a prescribed time and place. On the other hand, Internet meetings can solve this problem because participants can attend the meeting from personal computers at any location.

Regarding the provision of planning information, officers and planners can provide an explanation and additional handouts in conventional meetings. However, in the case Internet meetings, the participants have to read the URL site carefully in order to determine the contents of the plan, which requires some knowledge of computer operations, and participants will occasionally overlook a

Table 9.9 Comparison of the conventional meeting and the virtual meeting

	Conventional meeting	Virtual meeting
Discussion	Since the place and agenda of the meetings are fixed, participants have difficulty in attending all of the meetings	If participants can use the Internet, they can freely access the web site at favorite places within a given period of time without spatial and temporal limitations
	It is difficult to hold long meetings (over approximately 2 h) because of the restriction of participants	The holding period can be longer because the participants are not restricted during the holding period (several days to several weeks)
	Exchanging opinions face to face in a meeting room	Exchanging opinions remotely in a chat room
	Questions that cannot be answered during the meeting will be carried over to the next meeting	If it is difficult to reply immediately, the participants can reply after confirmation and determination over the Internet
	The decisions should be made in the meeting room every time. It is difficult to gather external reference information during meetings	Participants can make decisions within a period prescribed by the commission. It is possible to gather reference information before making a decision
Media	The secretary presents materials and alternative plans	Each participant operates a PC and reads the web site by himself/herself. Participants need to know how to operate a PC
	Paper documents and drawing materials are separately presented. The PC can be used to present alternative plans	Drawings and multimedia materials are open to the public over the Internet. Hyperlinks provide reference information. A record of previous information is available for review
	Participants can review models	Participants can experience and walk through the virtual space

portion of the information. In addition, information about the project area and printed documents in conventional meetings will be distributed to regular attendees, which can be read by all local citizens if made available over the Internet. In conventional meetings, planning information is always provided in the form of paper documents and photographs. Because of the numerous separate pages of information, it is difficult to understand the how the articles or pages are interrelated. In particular, when presenting relatively large-scale designs and planning of public space in a city, it is necessary for planners to present a lengthy oral explanation to members using paper documents. Even though an oral explanation is not available over the Internet, participants can read the website carefully and check the links to each page, which takes advantage of the use of voice, VR, and other multimedia materials to present a variety of alternative plans.

Internet chat meetings can be held for a longer period (e.g., a week or a month) than conventional meetings, and members can freely access a database of the meetings to review the design scheme, make suggestions, and present responses when necessary during a meeting. Deliberations between members are not necessarily conducted in a face-to-face manner, and members can present opinions and suggestions on the web. Thus, it is possible for planners and officers to answer questions after sufficient preparation and properly preparing responses to various questions during the Internet meetings. In conventional meetings, officers and planners should prepare responses to questions raised by planning committees for the next meeting if there are issues that cannot be answered during the current meetings.

Moreover, in the conventional meetings, deliberations are usually conducted face to face, and there are always difficult problems that cannot be solved during the meetings. Thus, it is possible for members to summarize these problems after the meeting if the meeting is conducted or continued over Internet. The Internet can also be used to improve the quality of the discussion instead of just rapidly exchanging information and ideas, as in the conventional meetings, because planners will have time to investigate and prepare necessary documents for supporting their ideas. It is also helpful for planning committees to consider the opinions of other participants after reviewing the chat logs over the Internet.

Conclusion

In this chapter, an online multimedia tool is developed to support local planning committee meetings for green corridor design. An experimental virtual meeting was carried out to examine the effectiveness and applicability of the tool.

Experiments were conducted over the Internet for comparison with conventional paper drawings distributed during the planning committee meetings. Presenting planning information using VRML and audio files over the Internet received a relatively high evaluation. The combined online multimedia including VRML, audio, and text files revealed that providing the context for program-related information is effective.

Virtual meetings for different planning stages, including the learning and reviewing process, are available for participants to access from various times and locations. Participants posted their opinions over the Internet during discussions. Thus, it is possible to expand the range of participants as compared with the conventional planning committee. Virtual meetings over the Internet will be enable longer meetings than conventional meetings. Therefore, participants can post more opinions and have a more in-depth discussion on the details of each step during the planning process. However, it is important to set up rules for participants to read and post their opinions during discussions over the Internet. If a facilitator can carefully check who posts opinions, it will be helpful to interactively promote the discussion by clearly mentioning with the name of the person.

Acknowledgments We would like to thank the financial support of the Grants-in-Aid for Scientific Research (No. 22560602C), Japan Society of the Promotion of Science.

References

Atav, A. and Ezgi Haliloğlu Kahramana, Z. (2009), Constructing collaborative processes through experiential learning: Participatory planning in Kaymakli, Turkey, Habitat International, Volume 33, Issue 4, 378–386.

deVries, B., Tabak, V. and Achten, H. (2005), Interactive urban design using integrated planning requirements control, Automation in Construction, Volume 14, Issue 2, 207–213.

Eppler, Martin J. and Platts, Ken W. (2009), Visual Strategizing: The Systematic Use of Visualization in the Strategic-Planning Process, Long Range Planning, Volume 42, Issue 1, 42–74.

Fraser, E.D.G., Dougill, A.J., Mabee, W. E., Reed, M. and McAlpine, P. (2006), Bottom up and top down: Analysis of participatory processes for sustainability indicator identification as a pathway to community empowerment and sustainable environmental management, Journal of Environmental Management, Volume 78, Issue 2, 114–127.

Germain, Reně H., Floyd, Donald W. and Stehman, Stephen V. (2001), Public perceptions of the USDA Forest Service public participation process, Forest Policy and Economics, Volume 3, Issues 3–4, 113–124

Groome D. (1990) "Green corridors": a discussion of a planning concept, Landscape and Urban Planning, Volume 19, Issue 4, 383–387

Jepsen, J.U., Baveco, J.M., Topping, C.J., Verboom, J., Vos, C.C. (2005), Evaluating the effect of corridors and landscape heterogeneity on dispersal probability: a comparison of three spatially explicit modelling approaches, Ecological Modelling, Volume 181, Issue 4, 445–459

Manabe, R., Murayama, A., Koizumi, H. Okata, J. (2003), The New Turn of the Internet Mapped Information Board System -Kakiko Map-, City planning review. Special issue, Papers on city planning, 38(3), 235–240

Moon, T.H. (2003), Development of Web-based 'Public Participation and Collaborative Planning System (PPCPS)', CUPUM'03 Sendai (The 8th International Conference on Planning and Urban Management), 3a2, 2003

Oura, M., Arima,T., Hagishima, S., Sakai, T. (2001), Development of the multimedia town planning support system using WWW (2), Proceedings of the 24th symposium on computer technology of information, systems and applications, Architectural Institute of Japan, Vol. 24, 61–66

Richard, Register (1987) Ecocity Berkeley: building cities for a healthy future. North Atlantic Books, Berkeley.

Shen, Z., Kawakami, M., and Kishimoto, K. (2002), "VRML applied research on the possibility of cooperation projects using the design system", Journal of the City Planning Institute of Japan, Vol. 37, 73–78

Shen, Z. and Kawakami, M. (2010), An online visualization tool for Internet-based local townscape design, Computers, Environment and Urban Systems, Vol 34, 104–116

Tippett, J., Handley, J. F. and Ravetz, J. (2007), Meeting the challenges of sustainable development —A conceptual appraisal of a new methodology for participatory ecological planning, Progress in Planning, Volume 67, Issue 1, 9–98

Ulla Mörtberg, Hans-Georg Wallentinus (2000), Red-listed forest bird species in an urban environment — assessment of green space corridors, Landscape and Urban Planning, Volume 50, Issue 4, 215–226

Chapter 10
Online Cooperative Design for the Proposal of Layouts of Street Furniture in a Street Park

Zhenjiang Shen, Dingyou Zhou, Mitsuhiko Kawakami, Kazuko Kishimoto, and Seitaro Imai

Introduction

In the last decade of the twenty-first century, the rapid development of Internet and computer technology has enabled conduct cooperative design through the Internet. Al-Douri (2010) argued that conventional urban planning may lack adequate coverage of the essential design aspects of the built environment, which can be corrected by digital models and information technology tools that may help designers to visualize and interact with design alternatives and correct this problem. Following the work of Carmona and Punter (2002), Ai-Douri classified urban design aspects, such as urban form, townscape, public realm, conservation areas, land use, and landscape architecture. Under the multi-user architecture for team environments, the technical conditions for the implementation of cooperative design comprises both a visual representation and a semantic representation, as well as a shared workspace for shared understanding (Saad and Maher 1996). Meanwhile, research on online design tools has recently appeared in literature regarding coordination, collaboration, and communication in cooperative design (Klein 1995; Nurcan 1998; Okukuni et al. 2000; Wang et al. 2009) using virtual reality (VR). It is worthwhile to explore the prospect of online cooperative design as an application and extension of VR technology. In this chapter, we focus on a shared visual representation and computer-mediated communication (Miller and Brunner 2008) between participants for reaching a consensus on the design of a street park.

For design cooperation regarding the representation of public perception, such as public space usage and street furniture, virtual representation can be used to design

---

Z. Shen (✉) • M. Kawakami • K. Kishimoto • S. Imai
School of Environmental Design, Kanazawa University, Kanazawa, Japan
e-mail: shenzhe@t.kanazawa-u.ac.jp

D. Zhou
School of Architecture, Dongnan University, Nanjing, China

Z. Shen, *Geospatial Techniques in Urban Planning*, Advances in Geographic
Information Science, DOI 10.1007/978-3-642-13559-0_10,
© Springer-Verlag Berlin Heidelberg 2012

concept development (Hoskins 1979) in the planning process. Since the 1990s, a number of universities have used virtual reality to build three-dimensional (3D) urban models for presenting urban planning and design. For instance, the preliminary stage of research on 3D modeling for urban planning and design was started by Batty (2001) at University College London and Day (1994) at Bath University, among others. As the models built by Bath University (Day 1994), which include more than 150 model blocks, the 3D information system was used to evaluate the physical transformation of cityscape after urban redevelopment. Day argued that a problem-finding approach using such models is generally applicable and, in particular, is appropriate for nonexpert users. Bulmer (2002) also discussed the possibility of design collaboration using VR for improving public participation. In Japan, a number of research projects have attempted to assess urban design using virtual reality modeling language (VRML) through collecting public opinion on the Internet. For instance, a group in Oita University (Yamataki et al. 1999) investigated the effectiveness of different proportions of tree and building in virtual reality. Nagoya Institute of Technology (Hammad et al. 1999) implemented an experiment on the effectiveness of virtual representation of urban design using VRML through psychological evaluation.

Generally speaking, with respect to design coordination, it is necessary for participants to execute design activities as individual tasks and to then develop a single design proposal together as a team based on the results of individual design activities (Tichem and Storm 1997). The RoomNavi tool (Kaga et al. 1996) is an example system that was used for individual tasks performed by inexpert designers. RoomNavi developed using the network open design environment (NODE) by Osaka University, which is based on the technology of Blaxxun and Shout 3D. Using this tool, house buyers can design room layouts so that real estate developers can understand the intentions of buyers, which can improve communication between designers and house buyers. With respect to cooperative design for developing a design proposal of public spaces used in a multi-user environment, each participant was asked to build his/her own space in the virtual community platform. The technical feasibility of cooperative design has been successfully proven by the research of Alpha World Project (Smith et al. 1998). Up to now, more than 30,000 participants have registered and built their own spaces in Alpha World. In addition, the project of Ryoanji (Soide et al, 1998; Okabe et al. 1999) is an attempt to practice collaboration design by positioning a stone in a Japanese garden in a temple located in Kyoto, Japan. Moreover, an online integrated multimedia-GIS tool was developed in the context of scenario planning to enable a community to actively explore different options for the use of public spaces (Pettit et al. 2004a, 2004b). In another study, participants were demonstrated to be able to share an online multi-user design tool to reach a consensus concerning public use of spaces scheduled at prescribed times through design coordination of a facilitator through the Internet (Shen and Kawakami 2010).

Accordingly, with respect to design collaboration regarding the use of public space, it is possible for participants to make design proposals as individual tasks and then develop a single design proposal together as a team. In this chapter, for the

10 Online Cooperative Design for the Proposal of Layouts of Street Furniture 207

completion of design tasks of a street park, an online design tool is developed for nonexpert designers to individually make design proposals in an Internet environment and to save these designs in a web database for comparison with the design of others. All of the individual designs will be reviewed by professional planners and designers. In the next section, we discuss the research background. After introducing the planning practice regarding street park designs in Japan in Sect. "System Development", a system framework of the online design tool is described in Sect. "System: Development". In Sect. "System: Application and Evaluation", we describe a case study carried out in Kanazawa City, and conclusions are presented in Sect. "Conclusion".

Research Background

In Japan, since the mid-1970s, the bottom-up decision making process of public space design has been widely used in public-participation-based communal management (PPBCM) systems (Matsushita 2003). The public has shown significant interest in improving the residential environment in built-up areas in which there are not sufficient public spaces, i.e., urban green spaces and street parks (Funabiki 2009). Generally, planning meetings regarding public spaces are scheduled by local government, and scheme designs are proposed by planners and designers of consultant companies. Then, local residents make comments and suggestions concerning scheme design during these meetings (Shen and Kawakami 2010). In this bottom-up process, the plan consulting company should collect the planning requirements of all participants that may be in the form of replies to questionnaires and coordinate designs through sketches and drawings created by participants in design workshops (Nishikizawa and Harashina 2004). Moreover, local governments hold design workshops using VR in order to actively involve all stakeholders and collect public opinions regarding public spaces, townscapes, and buildings (Fukushima et al. 2001; Koga et al. 2008). However, it is difficult for planners to understand the design concepts of stakeholders through their sketch drawings because untrained participants have difficulty in expressing their ideas regarding the use of public spaces based on their daily living experiences. For this reason, we developed an online design tool in order to facilitate the expression of public space designs by the participants.

In this chapter, we attempt to develop an online design tool for design collaboration between planners and nonprofessional participants, which is different than that for supporting professional planners to present their design work. For the representation of the design proposals of non-expert individual tasks concerning the layout of street furniture, we consider a possible system framework and the evaluation of the tool by the participants. Most of the participants were either housewives, children, or elderly people who lived in the neighborhood of interest (Shimizu et al. 2007). We intend to invite children to use the tool and make design proposals of public spaces design, which will allow planners and designers to obtain ideas from participants proposals. In the present research, we assume that VR has an advantage over conventional plan drawings in that the former can significantly improve the representation of design concepts in 3D form. In particular, it is attractive for participants to share virtual representations of public spaces. In the absence of professional-quality design drawings, an online design tool would be helpful for design coordination and communication between nonprofessional individuals and planners in the process of design collaboration.

Accordingly, an online tool for street park design is expected to aid nonprofessional individuals in expressing their proposals in 3D virtual space. The design content in their proposals is related to the use of public space (Goličnik and Thompson 2010) and the requirements of street furniture, so that individuals should provide their design intensions and information concerning the design of street furniture and site layout planning using this design tool. Therefore, a simple design tool for expressing design intension is needed, and individuals are asked to submit their design information including coordination information of 3D objects' positions as their preference to web database in order to reconstruct 3D design alternatives. On the other hand, cooperative planning and design through public participation does not mean that design schemes proposed by professional designers are not needed. In order to improve the design content, professional designers perform their tasks while taking into account the proposals obtained through public participation. Professional designers also needed to help participants to finish their proposals using online tools for design and planning. In addition, tools for online communication, such as a chat system between public and professional designers, are also necessary. Next, we introduce an online cooperative design tool for park design.

System Development

Planning Practice of a Street Park in Kanazawa

Urban parks are classified into several types in the planning system of Japan. The street park, having a service radius of 250 m, is the smallest urban park, and, as of 2005, there are more than 72,000 street parks in Japan. Users of street parks are primarily neighborhood residents who live nearby. In the design process of these street parks, local governments consider the needs of neighborhood residents through public participation. It is expected that residents will make proposals based on their daily lives and will cooperate with each other to reach consensus on the design of street parks within local communities.

For collaborative design of a street park, local governments organize planning committee meetings, in which planners and designers from a consulting company present proposals and design schemes. A facilitator selected from among the stakeholders will coordinate the collaboration between the stakeholders and will make a final decision based on the deliberations in the planning committee meetings. The consulting company will obtain opinions on street park design through responses to questionnaires or through attendees in planning meetings within the neighborhood. Conducting questionnaires over the Internet will be more convenient for neighborhood residents and will be more efficient than conventional methods. If the opinions of stakeholders can be collected online, then neighborhood residents cannot only reply to questionnaires through the Internet but can also cooperate with others at home to present design proposals using VR without spatial–temporal limitations.

The process of participatory planning and design for street parks in Japan generally includes six steps, as summarized by the Setagaya Community Design Center (1998) based on experience in Setagaya District, Tokyo. The six steps are mutual understanding, proposal of planning objectives, investigation of the planning site, budget discussion, proposal of design schemes, and gaining consensus. In this chapter, a design tool is introduced to support the participatory planning and design of a street park in the last two steps: proposing design schemes and gaining consensus.

The planning section of urban parks and green spaces in Kanazawa City implements a six-step process similar to that proposed by the Setagaya Community Design Center (1998) and Sanoff (1993) in order to achieve street park design through a collaborative process involving neighborhood residents. In this chapter, we describe a case study involving Yamanoue Street Park, which is located in the area of low mountains and hills near downtown Kanazawa, as shown in Fig. 10.1. The area of Yamanoue Street Park is approximately 600 m^2. The land use zoning in Yamanoue Community is planned as a residential area, where most of the houses are traditional Japanese houses. Yamanoue Community consisted of 183

Fig. 10.1 Location of Yamanoue Street Park in Kanazawa City

Fig. 10.2 Method for participatory planning and design in Kanazawa City – icon design game

households, and the elderly population of Yamanoue Community accounts for 48% of the total population. Since all parcels were in private ownership, there were no public spaces in this community until a property owner sold 1,000 m^2 of land in 2000, and the local government decided to acquire 600 m^2 for the construction of a street park.

As shown in Fig. 10.2, the conventional cooperative design game, namely, the icon game, is usually conducted by local governments in the process of street park design for gathering public opinion on the use of urban park spaces. Instead of the conventional planning approach, we developed an online cooperative design tool to support the process of design collaboration involving neighborhood residents. A local planning commission was set up by the local government for the planning of Yamanoue Street Park. Together with local residents, planners from a consulting company proposed design alternatives while using a design tool developed by our team. The style of the street park and how to satisfy the needs of elderly individuals and children were considered during the planning meetings.

In the next section, the system framework of our cooperative design tool, which was developed to meet the needs of street park design through public participation, is discussed.

Key Considerations in System Development

With the rapid development of computer technology, VR is increasingly being used as a technological support tool in online design collaboration for improving public participation through the Internet. Methods for developing useful design tools and investigating the possible effects of using these design tools are topic for further study, and practical applications involving both system developers and urban planners are being developed. Most modeling software, which is based on CAD and CG software technology, is developed as professional software for designers and planners. However, using VR technology in order to improve design

10 Online Cooperative Design for the Proposal of Layouts of Street Furniture

collaboration between nonprofessional participants and professional planners remains a challenge. Researches on the application of VR indicate the existence of various opportunities for Internet participation (Doyle et al, 1998; Smith et al. 1998; Bulmer 2002). Thus, in this chapter, we attempt to support participants in generating design proposals using virtual reality directly over the Internet.

A number of VR systems have released web browser plug-in software for design presentation. The advantage of VR plug-in software is that a web browser can be used for design and plan evaluation from a remote terminal computer. Since VRML can provide an interactive environment, most software provides an export function for converting its file format to VRML format in order to publish designs and plan for interactive design presentation on a web site. For stakeholders to make design proposals at home, a web database is necessary in order to record all of the design schemes submitted from the Internet. Meanwhile, in order to review and discuss the design alternatives, it is necessary to store all of the design schemes for design collaboration.

Accordingly, we attempt to develop an online VRML cooperative design tool, which has the following features (1) VRML is used to create a virtual street park that can be experienced by anyone over the Internet. (2) A web database is necessary for storing the dataset reflecting the design proposals of the participants. (3) Each design proposal can be retrieved from the web database for representing a design proposal in a VRML world for comparison with the designs of other participants.

System: Development

System Framework

As mentioned earlier, a cooperative design tool and a web database are necessary for storing design proposals. The structure of this system is shown in Fig. 10.3, in which rectangles represent VRML files, circles and ellipses represent user interfaces, and the cylinder represents the web database. In addition, lines represent running processes. The web-based application is established using the HTTP service on a Windows server, and the client operating system can be Window2000/NT/XP/VISTA.

As in the case of the real ICON game, the size of urban furniture cannot be changed, and only the location and orientation of furniture can be adjusted. Thus, based on the Icon game, the tool can be developed as a two-dimensional design tool, while the design results must be displayed in a 3D environment.

Participants can select and arrange street furniture by moving a mouse over a JPEG image of the park site in the system. Thus, participants can arrange the furniture in the park, and the results, together with the personal information of the designer, can be saved to the database through Active Server Pages (ASP).

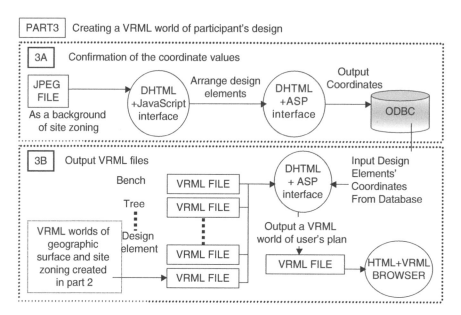

Fig. 10.3 Software framework of online cooperative design

The coordinate values of urban furniture can be output to database records through ASP and Open Database Connectivity (ODBC) and can be retrieved from the database to create a VRML file through ASP. The VRML file created based on the coordinate values of urban furniture is the design result of the participants. These VRML files can be created at anytime within the Internet environment and are important information for making a final decision regarding the street park.

User Interface of the Design Tool

The first page of the proposed system is a registration page, which is linked to the following content:

1. Registration page and design game pages

    The registration page is linked to a design game page, which is developed for arranging detailed urban furniture elements (Fig. 10.4). We use JavaScript to collect the design proposals of participants via the web page, where participants can decide the locations of trees, benches, and other items in a park site.

    Database access pages are prepared for the input of design information (Figs. 10.5 and 10.6), where we use ASP to record the design information of participants in the system. A Microsoft Access database (Fig. 10.7) is prepared for recording the coordinates and the direction of trees and urban furniture according to the recommendations of participants.

10 Online Cooperative Design for the Proposal of Layouts of Street Furniture 213

Fig. 10.4 Interface of the online design tool

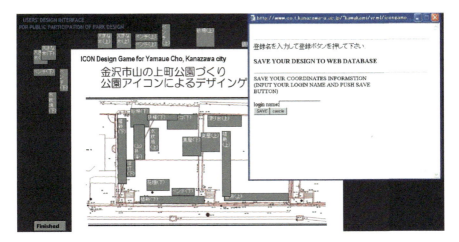

Fig. 10.5 Street park design interface for Yamanoue Machi, Kanazawa City

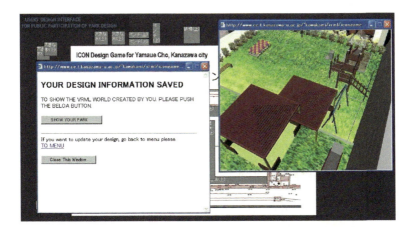

Fig. 10.6 Reloading a cooperative plan for a street park in Yamanoue Machi, Kanazawa City

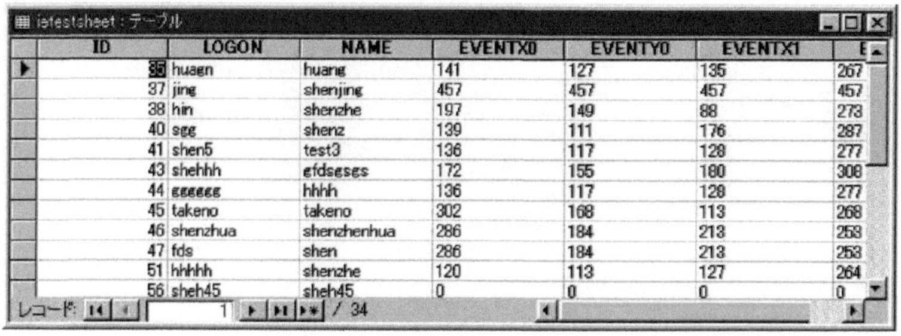

ID	LOGON	NAME	EVENTX0	EVENTY0	EVENTX1	E
35 huaen	huang	141	127	135	267
37 jing	shenjing	457	457	457	457
38 hin	shenzhe	197	149	88	273
40 sgg	shenz	139	111	176	287
41 shen5	test3	136	117	129	277
43 shehhh	efdseses	172	155	180	308
44 gggggg	hhhh	136	117	129	277
45 takeno	takeno	302	168	113	268
46 shenzhua	shenzhenhua	286	184	213	263
47 fds	shen	286	184	213	263
51 hhhhh	shenzhe	120	113	127	264
56 sheh45	sheh45	0	0	0	0

レコード: 1 / 34

Fig. 10.7 Coordinate values recorded in the Microsoft Access database

2. Reviewing page of the designs of other participants

   This page involves procedures such as confirmation of the participant's login name and the creation of a VRML file (Fig. 10.4). The database access pages will help participants to decide which design to review. First, the participant confirms his/her login name while accessing the database. Upon confirmation, a VRML file of the design of the participant will be output. When the participant like to review a design scheme from database and has chosen the participant's name, the VRML browser can be launched by a mouse click.

3. Questionnaire pages for mutual design evaluation

4. Pages for viewing the current park site before design

5. Pages showing a sample of the park design

We also prepared a chat page to support the exchange of opinions in the online cooperative design system.

Design Tool for Making Design Proposals

Arrangement of Urban Furniture

At the beginning of the design game, various types of street furniture are prepared on the furniture arrangement web page (as shown in Fig. 10.5). The types of furniture that are made available are based on the questionnaire results and the construction budget. Participants can move or change the orientation of furniture using a mouse. After arranging the furniture, participants can press the finish button to save the completed design. When the participants want to reference the design schemes of others, they must first select a participant through the database. After confirming the name of the participant from a list of participants, the system will generate a 3D VR space for assessing the design schemes (Fig. 10.6).

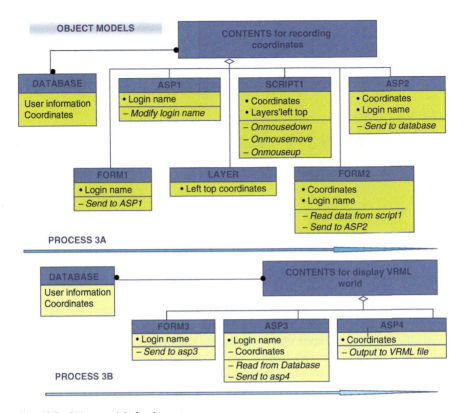

Fig. 10.8 Object models for the system

Development of the Design Tool Using ASP

Processes 3A and 3B, as shown as Fig. 10.8, are the two processes of the proposed system. These processes were developed using Dynamic Hyper Text Markup Language (DHTML). Object is a term used in programming, and, in Fig. 10.8, ASP1 and ASP2, for example, are objects. In the figure, ASP2 indicates the name of the object, and "coordinates" and "login name" are variables used in this object. Finally, "send to database" is the method or function used in this object. The object model makes a statement of the relation between the objects shown in Fig. 10.8 in two parts, namely, 3A "content for recording coordinates" and 3B "content for displaying the VRML environment".

Process 3A: Storage of the Coordinate Values of Urban Furniture

Process 3A (shown in Fig. 10.8) is composed of a content set of DHTML and ASP. In this set, Form 1 is used by participants to register their login names, which is necessary in order to modify their design information when they want to change

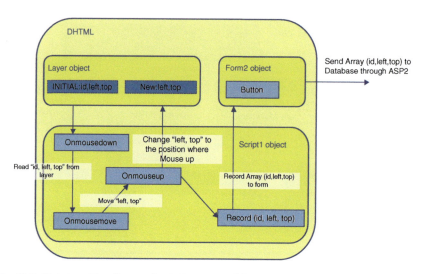

Fig. 10.9 State transition diagram for design process 3A

their designs. And the design information is the coordinates of the urban furniture elements that are decided according to the coordinates of the layer objects. Participants can move layer objects freely and can place urban furniture elements as Script 1 objects. Through Form 2, the coordinate values can be obtained from Script 1 objects and sent to the database as an ASP2 object.

The state transition diagrams of Processes 3A and 3B are shown in Figs. 10.9 and 10.10, respectively. In Fig. 10.9, we show how the layer objects can be used to record the coordinate values of the urban furniture elements. Each layer can be decided as a design element in the DHTML script. The properties of the layer object include layer position, which is indicated as left, top in the this example. This information can be used as coordinate values for urban furniture elements. In the Script 1 object, there are three methods for handing over the layer position to Form 2. The first is *onmousedown*, which is used to obtain active layer position. The second is *onmousemove*, which is used to calculate the new layer position when the mouse moves. Finally, the third method is the *onmouseup*, which is used to move the layer object to where mouse dragged, and the position of *onmouseup* is passed to Form 2 as the coordinate values of the urban furniture elements, which can be stored in the database at the server site through ASP2. Fig. 10.5 shows the interface of Process 3A, and Fig. 10.7 shows the coordinate values recorded in the Microsoft Access database.

Process 3B: Output VRML Files

In the Process 3B (shown in Fig. 10.8), we use a set of contents to obtain the coordinate values of participants from the database, and the coordinate values can

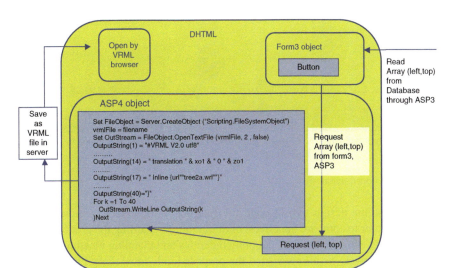

Fig. 10.10 State transition diagram for design process 3B

also be used to build a VRML world on the server site. In Form 3, the login name of a participant is input from the login name list and is sent to ASP3 and ASP4, which output the corresponding VRML file based on the coordinate values recorded in the database.

Figure 10.10 shows Process 3B. In this process, ASP3 can obtain coordinate values for urban furniture elements from the database, which are input by each participant in Process 3A. In addition, Form 3 transfers the coordinate values to ASP4. Through ASP4, a VRML file will be created at the server site according to the coordinate values. In the same manner, according to the design information of each participant, the design elements can be arranged correctly in the VRML environment.

Through Processes 3A and 3B, the arrangement of the urban furniture elements will emerge as a 3D VRML environment. That means that in the proposed system, the design information of each participant can be recorded and displayed for comparison.

In this section, urban furniture arrangement using DHTML is discussed. Participants can design a street park from home using only a mouse. Furthermore, participants can review the design schemes of other participants through a database.

System: Application and Evaluation

As a case study involving the Yamanoue Street Park design, two applications of the proposed system are tested. The first application is design coordination by a facilitator for achieving consensus in planning meetings, and the second application is free online participation, namely, the individual design proposals of participants

Free online participatio (Personal) Free online participatio (Group) Design coordinaion in a planning meeting

Fig. 10.11 Cooperative design

using the online design tool through the Internet. We conducted a questionnaire survey in order to compare the effects of deliberation regarding the proposed design schemes between conventional planning meetings and planning meetings using the system. Moreover, we also analyzed the proposed individual design schemes from the web, focusing primarily on the function of the online design tool in urban furniture arrangement. Figure 10.11 shows the two applications of this system.

Online Participation for Individual Design Proposals

Park construction began approximately 1 year from the introduction to the system to the public. During this period, 195 design schemes were proposed by 195 participants, and several hundred messages were exchanged. The locations of the participants were not revealed, because most of the participants were not residents of Yamanoue community and some of them were pupils or middle school students who were interested in the online design system. Based on the data collected from the online questionnaire respondents, 71% of the respondents disagreed with the statement, "The VRML can fully reflect their design concepts", but 83% of the respondents agreed with the statement, "The system is helpful for sharing design ideas with others". Although the proposed system is not sufficient for clearly reflecting the design concepts of participants, the system is very useful for communicating design concepts. In addition, although 56% of the respondents could not freely explore every corner of the VRML environment, 82% of the respondents reported that they could "understand the design intentions of other participants". In conclusion, the proposed system successfully presented the design intentions of nonprofessional designers.

Moreover, 77% of the respondents agree with the statement that their "interest in design collaboration increased as a result of using of the system", and 88% of the respondents reported that they would take part in design collaboration in the future. This indicates that the proposed system is useful not only for expressing design concepts and online communication but also for promoting public participation through online design collaboration.

Design Coordination in Planning Commission Meetings

This system was implemented in Yamanoue to support the process of achieving consensus of a street park design in the prescribed planning meetings in 2001. There were 11 residents from Yamanoue, three officers from the Kanazawa government, and two planners from a consulting company, and two staffers from Kanazawa University who attended the planning meetings. During the meetings, the planners from the consulting company explained their design scheme using conventional planning drawings and the proposed system were used by the staffers from Kanazawa University. In order to compare the conventional planning drawings with the VR representations, we presented both the conventional planning drawings and the VR representations at the meetings and recorded the discussions of participants using video and sound recorders so that analysis and comparison of the design drawings produced by the consulting company and the VR representations generated by the proposed system can be conducted.

In these meetings, on August 31, 2001, planners and officers emphasized the discussion of different design schemes proposed by participants through the Internet. A number of representative design schemes are shown in Fig. 10.12. These schemes primarily reflect differences in opinion regarding green spaces and zoning division for relaxation and play. In this section, we focus on the potential contribution of applying the system rather than the preparation of the design schemes.

Fig. 10.12 Representative design schemes proposed by participants

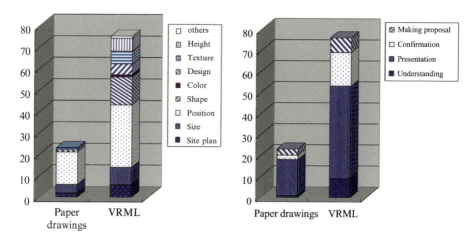

Fig. 10.13 Analysis of comments on VRML and on-paper drawings

During the planning meetings, there were 75 comments on the design schemes presented by VRML, which is more than three times the number of comments on the design schemes present on paper. This indicates that it is more effective for participants to discuss the VRML design schemes. These comments are classified according to topic and the effectives of VRML is compared with on-paper drawings. In Fig. 10.13, the two graphs indicate that most of the participants used VRML to discuss the design schemes, because the VRML environment is easier to understand and more effectively reflects the relationships among design elements and the surroundings. All of the design schemes can be checked carefully by moving around in the VRML environment, which can be done collectively by everyone in the planning meetings. On the other hand, in deliberating on the arrangement of water supply service pipes, garbage boxes, and evacuation roads in the community, a number of the participants used the on-paper drawings.

An interview survey was conducted after the planning meetings, in which the participants commented that the system made a connection between the VR representation and their daily life, which was helpful for the discussion of planning issues. Although the VRML is very useful in the deliberation process, the on-paper drawings tended to be more useful for discussing the entire planning site. This is because on-paper drawings allow the entire planning site to be viewed, whereas, in the VRML environment, participants have to review the entire design scheme from a top viewpoint. Although there are some drawbacks to the proposed system, the system was useful for individual design proposals over the Internet and design coordination in planning meetings. Officers of the local government recognized that the proposed system can promote public participation through design collaboration over the Internet, and so were more concerned with the expense of system development. The planners of the consulting company were eager to make the change from the conventional approach because the proposed system can visualize the demands from the local community as a VRML representation.

After several meetings, the consulting company presented a final design for the street park (as shown in Figs. 10.14 and 10.15) based on the discussions and the design proposals of participants. The final scheme was accepted during the final planning meeting, and the street park has been in use until the present.

In this section, we have not only concentrated on investigating a technical application, but also explore the possibility of the design tool with respect to public participation in planning practice. In Japanese cities, local planning committee meetings regarding public spaces in neighborhoods are always scheduled from 7 to 9 p.m. so that all interested stakeholders will have the opportunity to attend. An online design tool would be helpful to stakeholders who cannot attend the scheduled meetings.

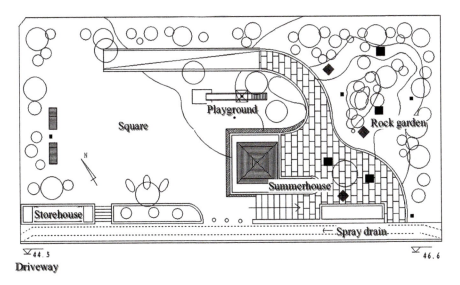

Fig. 10.14 On-paper drawings of the final scheme

Fig. 10.15 VR images of the final scheme

Conclusion

In this chapter, we developed an online design tool using VRML that would allow local residents to propose uses, sectorization (site planning), and street furniture items for street parks in their neighborhood. We considered an online design tool using VRML and an online communication tool for public participation. The system capabilities were considered from the viewpoint of practical use. Specifically, we investigated whether the system could satisfy the need for local public participation in the design of street parks in Japan. In the proposed system, ASP and JavaScript are used as the main development tools. As such, the proposed system can be installed at a very low cost, and anyone can use this system to create a VRML environment as his/her design proposal over the Internet.

In design collaboration, the design tool consists primarily of furniture arrangement, which involves arranging and adjusting the positions and orientations of urban furniture. Participants can login to the system to make their design proposals and save the completed proposals to a web database. Moreover, the participants can compare and comment on submitted design proposals. In addition, nonprofessional participants can use the design tool to quickly represent their design concepts in a 3D VR environment without the requirement for special training.

Participants can express their design conceptions by building their own virtual public spaces using the online cooperative design tool, and they can also comment on the design schemes of other participants using the online communication tool. We applied this system to the Yamanoue Street Park design in Kanazawa City. According to the results of a questionnaire and an interview survey after the planning meetings, the design tool for furniture arrangement was verified to be useful for practical use. The differences between the design tool and conventional on-paper drawings were investigated. According to the results, VRML is more effective for use in the deliberation process than conventional on-paper drawings. For instance, by VRML, design concepts are more easily represented, and depicting the relationships between design elements and their surroundings is easier. Participants can confirm their design schemes in detail from various viewpoints in the VRML environment, while considering how to satisfy their needs and requirements in their daily lives. Although the online design tool does not completely satisfy the needs of the participants in fully reflecting their design concepts, the tool was helpful for communication among participants as well as for clarifying their design intentions. On the other hand, on-paper drawings were more convenient for reviewing planning issues regarding the entire planning site.

Consequently, the proposed system succeeded in representing design proposals from nonprofessional participants so that professional planners can easily collect design suggestions from the local community. This system improved public participation through individual design proposals over the Internet and design coordination during the planning meetings.

10 Online Cooperative Design for the Proposal of Layouts of Street Furniture

Acknowledgments We would like to thank the financial support of the Grants-in-Aid for Scientific Research (No. 22560602C), Japan Society of the Promotion of Science.

References

Al-Douri, F. A. (2010), The impact of 3D modeling function usage on the design content of urban design plans in US cities, Environment and Planning B: Planning and Design, 37(1) 75–98

Batty, M., Chapman, D., Evans, S., Haklay, M., Kueppers, S., Shiode, N., Smith, A., & Torrens, Paul M.T. (2001). Visualizing the city: Communicating urban design to designers and design makers, Redlands, California, USA: ESRI Press, 405–443

Batty, M. (2001), Virtual Reality, Spatial Analysis, and the Performance of the Built Environment, Building Performance, 3, 24–30.

Bulmer, D. (2002), How can computer simulated visualizations of the built environment facilitate better public participation in the design process? On-line design Journal, http://www.onlineplanning.org/

Carmona, M. and Punter, J. (2002), "From Design Policy to Design Quality", London, Thomas Telford.

Day, A. (1994), From map to model: the development of an urban information system. Design Studies, 15:366–84.

Doyle, S., M. Dodge, and A. Smith. (1998). The potential of web based mapping and virtual reality technologies for modeling urban environment. Computers, Environments and Urban Systems, 22:137–55

Fukushima, R., Nakanishi, H., Imamura, K., Kashiwagi, M., Nakashima, R. & Sawada, K. (2001), Development of the Web3D-based VR system for living space – The data optimization for high quality and high compression, Proceeding of the twenty-fourth symposium on computer technology of information systems and applications (pp. 85–91), Tokyo, Japan.

Funabiki, T. (2009). A study on improvement of the system of landscape conservation by the publication of green-spaces value. Journal of the City Planning Institute of Japan, Vol. 44, 5–10

Golićnik, B., Thompson, C. W. (2010), Emerging relationships between design and use of urban park spaces, Landscape and Urban Planning, Vol 94, 38–53

Hammad, A., Sugihara, K., Matsumoto,N. and Hayashi, Y. (1999). Integrating GIS CG and the WWW for Public Involvement in Urban Landscape Evaluation. In, Proceeding of CUPUM '99. Computer in Urban Design and Urban Management (in CD-ROM).

Hoskins, E.M. (1979), Design development and description using 3D box geometries Original Research Article, Computer-Aided Design, Vol 11, Issue 6, 329–336

Kaga, A., Hosono, H., Sasada, T. (1996), Collaboration design system with network technology (1), An our way of thinking about system architecture. Proceeding of the nineteenth symposium on computer technology of information systems and applications (235–240), Japan.

Koga, M., Ikaruga, S., Tadamura, K., Ohgai A. and Matsuo, M. (2008). Study on image sharing support method in landscape planning using information technology, Journal of Architecture and Planning (Transactions of AIJ), Vol. 73, No. 633, 2409–2416.

Klein, M. (1995), Integrated coordination in cooperative design, International Journal of Production Economics, Vol. 38, Issue 1, 85–102

Matsushita, J. (2003), Study on Present Attainments and Required Components of Public- Participation-Based Communal Management Systems, Journal of the City Planning Institute of Japan, Vol. 38, 50–55

Michael B., David, C., Steve E., Scott F., Mordechai H., Stefan K., Naru S., Andy S., & Paul M.T. (2001). Visualizing the city: Communicating urban design to designers and design makers, (pp. 405–443). Redlands, California, USA: ESRI Press.

Miller, Michael D. and Brunner, C. Cryss (2008), Social impact in technologically-mediated communication: An examination of online influence, Computers in Human Behavior, Vol 4, Issue 6, 2972–2991

Nishikizawa, S., Harashina, S. (2004). The Role of Citizen Representatives in the Workshop Approach for Making Planning Scheme, Journal of the City Planning Institute of Japan, Vol. 39, 1–6

Nurcan S. (1998), Analysis and design of co-operative work processes: a framework, Information and Software Technology, Vol. 40, Issue 3, 143–156

Okabe, A., Sato, T., Okata, J., and Okunuki, K. (1999). A Study on an Internet-based Decision Support System for City Design: a Virtual Environment for Interactive Operations with Three Dimensional Objects. Center for Spatial Information Science, University of Tokyo, Paper #12.

Okunuki, K., Sato, T., & Nisikawa, S. (2000). The possibility of design collaboration. In Architectural Institute of Japan, & Cyber Master Design Project of Keio University (Eds.), The Situation and Challenge of Information Exchange in the Internet (http://cmp.sfc.keio.ac.jp/contents/aijwgpdf/wgindex.htm).

Pettit, C., Nelson, A. and Cartwright, W. (2004), Using On-Line Geographical Visualisation Tools to Improve Land Use Decision-Making with a Bottom-Up Community Participatory Approach Jewell Station Neighbourhood – Case Study, in Recent Advances in Design and Decision Support Systems in Architecture and Urban Planning, edited by Jos P. Van Leeuwen and Harry J. P. Timmermans, Dordrecht/Boston/London: Kluwer Academic publishers, 53–68.

Pettit, C. Nelson, A. and Cartwright, W. (2004b): Recent advances in design & decision support systems in architecture and urban planning, edited by J. P. van Leeuwen & H. J.P. Timmermans. Dordrecht/Boston/London: Kluwer Academic publishers, 53–68.

Saad, M, and Maher, M. L. (1996), Shared understanding in computer-supported collaborative design, Computer-Aided Design, Vol 28, Issue 3, 183–192

Setagaya community design center (1998), Design tools for public participation, Setagaya community urban redevelopment authority, Japan.

Sanoff, H. (1993). Design Games: playing for keeps with personal and environmental design decisions (Japanese translation). Tokyo: Japan UNI Agency, Inc.

Shen, Z., Kishimoto, K., Kawakami, M. (2007), Study on Possibility for Education Support using On-line Cooperative Planning and Design System : Case Study of Design Game System for Public Park, Reports of the City Planning Institute of Japan, Vol.6-3, 84–89

Shen, Z., Kawakami, M. (2010), An online visualization tool for Internet-based local townscape design, Computers, Environment and Urban Systems, Volume 34, Issue 2, 104–116

Shimizu, H., Sakai, A., and Ono, H. (2007), An approach of space planning by children for children basing on a "community salon", Journal of the City Planning Institute of Japan, Vol. 42, 32–38

Shiode, N., Okabe, A., Okunuki, K., Sagara, S. and Kamachi, T. (1998), Virtual Ryoanji Project: Imolementing a Computer-assisted collaborative working environment of a virtual temple garden, International workshop on "Groupware for urban design", Lyon, France.

Smith, A., Dodge M. and Doyle S. (1998), Visual communication in urban design and urban design, Center for advanced spatial analysis working paper series. http://www.casa.ucl.ac.uk/urbandesign.pdf.HTML

Tichem, M. and Storm, T. (1997), Designer support for product structuring—development of a DFX tool within the design coordination framework, Computers in Industry, Vol 33, 155–163

Wang, J. X., Tang, M.X., Song, L. N. and Jiang, S. Q. (2009), Design and implementation of an agent-based collaborative product design system, Computers in Industry, Vol. 60, Issue 7, 520–535

Yamataki, K., Sato, S., Kobayashi, Y., & Saotomi, T. (1999). Street proportion using virtual reality: The relation between height of building and street tree- part1-2. Summaries of technical papers of annual meeting, Architecture Institute of Japan, F-1, 457–460.

Chapter 11
Online Learning Tool for Repair of Traditional Merchant Houses: Machiya

Zhenjiang Shen, Mitsuhiko Kawakami, Masayasu Tsunekawa, and Eiichi Nishimoto

Introduction

In recent years, local governments have gradually come to accept the use of various visualization technologies for enhancing public participation in urban planning and design processes (Warren-Kretzschmar and Tiedtke 2005). For instance, stakeholders such as local residents can deliberate in planning committees in the process of making decisions about townscape design guidelines by sharing images of ideas for the future represented in a virtual world (Shen and Kawakami 2010). In this chapter, to reach consensus on townscape design guidelines in a historical district, we propose using a visualization tool for stakeholders that will allow them to gain the knowledge they need about traditional Japanese-style merchant houses, or Machiya, in order that building owners can know how to more correctly and effectively keep these buildings in good repair with the help of administrative officers and consultants.

Forms of Communication Between Stakeholders in Planning Meetings

In approaching public participation in local planning committee workshops, there are three dimensions: timing/phasing, a public dimension, and a form-of-communication dimension (Benwell 1979). To facilitate information exchange between associated participants in the different phases, professional planners introduce the objectives and contents of their planning and design schemes by providing paper drawings and supplementary planning documents as a form of communication (Hanzl 2007). Participants (stakeholders) can use these as a basis for discussion, which enhances

---

Z. Shen (✉) • M. Kawakami • M. Tsunekawa • E. Nishimoto
School of Environmental Design, Kanazawa University, Kanazawa, Japan
e-mail: shenzhe@t.kanazawa-u.ac.jp

Z. Shen, *Geospatial Techniques in Urban Planning*, Advances in Geographic
Information Science, DOI 10.1007/978-3-642-13559-0_11,
© Springer-Verlag Berlin Heidelberg 2012

the effectiveness of public involvement. To reach a consensus on townscape design, a local planning committee will meet several times until final agreement is reached. However, it is hard for residents to share images of the entire townscape just using design alternatives presented in the form of documents and paper drawings. In previous findings (Shen and Kawakami 2010), we found that local planners could enhance the level of communication among stakeholders by using a shared virtual world as a form of communication because such a shared virtual world allows participants to look around the entire townscape and check the details from different viewpoints. As a conventional form of communication, conventional committee meetings require that all participants take part in the discussion of design alternatives at a determined time and place. However, because of the spatial-temporal limitations of routine planning committee meetings, some participants must be absent, and the process of deliberation may become a mere formality, lacking real content. In the Woodberry Down experiment (Hudson-Smith et al. 2002), online participation, namely a new form of communication is regarded as a permanent part of a 10-year regeneration scheme process, and local resident and tenant committees are continually being reconstituted without spatial-temporal limitations.

In this paper, we are describe developing a learning tool for providing available forms of communication between stakeholders in order to improve residents' level of understanding of architectural style and townscape design with respect to the traditional merchant house.

Stakeholders' Learning Through Virtual Communication

Cheng (2003) explained the conceptual frameworks and standards that enable interdisciplinary exchange by envisioning and structuring interactive approaches to collaboration design, and analyzed how media affects teamwork. Ageron et al. (1994) suggested *images database management system* to manage all information about digitized images, giving some parties access to images and authority to modify, work on, and export those images to other information retrieval systems. As an example of online multimedia learning, Google was, at that time, already providing search tools for retrieving images and videos using keywords, which help users access photos or movie clips that aid in their understanding.

In this paper, we provide a server–client model of virtual reality in which users interact to communicate by such means as Q&A (question and answer) postings and discussions on design alternatives during online VR world meetings, in which interaction between participants and planners is expected to improve participants' understanding of design alternatives for traditional buildings.

Goul et al. (1997) argued that the server–client model is one choice on which to base proposing and justifying requirements for deploying and sharing a decision support system on the Internet. Monahan et al. (2008) presented a Collaborative Learning Environment with Virtual Reality (CLEV-R), a web-based system using virtual reality (VR) and multimedia for providing communication tools to support collaboration among participants. Virtual learning environments (VLE) can be a means of enhancing, motivating, and stimulating learners' understanding of difficult or abstract notions (Pan et al. 2006). Hernández-Serrano et al. (2009) argued that virtual spaces can be looked on as not only technological artifacts and contexts or environments of coexistence, but also as interactions and guides for personality development and construction of identities.

Considering examples of applied virtual communication, Zhou et al. (2010) argued the potential benefits of social virtual worlds (SVWs) in many real-life domains such as business and education. As for extant virtual worlds, including education-focused, theme-based, community-specific, children-focused, and self-determined worlds. Messinger et al. (2009) posited that collaborative design systems are multi-purpose platforms for promoting education, design themes, community learning, and participants' self-development processes. In townscape design practice in Japan, Shen and Kawakami (2007) focused on providing an online design tool to help participants obtain a better understanding of townscape design in the deliberation/consensus building process, reminding participants of the relevance of planning and design.

For improving communication in virtual world, Economou et al. (2000) reported the elicitation of requirements for virtual actors in collaborative virtual learning environments and a methodological approach to phased development of a series of learning environments. Teo et al. (2003) employed an experiment to examine the sustainability of virtual learning communities based on the impact of information accessibility and community adaptivity. Virtual worlds are providing new opportunities and a new range of questions regarding design creativity and mass personalization (Ward and Sonneborn 2009; Tseng et al. 2010; Wang et al. 2010). Virtual spaces are useful for cooperative learning in the context of social activity. How this can lead to new insights and understanding about creativity and personalization in general remains as further work. Thus, our learning tool based on the server–client model is expected to be a promising application of a virtual society for learning in the field of urban design and planning.

Motivation

In this work, we devote our efforts to visualizing the architectural style design guidelines using virtual reality in an area containing traditional merchant houses. This is a new challenge in urban planning and design: for stakeholders to use the Internet to study how to repair their traditional houses in order to preserve the traditional townscape via virtual communication.

Conservation and inheritance of Kanazawa Machiya (traditional merchant houses in Kanazawa City) is one of the city's historical preservation projects. One of the key issues in Kanazawa Machiya revitalization is townscape design that can achieve a harmonious urban landscape across the entire traditional area. With this as a goal, a nonprofit organization, the Kanazawa Machiya Research Institute, carried out a program whereby owners of Machiya could learn the elevation design of traditional Kanazawa Machiya. In this situation, a learning tool via virtual communication for those planning to repair their traditional wooden merchant houses would be helpful.

When repairing traditional houses, it is not possible to replace room layouts, and so the floor area ratio and building coverage rate are retained in their original form in most cases. The most difficult problem for building owners is what to do to repair damage to traditional wooden houses' elevations and structures. Work related to elevations accounts for about 60% of the total cost of building repairs. Accordingly, it is particularly important to discuss the design elements of building elevations and residents have to understand what traditional alternatives are applicable, such as lattices and roofs in the case of Kanazawa Machiya. Therefore, we supplied residents with a tool for learning about repair design alternatives. By employing this tool, different alternatives that mesh with the design elements in buildings' elevations could be visualized using a virtual reality modeling language (VRML).

In Sect. "Approach", we present our approach to developing a learning tool via virtual communication for learning about Kanazawa Machiya. In Sect. "System Framework Using VRML", we suggest a system framework. This is followed in Sect. "Visualization Tool for Learning Design Guidelines" with a discussion on how to visualize design guidelines using the learning tool. Finally, we present our conclusions.

Research Approach

This study considers a visualization tool capable of representing traditional house elevation design alternatives for traditional house elevations in a historical conservation area. Stakeholders can use this tool to experience the traditional townscape in a virtual world to study how to repair their Kanazawa Machiya structures via virtual communication. In this section, we focus on the approach to visualizing design alternatives for a traditional merchant house.

We analyzed system functions necessary to communicate what is required to conserve traditional houses, based on townscape design guidelines for Kanazawa Machiya dictated by the local planning committee. In planning practice, this online learning tool must allow visualization of planning alternatives in an online virtual classroom in order for participants to study design alternatives through the Internet. For this purpose, functions that allow representation of design elements, comparison of different alternatives, and online communication about repairs are needed.

1. Visualizing alternatives for repairing traditional merchant houses

3D objects representing all design elements should be created. The learning tool can be employed to combine design elements to compose different Kanazawa

Machiya elevations. It is necessary for residents to identify different alternatives they can apply to their building elevations and to understand how different design alternatives impact the entire townscape before they contract with an architect to repair their Machiya.

2. Learning through virtual communication

Internet-based tools help participants learn different townscape alternatives that they can apply when designing their homes. To host a training program for participants on the Internet, we anticipated using a system with communication function based on a server–client model. An avatar representing a lecturer (a member of the staff of the Kanazawa Machiya Institute) will deliver lecture presentations at scheduled times. Participants who attend a lecture can experience the virtual world of the entire townscape and alter design alternatives to check their designs and their impact on the townscape under the lecturer's supervision.

In order to complement the temporal-spatial limitations of scheduled training sessions stakeholders who cannot attend lectures need other information sources. Online learning is a possible solution for those who are absent from the classroom. It can help them acquire information about the Machiya design style and allow them to leave their comments and questions online. The lecturer can also use a chat room to respond to participants online. We expect this system to arouse participants' interests in learning traditional Kanazawa house design and make it possible for participants to consider the design of their own Machiya.

System FrameWork Using VRML

Visualizing Design Alternatives Using VRML

The function of coordinating design alternatives is necessary to allow participants to combine different design elements by themselves on the Internet. For this purpose, we employ VRML to develop our design coordination tool. The interchangeable design elements are created as VRML objects in advance. They are input using an "Inline" function to a VRML file that has all the definitions of VRML objects placed in the entire area. There is a "Switch" Node that can be employed to show or hide a VRML object using a script embedded in the VRML file. A "ROUTE" node controls the field value of the Switch. As shown in Fig. 11.1, there are buttons on the user interface that control which VRML objects representing design alternatives to show or hide. When button 1 is clicked the itemClicked 1 script will be launched to change the field value of VRML object "no. 1". The field value of "no. 1" will be changed to 1 so that design element "no. 1" can be shown to users while the field value of other VRML objects are set to 0 and are hidden from view.

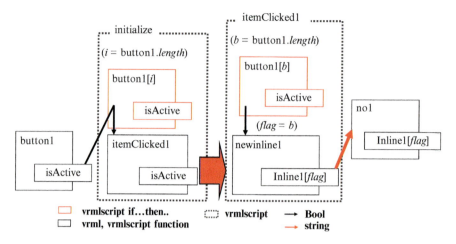

Fig. 11.1 Design alternative coordination function using Script

Online Learning Through Virtual Communication Using a Server–Client Model

As shown in Fig. 11.2, we use an extension of VRML called ShareEvent in the application of the server–client model in this case study. ShareEvent is used to build the VRML world, for which definitions of EXTERNPROTO, BlaxxunZone, and ShareEvent should be declared first. Using this tool, any design alternative information represented by VRML objects in client sites that has been changed can be retrieved and delivered to all clients in real time, and thus stakeholders can use this tool to coordinate design alternatives when considering how to repair Machiya while communicating with an online teacher on the Internet.

Concretely speaking, the field values for ShareEvent, which are used to control the parameters of the VRML objects that stand for design elements represented in the VRML world, need to be defined. To share changes made to information about design alternatives in real time on the Internet, we need to define two types of field values to share one certain event. These are EventIn and EventOut. For instance, a client's Event can be sent to the application server through the field "SetType" for EventIn and "TypeToServer" for EventOut. Conversely, the field "Typechanged" for EventOut can be used to send information about changed design elements to all clients from the application server in real time. For the event mentioned above, a node, BlaxxunZone, must be defined which includes all necessary fields and nodes for sending information to the server and obtaining information from the server.

For stakeholders to be able to learn the Machiya design on the Internet, when design alternatives are changed, they should be shown to all clients in real time. Therefore, a tool based on a server–client model can be employed to get requirements from client sites for coordinating design elements in a VRML world

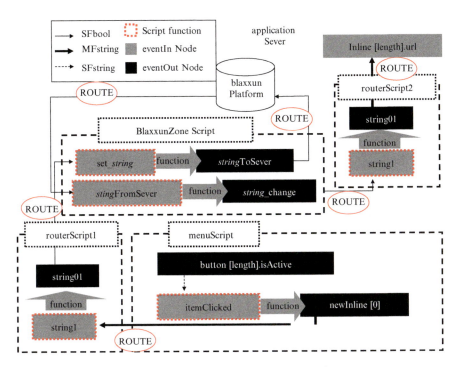

Fig. 11.2 System structure of the online learning tool on the server site

and respond to all clients from the server by delivering the new design element information in that world.

Visualization Tool for Learning Design Guidelines

As mentioned above, building owners can consider how to repair Machiya structures while utilizing the proposed learning tool. As a precondition, it is necessary to visualize the different design alternatives reflecting the possible combinations of design elements such as windows and lattices in building elevations. For this, a dataset of all design elements must be prepared. Accordingly, design elements in building elevations should be carefully classified.

Design Guidelines and 3D Dataset

Table 11.1 shows how design elements necessary for repairing Machiya structures are classified. Although floor plan changes may alter window positions in the building elevation, repairs will not result in major floor plan changes. Owners

Table 11.1 Design guidelines

Design elements			
Building roof	Roof materials	Roof tile	
		Totan	
		Itabuki	
	Eave boards inside the roof	Upper beam	
		Bracket	
		Segai	Singlet
			Double
Second-floor walls	Wall	White-plaster	
		Black-plaster	
		Itabuki	
Second-floor windows	Window	Flat lattice	Old lattice
			Coarse lattice
			Fine lattice
			Grass crates
			Aluminium sash
		Child elative	Old lattice
			Coarse lattice
			Fine lattice
			Grass crates
			Aluminium sash
	Uwa-shita Nageshi		
	Wall sleeve		
Eave boards on the first floor	Roof tile		
	Totan		
	Itabuki		
	Double eaves		
	Eaves		
First floor	Wall	White-plaster	
		Black-plaster	
		Itabuki	
	Window	Flat lattice	Old lattice
			Coarse lattice
			Fine lattice
			Grass crates
			Aluminium sash
		Child elative	Old lattice
			Coarse lattice
			Fine lattice
			Grass crates
			Aluminium sash

should make design decisions for their building elevations when repairing their Machiya, choosing alternatives that will contribute to a harmonious urban townscape. For this, we classified only the design elements that appear in building elevations.

11 Online Learning Tool for Repair of Traditional Merchant Houses: Machiya

Fig. 11.3 Creation of 3D dataset

In repair practice, deliberation will be focused on design alternatives composed of a combination of design elements such as roofs, windows, eave boards, and exterior walls. All design elements should be edited and classified into different groups so that users can choose from among them to compose the building elevation of a possible alternative for repairing Machiya buildings. 3D data representing building elevations should be created separately. As shown in Fig. 11.3, whole buildings are divided into different parts. In addition, the design elements and classifications are kept consistent with design guidelines for the traditional Machiya area townscape, about which consensus of local stakeholders has already been reached. For instance, the 3D roof dataset was prepared on the basis of materials and slopes regulated in the design guidelines. Figure 11.3 shows Machiya-related elements of the townscape. The elements were divided into four groups as organized in a pamphlet published by The *Kanazawa Machiya Research Institute*. The four groups are first floor exterior walls, eave boards, second floor exterior walls, and building roofs. Lattices, materials, and other design elements are also listed in Table 11.1.

To prepare the 3D database, 3DMAX is employed to build the 3D models representing the traditional Machiya design elements. These are converted to VRMl files and made available to the public on the Internet.

Online Learning Tool Interface

This research is focused on the design and implementation of a learning tool for representing the design alternatives for traditional Machiya houses. The intent is to

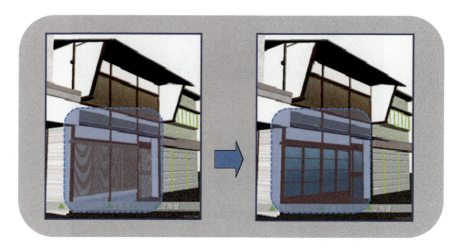

Fig. 11.4 Design alternatives for first-floor building elevations in Machiya

show that this tool can be used to facilitate a dynamic learning process between advisors and stakeholders over the Internet.

A 3D dataset representing the design alternatives was created based on the classification of design elements in Kanazawa City as shown in Table 11.1. Stakeholders can learn about repairing traditional Machiya houses Machiya by using the design element represented in the 3D dataset.

For presentation on the Internet, different design alternatives should be displayed for users so that they can learn how to correctly repair their traditional Machiya houses. As shown in Fig. 11.4, there are two first-floor elevation alternatives. They are made from different types of eave boards, windows, and walls. For online learning, users should be able to compare all possible combinations of these three design elements. Thus, the system should have a useful tool for comparing the different design elements. For presenting design alternatives, the buttons designed in the VRML world can represent the various combinations of the five types of design elements as shown in Fig. 11.5.

To create different design alternatives based on the five types of design elements, we set up a layout of five buttons in the virtual world that let users choose each type of design element. Each design element has three to five options. For instance, a detailed example of design elements prepared for online learning with respect to the first floor is presented in Table 11.2. Using this sort of display, it is possible to discuss several design alternatives that combine the five design elements to be used when repairing Machiya structures.

We have developed a learning tool featuring Kanazawa Machiya for use in townscape design in a traditional preservation area. The tool's interface layout is designed as shown in Fig. 11.6. The buttons embedded in the VR world are set with sensors so that the JavaScript will show or hide the chosen design elements when the sensors are pushed by users. This feature makes it possible for the system to represent building changes within the VRML, show different design alternatives,

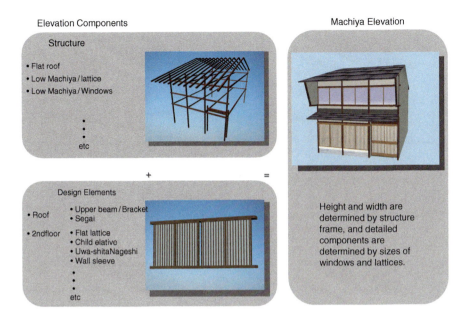

Fig. 11.5 System function representing design elements

and display the differences between the alternatives. As shown in Fig. 11.7, at the bottom left of the tool there is a chat system that participants can use to communicate with each other on the Internet. Next to the chat tool is a list of participants. Each person can record his or her chats regarding questions, answers, or advice received from online advisors. An explanation of the system resides in the top left of the interface. Thus, the developed learning tool can be used to coordinate design alternatives by the sharing of a VR image through the Internet. In addition, when a design alternative is changed, the changes can also be delivered to all clients in real time and participants and an online teacher can thus share the VR world for virtual communication.

Discussion and Practical Application

This learning tool is used to study the design alternatives for Machiya structures located in the downtown area of a particular Japanese city. It allows the entire area, including original Machiya, reconstructed buildings, and repaired Machiya, to be represented in one integrated VR world.

Unlike face-to-face group activities in the real world, ambient conditions in a social virtual world are largely the same and participants are usually from different places (Ward and Sonneborn 2009). However, in our case study, stakeholders who join our training program almost all live in the same community, and so the ambient

Table 11.2 Examples of 3D datasets for the first floor

Coarse lattice

Grass crates

Back wall

Child elative

conditions in participants' living environments are almost identical. It is possible for residents to cooperate with a virtual lecturer or to engage in individualized learning. In most of the cases, children accessed the learning tool, playing with the tool as an online game together with their grandparents. Thus, the tool is useful in delivering design knowledge about traditional merchant houses to younger generations. The participants also made comments, such as "this tool helps us understand the design guidelines that had been agreed upon in the local community" and "it is amazing to communicate with an online teacher to learn something about Kanazawa Machiya." From these comments, we learn that this tool is a good form of virtual communication that can be expected to improve the level of understanding of the traditional architectural design style in Kanazawa.

In planning practice, 3D representations of design alternatives for traditional merchant houses has the potential to allow participants to check if a particular

11 Online Learning Tool for Repair of Traditional Merchant Houses: Machiya

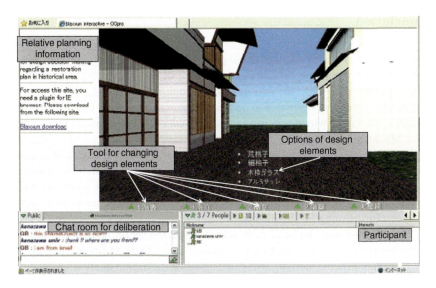

Fig. 11.6 System interface

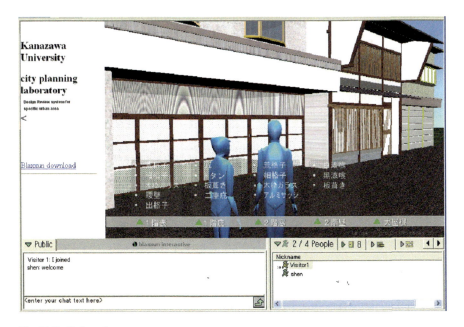

Fig. 11.7 Online classroom

alternative Machiya elevation is harmonious with the overall townscape from different viewpoints. However, at this stage, the tool is still available for public use for learning but not for design review board committee meetings.

Conclusion

This research focused on developing and implementing a learning tool for participants to use to aid their understanding of design alternatives for traditional Machiya houses in order to maintain a harmonious townscape. Using this tool, alternative elevation designs for traditional merchant houses can be visualized within a VR world, which is helpful for stakeholders as they consider how to repair their Machiya buildings.

In this chapter, we discussed a methodology for visualizing design alternatives. The design elements are classified based on suggestions from the Kanazawa Machiya Research Institute. The learning tool was developed based on VRML technology that supports a server–client model for virtual communication and was designed for learning on the Internet.

This learning tool can be directly accessed without spatial-temporal limitations. Participants can use it to learn about design alternatives and communicate with advisors. Because the learning tool can display all the design alternatives in the VR world, those alternatives become easier to understand for all the participants, whether or not they are professionals. The online design coordination of this tool overcomes the drawbacks of traditional ways of learning design alternatives for Machiya structures and makes the cooperative design of refurbished Machiya houses more attractive and practical. Some researchers (Zhou et al. 2010) have argued that no significant inter-group difference of any experiential motivation has been found regarding education and experience, and in our case study, even though we did not investigate the effectiveness of the learning tool in the field, we expect that learning tools like this one *can be appropriate for* all kinds of participants for learning traditional building design.

Acknowledgments We would like to thank the financial support of the Grants-in-Aid for Scientific Research (No. 22560602C), Japan Society of the Promotion of Science.

References

Ageron, P., Besson, F. and Desfarges, P. (1994), Images database management system: A "Server–client producer" system on a local network and on the internet, Computer Networks and ISDN Systems, Volume 26, Supplement 2, S101-S106

Benwell, M. (1979), Four models of public participation in structure planning. Cranfield CTS report No. 15

Cheng, N. Y. (2003), Approaches to design collaboration research, Automation in Construction, Volume 12, Issue 6, 715–723

Economou, D., Mitchell, William L., Boyle,T. (2000), Requirements elicitation for virtual actors in collaborative learning environments, Computers & Education, Volume 34, Issues 3–4, 225–239

Goul, M., Philippakis,A., Kiang, M.Y., Fernandes, D. and Otondo, R. (1997), Requirements for the design of a protocol suite to automate DSS deployment on the World Wide Web: A client/server approach, Decision Support Systems,Volume 19, Issue 3, 151–170

Hanzl, Malgorzata (2007), Information technology as a tool for public participation in urban planning: a review of experiments and potentials, Design Studies, Vol. 28, Issue 3, 289–307

Hudson-Smith, A., Evans, S., Batty, M. and Batty, S. (2002), Online Participation: The Woodberry Down Experiment http://www.casa.ucl.ac.uk/working_papers/paper60.pdf, CASA Working Paper 60.

Hernández-Serrano, M.J., González-Sánchez, M., Muñoz-Rodríguez, J. (2009), Designing learning environments improving social interactions: essential variables for a virtual training space, Procedia - Social and Behavioral Sciences, Volume 1, Issue 1, 2411–2415

Messinger, Paul R., Stroulia, E., Lyons, K., Bone, M., Niu, Run H., Smirnov, K. and Perelgut, S. (2009), Virtual worlds – past, present, and future: New directions in social computing, Decision Support Systems, Volume 47, Issue 3, 204–228

Monahan, T., McArdle, G., Bertolotto, M. (2008), Virtual reality for collaborative e-learning, Computers & Education, Volume 50, Issue 4, 1339–1353

Pan, Z., Cheok, A.D., Yang, H., Zhu, J., and Shi, J. (2006), Virtual reality and mixed reality for virtual learning environments, Computers & Graphics, Volume 30, Issue 1, 20–28

Shen, Z., and Kawakami, M. (2010), An online visualization tool for Internet-based local townscape design, Computers, Environment and Urban Systems, Volume 34, Issue 2, 104–116

Shen, Z., and Kawakami, M. (2007), Study on Visualization of Townscape Rules Using VRML for Public Involvement. Journal of Asian Architecture and Building Engineering, Vol. 6, No. 1,119–126.

Teo, H.H., Chan, H.C., Wei,K.K., Zhang, Z. (2003), Evaluating information accessibility and community adaptivity features for sustaining virtual learning communities, International Journal of Human-Computer Studies, Volume 59, Issue 5, 671–697

Tseng, M.M., Jiao R.J., and Wang, C. (2010), Design for mass personalization, CIRP Annals Manufacturing Technology, Volume 59, 175–178

Wang, R., Wang, X. and Kim, M.J. (2010), Motivated learning agent model for distributed collaborative systems, Expert Systems with Applications, in press

Ward, Thomas B., and Sonneborn, Marcene S. (2009), Creative Expression in Virtual Worlds: Imitation, Imagination, and Individualized Collaboration, Psychology of Aesthetics, Creativity, and the Arts, Volume 3, Issue 4, 211–221

Warren-Kretzschmar, B., and Tiedtke, S. (2005), What role does visualization play in communication with citizens? – A field study from the interactive landscape plan. In: E. Buhmann, P. Paar, I. Bishop and E. Lange, Editor, Trends in real-time landscape visualization and participation, Wichmann Verlag, Heidelberg, 156–167.

Zhou, Z., Jin, X.-L., Vogel, Douglas R., Fang, Y. and Chen, X. (2010), Individual motivations and demographic differences in social virtual world uses: An exploratory investigation in Second Life, International Journal of Information Management, in Press

Chapter 12
Historical Landscape Restoration Using Google Technology in a Traditional Temple Area, Kanazawa, Japan

Zhenjiang Shen, Mitsuhiko Kawakami, Zheyuang Chen, and Linqian Peng

Introduction

Google Earth has caught the attention of researchers from all countries around the world since it was released in 2005. In this chapter, we attempt to use Google technology to visualize the historical landscape in a traditional temple area in order to create design guidelines for repairs and landscape work around the temple buildings. The simulation reproduces the historical landscape not only as survey material, but also to provide an effective means of deepening residents' understanding of the historical area. We use Google technology to grasp important points of the historical landscape in the temple area and verify the use of Google SketchUp and Google Earth to represent the historical landscape via a questionnaire.

Before the release of Google Earth, virtual reality (VR) had gradually become quite popular as an architectural presentation technique, making it likely that from now on, architects will have to learn to produce VR-models. Google Earth, a "virtual globe" software, is growing rapidly in popularity as a way to visualize and share three-dimensional (3D) environmental data. Scientists and environmental professionals, many of whom are new to 3D modeling and visual communications, are beginning to routinely use such techniques in their work (Sheppard and Cezik 2009). Pearce et al. (2007) completed an analysis of the imagery and 3D data depicting the entire earth that Google Earth provides as a novel information service to use in navigating information about conserving travel-related fuels.

Online tools, such as those pioneered by Google Earth, are changing the way scientists and the general public interact with 3D geospatial data in virtual environments (Wu 2007; Chen et al. 2009). There are many case studies about applications of Google Earth. For instance, Shaw et al. (2009) applied a participatory capacity building approach to the use of a multi-scale visualization consisting of

---

Z. Shen (✉) • M. Kawakami • Z. Chen • L. Peng
School of Environmental Design, Kanazawa University, Kanazawa, Japan
e-mail: shenzhe@t.kanazawa-u.ac.jp

Z. Shen, *Geospatial Techniques in Urban Planning*, Advances in Geographic
Information Science, DOI 10.1007/978-3-642-13559-0_12,
© Springer-Verlag Berlin Heidelberg 2012

global and local climate change scenarios, finally resulting in 3D visualizations of alternative local climate scenarios. Yamagishi et al. (2010) developed a visualization system for multidisciplinary geoscience data that visually displays different datasets by exploiting Google Earth technologies. In Lammeren et al. (2010), the authors discussed the affective appraisal and affective response of users to three different visualization types: colored raster cells, 2D-icons and 3D-icons. In that chapter, Lammeren developed a dedicated multi-layered visualization of current and future land uses in the Netherlands, that may allow policy-makers to assess and compare land use scenarios. For public participation, Wu et al. (2010) argued that the emerging technology of virtual globe-based 3D visualization is a unique opportunity to facilitate public participation in urban planning projects by promoting intuitive 3D interaction, with instant interoperability and seamless integration of 3D visualization and other traditional text and multimedia information channels.

The creation of building and facility models is a tedious and complicated task. As a solution to this problem, Holham et al. (2010) propose a combination of a procedural approach with shape grammars using Google SketchUp. In Sugihara and Hayashi (2008), the authors propose a new scheme using ArcGIS and 3dsmax for partitioning building polygons and show the process of creating a basic gable roof model and restoring an ancient Japanese temple and a pagoda, which are also possible using the Ruby tool of Google SketchUp. Furthermore, it is possible to investigate whether open source map (OSM) data can be used to create virtual city models (Over et al. 2010) because of increasing data supplements from governmental and commercial map agencies. In our work, it is necessary to create virtual 3D models of different traditional temples, but the procedural approach for programming buildings is not efficient for historical landscape restoration. Moreover, Google Map data is based on current spatial features; the historical dataset should be prepared after soliciting information from historical documents and gathering information from field surveys.

In our work, a VR model representing the historical landscape was created early in the planning process. It is employed as a planning tool for obtaining a consensus on design guidelines in our case study area. Sunesson et al. (2008) shows that clear information about how to produce and interpret VR models should be provided to stakeholders early in the planning process. Moreover, Zook and Graham (2007) focus on how people interact with place, and on the construction, access, and use of Google to deploy geo-reference information in the physical environment. In this chapter, we argue that Google Earth has technological and performance advantages in integrating the ground surface and historical 3D objects such as traditional buildings for historical landscape restoration.

In the following sections, we discuss the methodology and application of Google technology to represent historical landscapes. According to ancient paintings, historical records, drawings, and other materials, we suggest modeling traditional buildings using Google SketchUp and uploading them to Google Earth in order to synthesize a continuous, virtual historical landscape. In addition to providing this historical landscape simulation to the local government as survey material, the virtual landscape can be opened to the public through Google Earth to promote people's understanding of the historical landscape.

Research Approach

As mentioned above, we employed Google SketchUp and Google Earth to create a virtual historical representation of the landscape of the traditional temple area in Kanazawa City. We also made use of Google Earth to achieve net participation.

On April 27, 2006, Google announced a free, downloadable version of Google SketchUp. The public release of Google Earth in June 2005 caused media coverage about virtual globes to increase more than tenfold between 2005 and 2006, and is still driving public interest in geospatial technologies and applications. In this chapter, we deem Google technology to be very effective in making public and sharing planning information between officers, planners and citizens. SketchUp is designed to be easier to use than other 3D CAD programs for architects, civil engineers, and related professions. It has a very simple design and clear processes for using points and lines to create plane surfaces, and then extrude the planes as 3D models. Moreover, SketchUp's compatibility with other CAD software formats is excellent and it runs faster than most other CAD software. Although the time consumed by rendering and running animations is relatively short, SketchUp's rendering is still rough and imperfect.

Even though the free version of SketchUp is not as capable as the Pro version, it includes integrated tools for uploading content to Google Earth and to the Google 3D Warehouse, which is a repository of models created in SketchUp. For making public planning information and sharing planning experiences on different sites, the 3D Warehouse, which lets SketchUp users search for models made by others and contribute their own, is amazingly useful for planners. It allows them to compare similar urban projects in different locations because Google Earth also includes features to facilitate the placement of models stored in the 3D Warehouse. Moreover, using Google Earth, which is currently available for use on personal computers running Windows 2000 and above, planners and stakeholders can walk around, see things from a person's point of view, and check information linked to models. The satellite photos, terrain data, 3D building models, and building exterior photo datasets can be integrated on the platform of Google Earth.

As shown in Fig. 12.1, we retrieved the topography of the traditional temple area from Google Earth and imported it into SketchUp for making 3D models of temples

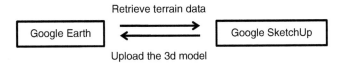

Fig. 12.1 Google SketchUp and Google Earth

and roads in the correct locations. After the 3D models are created, they can all be uploaded to Google Earth from SketchUp to visualize the historical landscape and let various stakeholders participate over the Internet (Fig. 12.1). After finishing the temple modeling in SketchUp and locating the temples correctly in the traditional temple area in Kanazawa City in Google Earth, we administered a questionnaire to investigate the effectiveness of the virtual historical landscape to the members of our research group. The questionnaire focused on the effectiveness and possibilities offered by Google Technology for visualization and net participation.

Visual Representation and Accuracy of Historical Temple Area

Historical Data and Information Gathering

In this chapter, we attempt to simulate the historical landscape of Kanazawa to conduct a case study using Google Earth. Kanazawa City is located on the Sea of Japan coast of Ishikawa Prefecture. It is the central city of the Hokuriku region. The city was originally formed in 1546, but its rise can be traced to 1583, when the local ruler of the Maeda family took up residence in Kanazawa castle. Since the formation of the Kanazawa castle town in the Edo period, its appearance has changed, but its urban geography, historical urban form, and other characteristics have been preserved. The traditional temple district has evolved over several centuries with the transformation of monastic communities in Kanazawa City. The traditional temple area in Kanazawa is one of the districts where many traditional buildings remain and have been preserved. Japan has been gradually modernizing during its rapid economic growth after World War II. This has been accompanied by rapid urban development. Because many historical buildings are decrepit, destruction of the historical landscape is occurring at an alarming rate. In order to better communicate to younger generations the importance of preserving the traditional temple landscape, Kanazawa City promulgated a bylaw called the "Action on preservation of temple landscapes". To protect the historical landscape, both experts' earnest investigation and the collaboration of local residents are necessary. Japan's Agency for Cultural Affairs and Kanazawa City began a project called "Preservation Districts for Groups of Historical Buildings", for which a traditional building survey was conducted and a planning committee formed in 2008. This project is supported by the historical heritage preservation unit of the History section of Kanazawa City's Bureau of Urban Policy. Members of the project include local officers, professors, historians, planners, and students. The representation of the historical landscape of the temple area is one of the tasks assigned to this planning committee.

12 Historical Landscape Restoration Using Google Technology

Table 12.1 Historical information and data employed for visual representation

	Planning drawings	Documents
Entire area	Land use in different time periods	Materials used by local custom, old landscape drawings
Buildings	Elevations, plan drawings, site layouts, roof layouts, measured building drawings	Historical documents carrying building information (construction year and style)
Trees	Preserved trees and preserved forests (species, height and position)	Photographs

One of the primarily tasks of 3D modeling intended to represent the historical landscape using CG is to determine the historical period. We gathered data and information regarding the formation and changes made to the historical temple area, and investigated landscape elements related to historical time periods, such as building styles, garden pieces, and trees, as well as investigating local customs. The landscape elements on the surface of the ground include roads, waterways, rivers, building sites, and green spaces. The traditional buildings are classified into temple buildings, merchant houses and samurai houses. The garden pieces are composed of stone stele and stone monuments, and the tree plantings include preserved trees, preserved forests, and flowerbeds. As shown in Table 12.1, old related drawings, temple plans, photographs, and other historical information reflecting the historical landscape of the temple area are employed as foundation materials for creating the virtual representation. Terrain is downloaded from Google Earth to confirm the positions of buildings, roads, and garden pieces.

Modeling Using SketchUp

The investigating committee gathered historical information using a field survey and documents in public libraries. As shown in Figs. 12.2 and 12.3, we modeled the traditional buildings using the materials gathered by the committee. The elevation drawings and ichnographies were imported to SketchUp and sized to match the scale of the drawings. The textures are mapped from pictures taken in our field survey. Some of the elevations were fixed but most of them were kept in their original state, except for burned and reconstructed buildings. We uploaded a map to Google Earth and overlay it on the aerograph map, which was edited based on the old map drawings, in order to place the traditional buildings and visualize the historical landscape. To locate the positions and orientations of those buildings, aerograph maps were downloaded from Google Earth. The old maps can be overlaid correctly in SketchUp. Modelers could then refer to the old maps to confirm the correct positions of traditional buildings on the sites.

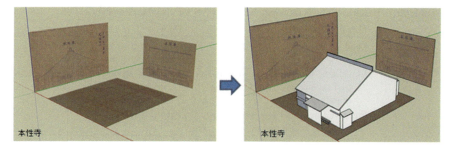

Fig. 12.2 Modeling based on elevation, side elevation, and ichnography

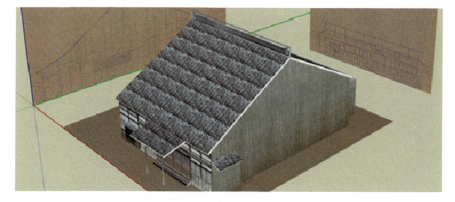

Fig. 12.3 Mapping photo textures to buildings

Accuracy from Different Usage Perspectives

The required accuracy of virtual representations differs for different purposes. When visual information is open to the general public, the atmosphere and surroundings of the place and characteristics of historical buildings will be crucial in the work of creating 3D models. The overall landscape simulation will create a more impressive and effective image than individual building modeling. However, if a model is to be used in planning meetings to display the basis of design guidelines, highly accurate models of architecture styles and details will be needed. The 3D models should be useable as a standard for repairs and landscaping in the historical temple area to be helpful to designers and residents who are learning the architecture style and structural components for repair and landscape projects. Therefore, precise 3D modeling of the components is essential. If, on the other hand, the virtual presentations are used as historical information rather than planning meeting materials and public access information, even more accurate 3D models will be needed because they will be stored in libraries as literature for researchers to use in their investigations.

12 Historical Landscape Restoration Using Google Technology

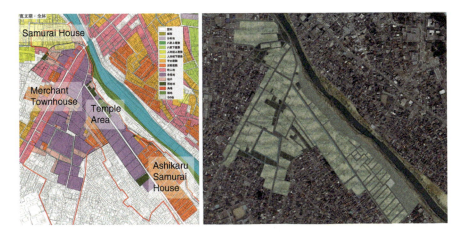

Fig. 12.4 Map of Kanbun era (*left*) and overlaid map with present (*right*)

Table 12.2 Landscape elements

Landscape elements	
Ground surface	Waterway, river
Buildings	Temple, samurai house, merchant house
Garden pieces	Sunk fence, stone monument, stone stele
Trees	Preserved tree, preserved tree forest, flowerbed, espalier

Creating the 3D Model Using Google Technology

As shown in Fig. 12.4, the Teramachi temple area in Kanazawa City is a typical historical area, where many temple buildings have been moved to their current locations since the Sixteenth Century. The Teramachi Temple Investigation Committee organized by Kanazawa City has carried out a comprehensive investigation, including searches of historical documents, building measurement surveys, site investigations, and fire protection research. According to the historical documents in the public records office in Kanazawa City, the Teramachi temple area was begun in the latter half of the 1600s. By analyzing historical documents, the research team decided to simulate the Kanbun Era of the Edo period (1661–1672).

The landscape elements shown in Table 12.2 were considered to be necessary in a virtual representation of the historical landscape. An explanation of how to create 3D models of buildings and trees follows.

Creating 3D Models of Traditional Temple Buildings

Old planning drawings kept by the owners of temple buildings were gathered in the investigation preparatory to creating 3D model of the temple buildings. The models

Fig. 12.5 Myaoden temple

we created of the temple buildings are shown in Figs. 12.5–12.7 and 3D model was created according to drawings of planes and elevations. The front and site elevations can be used as basis for modeling temples, however only two site elevations can be referred. Thus, the performance is partial and not complete; the other two sides are relatively ambiguous. For texture mapping on the elevations, the pictures taken in the field survey are employed because the materials about the appearance of buildings have not been found. The materials used for creating the 3D model are listed as Table 12.3.

Creating 3D Models of Samurai Houses

Samurai houses were distributed widely in the Teramachi temple area, and consisted of two classes, the Hitomochi house and the Ashikaru house. Ashikaru houses can be further divided into two subclasses: Kokashira Ashikaru houses and Hirajya Ashikaru houses. In the Teramachi temple area, the ratio of the number of Kokashira to Hirajya houses is 1:9. Kokashira Ashikaru houses can be further differentiated according to the position of their entrance into those with Tumaire layouts and those with Heraire layouts, which existed in approximately even numbers in the area. The historical landscape of the model was formed on the basis of these proportions of Samurai houses. The materials available for use in creating the 3D models of Samurai houses are listed in Table 12.4.

Fig. 12.6 Kougan temple

Fig. 12.7 Honjyo temple

The height, color, proportion, and scale of Samurai houses were set on the basis of the remaining Samurai buildings. Building materials were chosen to emulate remaining buildings and existing photographs printed in historical documents from their time of construction. For some buildings, those for which plan drawings of roof layouts exist, the created roof shapes are based on the drawings. Figure 12.8

Table 12.3 Materials for creating temple building model

	Plane	Elevation	Roof layout	Measured drawings	Survey photo	Ancient drawings
Myoden temple	○	Δ	×	×	○	×
Kougan temple	○	Δ	×	×	○	×
Honjyo temple	○	Δ	×	×	○	×
Gonkou temple	○	Δ	×	×	×	×

Data availability: ○ available, Δ some available, × none

Table 12.4 Materials used for creating Samurai houses

	Plane	Elevation	Roof layout	Measured drawings	Photos taken in field survey	Ancient drawings
Hitomochi Samurai house	○	×	○	×	×	×
Kokashira Ashikaru house	○	×	○	×	×	×
Hirajya Ashikaru house	○	×	○	×	×	×

Data availability: ○ available, × none

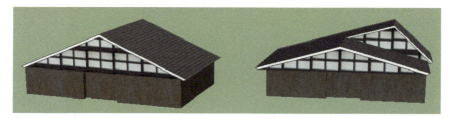

Fig. 12.8 The two types of Samurai house

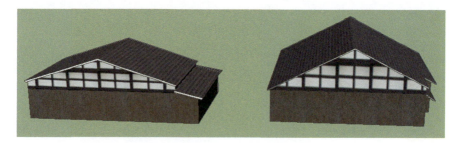

Fig. 12.9 The two types of Kokashira Ashikaru Samurai house

shows the Ashikaru and Hitomachi Samurai house subgroups and Fig. 12.9 shows the Kokashira and Hirajya subdivisions of the Ashikaru house type. Kokashira houses can finally be divided into Tsumaire and Heraire styles according to their roof shape as shown in Figs. 12.10 and 12.11. To represent the general area, all the

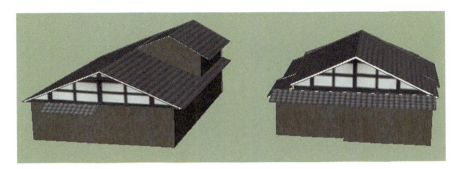

Fig. 12.10 The two types of Hirajya Ashikaru Samurai houses (Tumaire roof)

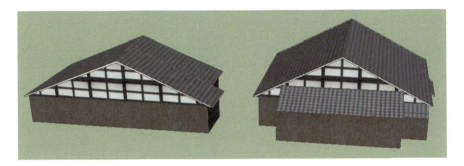

Fig. 12.11 The two types of Hirajya Ashikaru Samurai houses (Heraire roof)

Table 12.5 Materials for modelling Merchant townhouses

	Plane	Elevation	Roof layout	Measured drawings	Photos taken in field surveys	Ancient drawings
Merchant house	○	Δ	×	×	×	×

Data availability: ○ available, Δ some available, × none

types of Samurai house are arranged according to the locations and ratios of the different types of Samurai houses reported in available historical information.

Creating 3D Models of Merchant Townhouses

The materials available for creating merchant houses are shown in Table 12.5. For creating the 3D models, only floor plans and elevations were used. The 3D model is shown in Fig. 12.12. As with the Samurai house, we located the three types of merchant houses randomly in the original location of the traditional commercial street.

Fig. 12.12 Three types of Merchant townhouses

Table 12.6 Materials for modelling trees

	Height	Photo taken in field survey	Position
Tree	×	Δ	Δ
Preserved tree	○	○	○
Preserved forest	Δ	○	○

Data availability: ○ available, Δ some available, × none

Creating 3D Models of Preserved Trees

The Teramachi Temple area is located in downtown Kanazawa city, and there are continuous green spaces around the temple and associated houses. This creates a quiet and peaceful monastery landscape. The materials used for tree modeling are listed in Table 12.6. According to relevant local bylaws regarding preserved trees, we created 3D models of trees including regular trees, preserved trees, and preserved forest areas. It is relatively easy to model forest areas if information about the species, heights, and locations of trees are available, and textures for the 3D tree models can employ existing photos of different species of trees. Preserved tree forests are composed of a mix of tree types, as shown in Fig. 12.13.

On-line Visualization of Simulated Historical Landscape Using Google Earth

We edited the terrain data downloaded from Google Earth for shaping roads, rivers, and waterways as portrayed on old maps. Meanwhile, sunk fences, stone monuments, garden pieces, and trees were placed around the temple buildings along with samurai houses and merchant houses as shown in Fig. 12.14. All the

Fig. 12.13 Preserved forest

Fig. 12.14 Configuration of landscape elements

3D models discussed above, integrated with terrain data from Google Earth, were uploaded to Google Earth to complete the representation of the historical landscape in the Kanazawa City temple area. The models and data will allow stakeholders and investigation committee members to experience and share the historical landscape over the Internet (Fig. 12.15).

Evaluating the Virtual Presentation of the Teramachi Temple Area

Here, we assess the effectiveness of the virtual representation to verify the validity of the Google technology. We administered a questionnaire to members of the investigation committee on March 11, 2010 after a commission meeting. In the

Fig. 12.15 Historical landscape on Google Earth

meeting, we presented the virtual representation methodology and played movie clips of the simulated virtual world created using Google Earth. There were 13 respondents: seven researchers, four officers, and two historians.

Accuracy of the Virtual Representation

First, half of the respondents said that visualization of the historical landscape was possible to some extent. Because of the lack of relevant information, the shape and layout of temple building models were determined from old planning drawings, which showed only floor plans and elevations. This was successful even though only ambiguous information was employed. However, there were also negative comments about the virtual representation, such as "the building density looked incorrect", "in fact, there are no materials that show the style of temple buildings at that time", and "roof slopes are not correct and stone roofs are not shown in the models".

Second, many respondents expressed that "it is difficult (or impossible) to grasp the urban structure". The reason for this is inaccuracy of the virtual representation with respect to "the shape and density of buildings", the "layout and location of housing", and "the color of the road", in addition to other issues. Some respondents observed that, "because the photos of temple buildings taken in the field survey are mapped onto the ancient temple buildings, the circumstances of the modeled time period cannot be visualized correctly". Some respondents also expressed that, "because historical materials and plan drawings of the buildings in the modeled time period cannot be identified and are almost non-existent, the only way to judge the representation is in accordance with modern photographs and paintings". If only movie clips are presented to help viewers understand the overall structure, it is difficult for audiences to grasp the entire image because movie clips are composed of fragments. However, a respondent offered the view that "Google Earth is helpful in understanding the overall streetscape because users can freely change their position and view", indicating that Google Earth is better than pre-recorded animations.

Based on these results, we feel that virtualization of the Teramachi temple area based on insufficient information is, as yet, unsatisfactory. On the other hand, feedback from the questionnaire also showed that although information was insufficient, active use was feasible to some extent. Even though it is relatively difficult to make highly accurate 3D models, as long as accuracy can be improved, constructing the virtual representation is worth the effort.

Possibility of Utilizing the Virtual Representation as an Appendix to an Investigative Report

More than half of the respondents (61.1%) (Fig. 12.16) regarded using the virtual representations as appendices to investigative reports chose the options "fully possible" or "possible". However, there are many varying opinions on the correctness and accuracy of temple roofs, elevations, and other details. The main reason for the insufficient model correctness is a lack of necessary architectural survey data and historical documents detailing architecture style. Another reason is that the size of data that can be made open to the public over the Internet is limited, and so it was necessary to simply the 3D models to reduce the data file size. Even though there were negative comments about model precision, more than 60% of the members of the investigative committee agreed that virtual representation of traditional

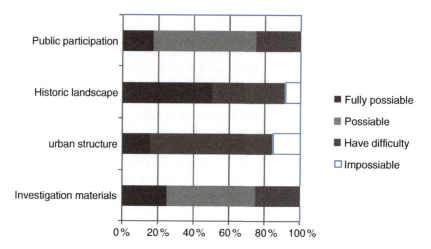

Fig. 12.16 Effectiveness of the virtual representation

buildings is feasible if used as an appendix to an investigative report. Therefore, the investigating committee expects to be able to use a virtual representation made with Google Earth.

Effectiveness of the Virtual Representation for Public Participation

Nine (75%) of the respondents thought that making planning information available to the public was "fully possible" or "possible". We therefore deem that the potential for using this virtual representation for public participation purposes has been verified as positive. The visual representation was effective as a learning aid, helping residents gain an impression of the historical Teramachi temple area. However, those respondents who thought making planning information public was "difficult" or "impossible" expressed opinions such as "it is necessary to improve the presentation of elevations" and "improper expression will convey a false historical impression to the public". The opinion "If more historical information and data were gathered, it would be easier to capture the historical characteristics of this area and increase the value of the virtual representation" was also presented. Although the virtual world was edited without sufficient historical information, it was not necessary to make highly accurate 3D models just to encourage residents to preserve the historical temple area. A more important reason for acceptance of the model is that the visualized landscape communicates a richer representation to the local community than words and photographs can convey.

12 Historical Landscape Restoration Using Google Technology

Table 12.7 Subsidy system for repairs and landscaping

	Assistance ratio (%)	Maximum (10,000 Japanese yen)
Landscape (building elevation)	70	200–300
Repair of preserved buildings	70	500
Fences and garden pieces	70	100–300
Landscape (windows)	70	–
Building design	30	30
Fire facility improvements	70	300
Fire structure improvements	70	300

Conservation Planning, Design Guidelines, and Standards for Repair and Landscape Work

Since the 1960s, the concept of preserving historical areas in Japan has been widely accepted and seriously emphasized. Preservation of the traditional temple area in Kanazawa City began with individual buildings and expanded to the entire block. The concept has been extended to restoration of urban history and culture from mere preservation of building property. For urban cultural and environmental preservation, Kanazawa City (1992) formulated a set of "Kanazawa traditional environmental preservation regulations". The Teramachi temple area, Utatsu mountain temple area, and Nagamachi samurai house area were chosen for preservation. Through public participation, supported by a virtual representation, consensus will be reached on design guidelines and subsidy regulations regarding the historical Teramachi temple area as shown in Tables 12.7 and 12.8. Thus, decisions on the spatial extent of the preservation area, design guidelines, and the subsidy system are possible outcomes of our investigation and planning work.

Conclusion

As a case study, we achieved technological success in building a 3D model of the Teramachi temple area in Kanazawa City. We conducted a questionnaire, the results of which verified that virtual representations using Google SketchUp and Google Earth are suitable for improving residents' understanding of the historical landscape. The idea of opening the model to public view on the Internet was evaluated positively by respondents to a questionnaire.

Information gathering is extremely important in historical landscape restoration. We constructed our virtual representation on the basis of an investigation of historical documents, old drawings, and building blueprints. The ground surface and the layout of the Temple area faithfully reflect the historical landscape based on old historical drawings. Even though we used historical documents and paper drawings as much as possible, we had to use photographs taken in field surveys to map the texture of buildings. Actually, building layouts and materials used to

Table 12.8 Design guidelines

			Design guidelines		Design review
Building	Position		Keep the same building line and fit in with surrounding buildings in the townscape		Design review is necessary for reconstruction and repair, of which local landscape commission in Teramachi area is in charged.
	Height		Keep building height close to that of surrounding buildings		
	Design	Roofs	Traditional roofing tile		
			Roofs with slopes using traditional building style		
		Exterior walls	Stable colors with dominant hues such as brown and gray		
			Traditional style		
		Other	Recycle used construction materials		
	Equipment		Keep equipment and instrument locations out of sight		
Street furniture	Gates		Preservation and repair of building gates and walls	Harmonious with traditional temple landscape	
	Sunk fences		Stone and wood are to be used to maintain a harmonious townscape		
	Advertisements		No commercial advertisements are allowed on buildings		
	Other		Interior design and repairs		
Parking spaces	Fences and walls should hide cars in parking spaces to maintain a continuous townscape				

		Consideration of demand for green space
Trees and green space	Tree plantings and green pavement should be used around parking spaces	
	Preservation of trees and gardens in the temple area	
Public facilities	Consideration of demand for overall community design in harmony with the historical landscape	
	Consideration of the historical atmosphere when approaching the Maeda Grave in road network design	
Other	Urban spaces for traditional events and customs	
	Rules on the positioning of vending machines	
	Other, including preservation of the landscape of the traditional temple area	

create building elevations in the Kunbun Era cannot be obtained or confirmed using existing historical documents, drawings, or photos, and so the virtual world presented in this chapter is only an imaged landscape of that time period. For this reason, the investigating committee voiced many arguments about the accuracy of building forms. Thus, we need more evidence-based information, such as building layout, elevations, and construction materials, to create a more reliable virtual representation.

Google SketchUp and Google Earth were verified as powerful tools through which planning information can be opened for public access over the Internet. We found that, if possible, the effort of gathering more historical information and integrating historians' opinions would improve the accuracy of virtual representation and be worthwhile.

Acknowledgments We would like to thank the financial support of the Grants-in-Aid for Scientific Research (No. 22560602C), Japan Society of the Promotion of Science.

References

Chen, A., Leptoukh, G., Kempler, S., Lynnes, C., Savtchenko, A., Nadeau, D., Farley, J., (2009), Visualization of A-Train vertical profiles using Google Earth, Comput. Geosci. 35, 419–427

Hohmann, B., Havemann, S., Krispel, U., Fellner, D., (2010), A GML shape grammar for semantically enriched 3D building models DOI:dx.doi.org, Computers & Graphics, Volume 34, Issue 4, 322–334

Kanazawa City, "The historical landscape and cultural assets preservation - Teramachi temple area" Kanazawa City Board of Education (1992), "The merchant house and historical building in Kanazawa" Kanazawa city, http://www4.city.kanazawa.lg.jp/29020/keikan/jourei/jisha/ji_guid901.html

Over M., Schilling A., Neubauer S., Zipf A. (2010), Generating web-based 3D City Models from OpenStreetMap: The current situation in Germany, Computers, Environment and Urban Systems, In Press, Corrected Proof, Available online 10 June 2010

Pearce, Joshua M., Johnson, Sara J., Grant, Gabriel B. (2007), 3D-mapping optimization of embodied energy of transportation, Resources, Conservation and Recycling, Volume 51, Issue 2, 435–453

Shaw, A., Sheppard, S., Burch, S., Flanders, D., Wiek, A., Carmichael, J., Robinson, J., Cohen, S., (2009), Making local futures tangible—Synthesizing, downscaling, and visualizing climate change scenarios for participatory capacity building, Global Environmental Change, Volume 19, Issue 4, 447–463

Sheppard, Stephen R.J., and Cizek, P. (2009), The ethics of Google Earth: Crossing thresholds from spatial data to landscape visualization, Journal of Environmental Management, Volume 90, Issue 6, 2102–2117

Sugihara, K., Hayashi, Y. (2008), Automatic Generation of 3D Building Models with Multiple Roofs, Tsinghua Science & Technology, Volume 13, Supplement 1, 368–374

Sunesson, K., Allwood, C.M., Paulin, D., Heldal, I., Roupé, M., Johansson, M., Westerdahl, B. (2008), Virtual Reality As a New Tool in the City Planning Process, Tsinghua Science & Technology, Volume 13, Supplement 1, 255–260

van Lammeren R., Houtkamp J., Colijn S., Hilferink M., Bouwman A. (2010), Affective appraisal of 3D land use visualization, Computers, Environment and Urban Systems, In Press.

Wu, H., He Z., and Gong, J.,(2010),A virtual globe-based 3D visualization and 391 interactive framework for public participation in urban planning processes, Computers, Environment and Urban Systems, Volume 34, Issue 4, 291–298

Wu, X. (2007), Stakeholder identifying and positioning (SIP) models: From Google's operation in China to a general case-analysis framework, Public Relations Review,Volume 33, Issue 4, 415–425

Yamagishi, Y., Yanaka, H., Suzuki, K., Tsuboi, S., Isse, T., Obayashi, M., Tamura, H., Nagao, H. (2010), Visualization of geoscience data on Google Earth: Development of a data converter system for seismic tomographic models, Computers & Geosciences, Volume 36, Issue 3, 373–382

Zook, Matthew A., Graham, M. (2007), The creative reconstruction of the Internet: Google and the privatization of cyberspace and DigiPlace, Geoforum, Volume 38, Issue 6, 1322–1343

Part III
Geography Information System and Planning Support

Various customized tools for urban planning and design.
Supporting decision-making and public participation.
Integration of VR, MAS and GIS.

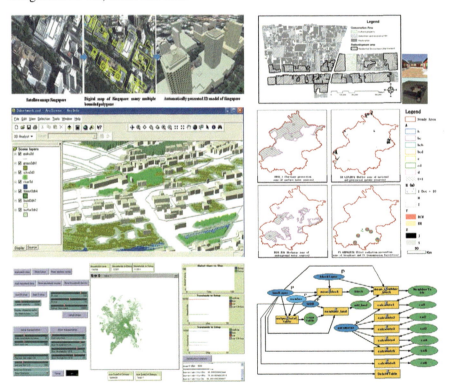

Chapter 13
Automatic Generation of Virtual 3D City Models for Urban Planning

Kenichi Sugihara and Zhenjiang Shen

Introduction

In the past 10 years, although planners and designers used a lot of paper drawings and planning documents in local planning committee, they have started to use VR technology gradually to produce a virtual world for citizens understanding the content of planning and design. Urban Planning Exhibition Hall in Expo' 2010 at Shanghai applied their unique VR system to show the stakeholders with the Urban Design Guideline in the city. Consequently, stakeholders can discuss with each other about the future of the city using the virtual 3D city model. On the one hand, designers and planners can get information feedback from the stakeholders to modify their designs; on the other hand, the stakeholders can better learn the content of the urban plan.

How to build a virtual 3D city model? Based on building polygons or building footprints on a digital map shown in Fig. 13.1 left, we propose a GIS and CG integrated system that automatically generates 3D building models. A virtual 3D city model as shown in Fig. 13.1 right is an important information infrastructure that can be utilized in several fields, such as, urban planning and landscape evaluation, disaster prevention simulation. However, enormous time and labour has to be consumed to create these 3D models, using 3D modeling software such as 3ds Max or Sketch Up. For example, when manually modeling a house with roofs by Constructive Solid Geometry (CSG), one must follow these laborious modeling steps: (1) generation of primitives of appropriate size, such as box, prism or polyhedron that will form parts of a house, (2) Boolean operation among these primitives to form the shapes of parts of a house such as making holes in a building

---

K. Sugihara (✉)
Faculty of Business Administration, Gifu Keizai University, Gifu, Japan
e-mail: sugihara@gifu-keizai.ac.jp

Z. Shen
School of Environmental Design, Kanazawa University, Kanazawa, Japan

Z. Shen, *Geospatial Techniques in Urban Planning*, Advances in Geographic
Information Science, DOI 10.1007/978-3-642-13559-0_13,
© Springer-Verlag Berlin Heidelberg 2012

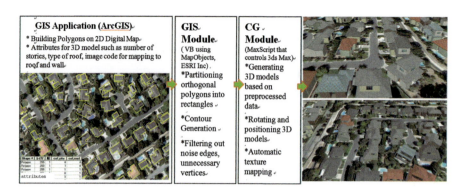

Fig. 13.1 An automatic generation system for 3D building models

body for doors and windows, (3) rotation of parts of a house, (4) positioning of parts of a house, (5) texture mapping onto these parts.

In order to automate these laborious steps, the GIS and CG integrated system that automatically generates 3D building models was proposed, depending on building polygons on a digital map (Sugihara 2005). As shown in Fig. 13.1 left, most building polygons' edges meet at right angles (orthogonal polygon). A complicated orthogonal polygon can be partitioned into a set of rectangles. The integrated system partitions orthogonal building polygons into a set of rectangles and places rectangular roofs and box-shaped building bodies on these rectangles. In order to partition an orthogonal polygon, it is proposed that a polygon will be expressed in RL (Right and Left turn) expression, and the partitioning scheme for deciding from which vertex a dividing line (DL) is drawn is also proposed (Sugihara 2006).

In the digital map, however, not all building polygons are orthogonal. As for non orthogonal polygons, the integrated system places parts of a building, such as walls and windows, doors, shop façade along the inner contour which is set backed from the original building polygon with the setbacked edges parallel to the original edges by straight skeleton computation (Aichholzer et al. 1995, 1996).

In this chapter, we propose a new scheme for partitioning complicated orthogonal building polygons. For non orthogonal polygons, we propose a new scheme for creating a complicated shape of building models based on a building polygon bounded by outer polygons (a multiple bounded polygon). For automatic generation of virtual 3D city model, we will carefully describe the new scheme before its application in Nagoya city.

Literature Review

Since virtual 3D city models are important information infrastructure that can be utilized in several fields, the researches on creations of virtual 3D city models are in full swing. Various types of technologies, ranging from computer vision, computer graphics, photogrammetry, and remote sensing, have been proposed and developed for creating virtual 3D city models. Using photogrammetry, Gruen and Wang (1998, 2002) introduced a semi-automated topology generator for 3D building models: CC-Modeler. Feature identification and measurement with aerial stereo images is implemented in manual mode.

During feature measurement, measured 3D points belonging to a single object should be coded into two different types according to their functionality and structure: boundary points and interior points. After these manual operations, the faces are defined and the related points are determined. Then the CC-Modeler fits the faces jointly to the given measurements in order to form a 3D building model. Suveg and Vosselman (2002) presented a knowledge-based system for automatic 3D building reconstruction from aerial images. The reconstruction process starts with the partitioning of a building into simple building parts based on the building polygon provided by 2D GIS map. If the building polygon is not a rectangle, then it can be divided into rectangles. A building can have multiple partitioning schemes. To avoid a blind search for optimal partitioning schemes, the minimum description length principle is used. This principle provides a means of giving higher priority to the partitioning schemes with a smaller number of rectangles. Among these schemes, optimal partitioning is 'manually' selected. Then, the building primitives of CSG representation are placed on the rectangles partitioned.

These proposals and systems, using photogrammetry, will provide us with a primitive 3D building model with accurate height, length and width, but without details such as windows, eaves or doors. The research on 3D reconstruction is concentrated on reconstructing the rough shape of the buildings, neglecting details on the façades such as windows, etc. (Zlatanova and et al. 2002). On the other hand, there are some application areas such as urban planning and game industries where the immediate creation and modification of many plausible building models is requested to present the alternative virtual 3D city models. Procedural modeling is an effective technique to create 3D models from sets of rules such as L-systems, fractals, and generative modeling language (Parish and Müller 2001; Müller and et al. 2006).

Müller et al. (2006) have created an archaeological site of Pompeii and a suburbia model of Beverly Hills by using a shape grammar with production rules. They import data from a GIS database and try to classify imported mass models as basic shapes in their shape vocabulary. If this is not possible, they use a general extruded footprint together with a general roof obtained by a straight skeleton computation (Aichholzer et al. 1995).

The straight skeleton can be used as the set of ridge lines of a building roof, based on walls in the form of the initial polygon (Aichholzer and Aurenhammer 1996). The roofs created by the straight skeleton are limited to hipped roofs or gable roofs with their ridges parallel to long edges of the rectangle into which a building polygon is partitioned. However, there are many roofs whose ridges are perpendicular to a long edge of the rectangle and these roofs cannot be created by the straight skeleton since the straight skeleton treats a building polygon as a whole and forms a seamless roof so that it cannot place individual roof independently on partitioned polygons. To create a various shape of 3D roofs, building polygons are to be partitioned by the different partitioning schemes; such as, separation prioritizing or shorter DL (dividing line) prioritizing.

Laycock and Day (2003) have combined the straight skeleton method and polygon partitioning in the following steps; (1) Partition the polygon into a set of rectangles by horizontal and vertical lines from all reflex vertices. (2) Construct the straight skeleton and grow an axis aligned rectangle (AAR) from each of the lines. (3) For each AAR, collect the rectangles which are interior to AAR and union them to obtain an exterior boundary. (4) Assign a roof model to each exterior boundary. Merge the roof models.

This method seems effective in independently choosing a roof model for each rectangle and merging the roof models for polygons with a small number of vertices. However, for polygons with a large number of vertices, implementation of partitioning along all DLs (dividing lines from all reflex vertices) often results in an unnecessarily large number of rectangles and collecting and merging steps become so cumbersome that they don't succeed in doing this. In our system, one has an option to choose partitioning scheme; prioritizing separation or prioritizing shorter DL. Our system tries to select a suitable DL for partitioning or a suitable separation, depending on the RL expression of a polygon, the length of DLs and the edges of a polygon.

More recently, image-based capturing and rendering techniques, together with procedural modeling approaches, have been developed that allow buildings to be quickly generated and rendered realistically at interactive rates. Bekins et al. (2005) exploit building features taken from real-world capture scenes. Their interactive system subdivides and groups the features into feature regions that can be rearranged to texture a novel model in the style of the original. The redundancy found in architecture is used to derive procedural rules describing the organization of the original building, which can then be used to automate the subdivision and texturing of a novel building. This redundancy can also be used to automatically fill occluded and poorly sampled areas of the image set. Aliaga et al. (2007) extend the technique to inverse procedural modeling of buildings and they describe how to use an extracted repertoire of building grammars to facilitate the visualization and modification of architectural structures. They present an interactive system that enables both creating new buildings in the style of others and modifying existing buildings in a quick manner.

Vanega et al. (2010) interactively reconstruct 3D building models with the grammar for representing changes in building geometry that approximately follow the Manhattan-world (MW) assumption which states there is a predominance of three mutually orthogonal directions in the scene. They say automatic approaches using laser-scans or LIDAR data, combined with aerial imagery or ground-level images, suffering from one or all of low-resolution sampling, robustness, and missing surfaces. One way to improve quality or automation is to incorporate assumptions about the buildings such as MW assumption. However, there are lots of buildings that have cylindrical or general curved surfaces, based on non orthogonal building polygons. By these interactive modeling, 3D building model with plausible detailed façade can be achieved. However, the limitation of these modeling is the large amount of user interaction involved (Jiang et al. 2009).

Research Objective

When modeling a virtual 3D city model with 100 or 1,000 of building interactively, it will take an enormous time and labour to create them. When creating virtual 3D city model for urban planning or facilitating public involvement, virtual 3D city models should cover lots of citizens' and stakeholders' buildings involved, and the base map as a source for these virtual 3D city models is expected to be the digital map for urban planning that are stored and administrated by GIS.

In this chapter, we attempt to generate virtual 3D city model for representing alternatives of urban design for deliberation between stakeholders. After explanation of our approach in session 2, we propose the GIS and CG integrated system in session 3 that automatically generates virtual 3D city models based on a digital map including multiple bounded polygons. In session 4, a case study is conducted in Nagoya city for deliberation within a local planning group, where citizens and stakeholder can coordinate the urban design alternatives using generated 3D building models for gaining consensus. Finally, we make a conclusion that the proposed methods for generating virtual 3D city model based on multiple bounded polygons is quite effective for stakeholders to learn and experience alternatives in virtual world.

Approach of Automatic Generation for Urban Design Alternatives

In this work, firstly we need 2D site plans, namely urban design alternatives that should be edited in the 'shape format' of ArcGIS (GIS application, ESRI Inc.) by urban planners. Meanwhile, attributes of all building features, such as colour, stories and others, that are necessary for generating virtual 3D city model, should be included in attribute tables of shape files of alternatives. Secondly, building polygons and attributes of building features in each alternative can be inputted to generation process of virtual 3D city model in order to visualize the design alternative.

As shown in Fig. 13.1, the automatic generation system consists of ArcGIS, GIS module and CG module. GIS module is necessary in the first process of virtual 3D city model generation of site plan. Accordingly, the source of a virtual 3D city model, namely design alternative is a digital map that contains building polygons linked with attributes data such as the number of stories and the type of roof. In the automatic generation process, prior to 3D modeling by CG module, the GIS module "pre-processes" building polygons on the digital map. The "pre-process" includes filtering out an unnecessary vertex whose internal angle is almost 180°, partitioning orthogonal building polygons into sets of rectangles, generating inside contours by straight skeleton computation for positioning windows and façades of a building and exporting the coordinates of polygons' vertices and attributes of buildings. The attributes of buildings consist of the number of stories, the image code of roof, wall and the type of roof (flat, gable roof, hipped roof, oblong gable roof, gambrel roof, Mansard roof, temple roof and so forth). The GIS module has been developed using 2D GIS software components (MapObjects, ESRI).

Following GIS module, the CG module receives the pre-processed data that the GIS module exports, for generating virtual 3D city models as urban design alternatives. CG module has been developed using Maxscript that is the scripting language of 3ds MAX (3D CG software, Autodesk Inc). For visualizing design alternatives, the CG module follows these steps: (1) generation of primitives of appropriate size, such as boxes, prisms or polyhedra that will form the various parts of the buildings, (2) Boolean operation on these primitives to form the shapes of parts of the buildings, for examples, making holes in a building body for doors and windows, making trapezoidal roof boards for a hipped roof and a temple roof, (3) rotation of parts of the buildings, (4) positioning of parts of the buildings, (5) texture mapping onto these parts according to the attribute received, (6) copying the second floor to form the third floor or more in case of building higher than three stories.

In the following sessions, we will introduce the key issues for generation of 3D building models regarding to GIS module and CG module for visualizing urban design alternatives. We will focus on pitched roof building using orthogonal polygon and flat roof building using inner polygon and multiple bounded polygon.

Virtual 3D City Model Generation

Generation of 3D Building Models for Pitched Roof Building

Polygon Expression and Partitioning Scheme for Orthogonal Polygons

At map production companies, technicians are drawing building polygons manually with digitizer, depending on aerial photos or satellite imagery as shown in Fig. 13.1. This aerial photo and digital map also show that most building polygons are orthogonal polygons. An orthogonal polygon can be replaced by a combination of rectangles. When following edges of a polygon clockwise, an edge turns to the right or to the left by 90°. Therefore, it is possible to assume that an orthogonal polygon can be expressed as a set of its edges' turning direction.

We proposed a useful polygon expression (RL expression) in specifying the shape pattern of an orthogonal polygon (Sugihara 2005). For example, an

13 Automatic Generation of Virtual 3D City Models for Urban Planning

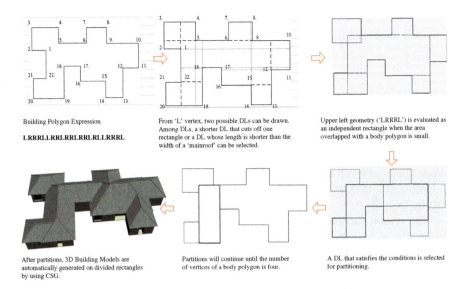

Fig. 13.2 Partitioning process of orthogonal building polygon into a set of rectangles

orthogonal polygon with 22 vertices shown in Fig.13.2 is expressed as a set of its edges' turning direction; LRRRLLRRLRRLRRLRLLRRRL where R and L mean a change of an edge's direction to the right and to the left, respectively. The number of shapes that a polygon can take depends on the number of vertices of a polygon.

The advantage of this RL expression is as follows.

1. RL expression specifies the shape pattern of a polygon without regard to the length of its edges.
2. This expression decides from which vertex a dividing line (DL) is drawn.

The more vertices a polygon has, the more partitioning scheme a polygon has, since the interior angle of a 'L' vertex is 270° and two DLs (dividing lines) can be drawn from a 'L' vertex. We proposed the partitioning scheme that gives higher priority to the DLs that divide 'fat rectangles' (Sugihara 2006). A 'fat rectangles' is a rectangle close to a square. Our proposed partitioning scheme is similar to Delaunay Triangulation in the sense that Delaunay Triangulation avoids thin triangles and generates fat triangles. However, our proposal did not always result in generating plausible and probable 3D building models with roofs. In our next proposal, among many possible DLs, the DL that satisfies the following conditions is selected for partitioning.

1. A DL that cuts off 'one rectangle'.
2. Among two DLs from a same 'L' vertex, a shorter DL is selected to cut off a rectangle.
3. A DL whose length is shorter than the width of a 'main roof' that a 'branch roof' is supposed to extend to.

A 'branch roof' is a roof that is cut off by a DL and extends to a main roof. A 'main roof' is a roof that is extended by a branch roof.

Partitioning Process for Orthogonal Polygons

Figure 13.2 shows the partitioning process of an orthogonal building polygon into a set of rectangles. The vertices of a polygon are numbered in clock-wise order. Stage 2 in Fig. 13.2 shows an orthogonal polygon with all possible DLs shown as thin dotted lines and with DLs that satisfy condition (1), shown as thick dotted lines. The example of a branch roof is shown as the rectangle formed by vertices 6,7,8,9 cut off by DL.

The reason we set up these conditions is that like breaking down a tree into a collection of branches, we will cut off along 'thin' part of branches of a polygon. Since each roof has the same slope in most multiple-roofed buildings, a wider roof is higher than a narrower roof and 'probable multiple-roofed buildings' take the form of narrower branch roofs diverging from a wider and higher main roof. Narrower branch roofs are formed by dividing a polygon along a shorter DL and the width of a branch roof is equal to the length of the DL. Therefore, we propose a scheme of prioritizing the DL that cuts off a branch roof, based on the length of the DL.

In the partitioning process as shown in Fig. 13.2, the DLs that satisfy the mentioned conditions are selected for partitioning. By cutting off one rectangle, the number of the vertices of a body polygon is reduced by two or four. After partitioning branches, the edges' length and RL data are recalculated to find new branches. Partitioning continues until the number of the vertices of a body polygon is four.

After being partitioned into a set of rectangles, the system places 3D building models on these rectangles. Figure 13.3 shows a variety of shapes of orthogonal building polygons with DLs implemented and 3D building models automatically generated from partitioned building polygons. The rectangle partitioned is extended to a wider and higher main roof so that it will form a narrower and lower branch roof. The vertices of a polygon are numbered in clock-wise order as shown in

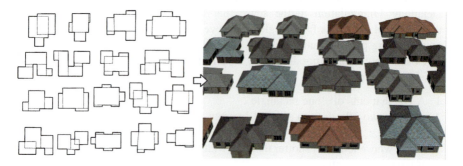

Fig. 13.3 Orthogonal building polygons with dividing lines and 3D models

Fig. 13.2. Here, how the system is finding 'branches' is as follows. The system counts the number of consecutive 'R' vertices (= n_R) between 'L' vertices. If n_R is two or more, then it can be a branch. One or two DLs can be drawn from 'L' vertex in a clockwise or counter-clockwise direction, depending on the length of the adjacent edges of 'L' vertex.

Generation 3D Building Model for Flat Roof Building

Inner Polygon and Straight Skelton

In the case of flat roof building, roof shape can be extruded simply. However, parts of a building, such as windows in façade, are placed on the inner polygon set backed by a fixed distance from the footprint or the original polygon of the building. This inner polygon receded by a fixed distance is computed by the straight skeleton (Aichholzer et al. 1995) which is defined by a shrinking process in which each edge of the polygon moves inwards parallel to themselves at a constant speed as shown in Fig. 13.4. Lengths of edges of the polygon might decrease or increase in this process. The edges incident to two reflex vertices will grow in length as shown by 'ed3' in Fig. 13.4. A reflex vertex is a vertex whose internal angle is greater than 180°. If the sum of the internal angles of two vertices incident to an edge is more than 360°, then the length of the edge increases, otherwise the edge will be shrunk to a point. Each vertex of the polygon moves along the angular bisector of its

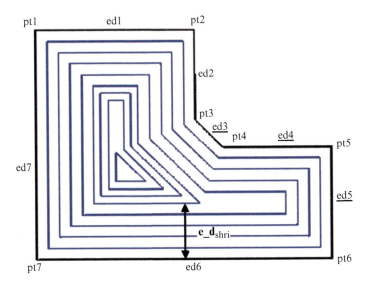

Fig. 13.4 Inner polygons shrunk by a distance of d_{shri} by straight skeleton computation. e_d_{shri} is the d_{shri} when an 'edge event' happens and ed4 and ed5 disappear. The edge (ed3) incident to two reflex vertices grows in length

incident edges. This situation continues as long as the boundary does not change topologically. There are two possible types of changes:

1. *Edge event*: An edge shrinks to zero, making its neighbouring edges adjacent now.
2. *Split event*: An edge is split, i.e., a reflex vertex runs into this edge, thus splitting the whole polygon. New adjacencies occur between the split edge and each of the two edges incident to the reflex vertex.

After either type of event, we are left with a new, or two new, polygons which are shrunk recursively if they have non-zero area (Aichholzer and Aurenhammer 1996).

Shrinking procedure is uniquely determined by the distance between the two edges of before and after shrinking procedure. We refer to this distance as d_{shri}, shown by e_d_{shri} in Fig. 13.4. e_d_{shri} is the d_{shri} when an edge event happens in the shrinking process.

The edge event d_{shri} is calculated as follows:

$$d_{shri} = \frac{L_i}{(\cot(0.5*\theta_i) + \cot(0.5*\theta_{i+1}))} \tag{1.1}$$

Where L_i is the length of ed_i (edge), and θ_i and θ_{i+1} are internal angles of vertices incident to ed_i.

Edge event will happen when $0.5*\theta_i + 0.5*\theta_{i+1} < 180°$, i.e., the sum of the internal angles of two vertices incident to an edge is less than 360°. After calculating d_{shri} for all edges and finding the shortest of them, when d_{shri} reaches the shortest found, an edge event happens and edges disappear for the first time in the process. In Fig. 13.4, when d_{shri} is e_d_{shri}, ed4 and ed5 are the edges that collapse into a point at the first edge event. After an edge shrinks to a point, the system makes its neighboring edges adjacent now and recalculates the length of each edge and internal angles of each vertex in order to find the shortest d_{shri} for next edge event. In Fig. 13.4, ed1 and ed2 will disappear in the next edge event. There remains a triangle that will be extinct by three edge events simultaneously. Inner polygons shrunk by straight skeleton computation are generated by the following algorithm in Fig. 13.5.

The straight skeleton, S(P), is defined as the union of the pieces of angular bisector traced out by polygon vertices during the shrinking process (Aichholzer and Aurenhammer 1996).

Figure 13.6 left shows non-orthogonal building polygons overlapped by satellite image. In Fig. 13.6 middle, inner building polygons are generated by the straight skeleton, and along these inner polygons, windows and set backed façade are placed and thus 3D building model are automatically generated as shown in the right of Fig. 13.6.

Algorithm: Shrink Polygon (**P**, **d**$_{shri}$)
  Input: n vertices of the polygon **P**, **d**$_{shri}$
  Output: polygon **P**$_s$ shrunk by **d**$_{shri}$
  The vertices of **P** *are numbered clockwise.*
1) *Let the previous vertex of* p_i *(the i-th vertex) to be* p_b *and the next vertex of the* p_i *to be* p_a.
2) *Normalize the vector from* p_i *to* p_b *and get* u_{ib}. *Normalize the vector from* p_i *to* p_a *and get* u_{ia}.
  // $(u_{ib}+u_{ia})$ *will be the angular bisector of i-th vertex* p_i.
  // *However,* $(u_{ib}+u_{ia})$ *does not always lie inside the polygon.*
3) *If the cross product of* u_{ib} *and* $u_{ia} > 0$
4) *then the inside bisector will be* $-(u_{ib} + u_{ia})$. // *The edge turns to the left.*
5) *Else the inside bisector will be* $(u_{ib} + u_{ia})$. // *The edge turns to the right.*
6) *End if*
7) *Each vertex of* **P**$_s$ *is positioned along the angular bisector of its incident edges. The offset of each vertex from original vertex is calculated as* $d\_bisect =$ **d**$_{shri}$ $/ sin(0.5 * angle(i))$ *where angle (i) is the internal angle of* p_i.
8) *Check if any edge shrinks to zero. If so, then make its neighboring edges adjacent now and recalculate the length of each edge and internal angles of each vertex.*

Fig. 13.5 Algorithm for shrinking a polygon by straight skeleton

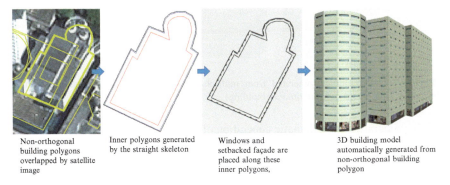

Non-orthogonal building polygons overlapped by satellite image

Inner polygons generated by the straight skeleton

Windows and setbacked façade are placed along these inner polygons,

3D building model automatically generated from non-orthogonal building polygon

Fig. 13.6 Automatically generated 3D building model with flat roof from non-orthogonal polygon

3D Building Models Using Multiple Bounded Polygons

In the shapes of many tall buildings, the third or more floors have the same shape as the second floor, as shown in Fig. 13.7 left. We call such a building a 'simple extruded building'.

On the other hand, there are some buildings whose floors take various shapes according to the number of stories as shown in Fig. 13.7 middle and right. We call these buildings a 'multilayer building'. These multilayer buildings are generated based on the building polygons bounded by outer polygons that we call 'multiple bounded polygon' as shown in Fig. 13.7. In this chapter, we propose a method for the automatic generation of 3D building models from multiple bounded polygons.

Fig. 13.7 Tall buildings generated from multiple bounded polygons (*left*: simple extruded building, *middle* and *right*; multilayer building)

In preparation for automatic generation of design alternative, building polygons are linked with attributes, such as the number of stories and the type of building structure, with the inner polygons in multiple bounded polygons being associated with the type of structure, such as a building floor or a roof or a fence. In multiple bounded polygons, we start with an inner polygon (ME_polygon) and search for outer bounded polygons that include ME_polygon and acquire the number of stories linked to bounded polygons. After getting and summing up the number of stories, we calculate the start height (the height at which the 3D model is built). Here is an algorithm shown in Fig. 13.8 for calculating the start height. In this algorithm, through knowing the type of the polygon found, the system does not need to sum up the number of stories of the polygon if the type of the polygon found is a roof or a fence placed at the top of the building which may include ME_polygon. Figure 13.9 is an example of virtual 3D city model automatically generated using multi bounded polygons. In this digital map in Fig. 13.9, there are many multiple bounded polygons from which multilayer buildings are generated. In this model, the start heights of penthouses, fences and roofs are calculated by searching for outer bounded polygons and summing up the number of stories of those bounded polygons, not having to calculate their start heights manually.

> (1) *For all polygons except ME_polygon, search for the polygon that includes one of the vertices of ME_polygon. If found, check if all the vertices of the ME_polygon are included in the polygon found.*
>
> (2) *After checking, acquire the type and the number of stories of the polygon found. Through knowing the type of the found polygon, do not sum up the number of stories of the polygon if the type of the found polygon is a roof or a fence placed at the top of the building that includes ME_polygon.*
>
> (3) *If there are several polygons (multiple bounded polygons) that include the ME_polygon, add up all the number of stories of these multiple bounded polygons.*
>
> (4) *THUS the start height of the 3D model from ME_polygon is the summation of the heights.*

Fig. 13.8 Algorithm for a multiple bounded

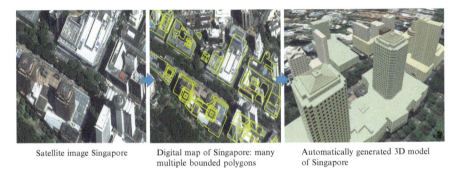

Satellite image Singapore   Digital map of Singapore: many multiple bounded polygons   Automatically generated 3D model of Singapore

Fig. 13.9 Automatically generated 3D building models from multiple bounded polygons

As discussed in this session, we present the partitioning theme for generation of 3D building model with pitched roof using orthogonal polygons. We also propose 'inner building polygon computation method' for placing windows and setbacked façade in case of dealing with flat roof building using non-orthogonal polygons. Depending on this method, we have presented multilayer buildings automatically generated based on the multiple bounded polygon. In the following session, we attempt to apply the above methods for automatically generating 3D urban design alternatives in Nagoya City.

Visualization of Urban Design Alternatives in Nagoya City

In our research, we aim at automatically generating 3D models for visualizing design alternative. To generate virtual 3D city models, the 3D shapes and material attributes of existing buildings and other objects will be reconstructed. In the reconstructing process, the image data will be acquired by taking photographs of the objects in the city. However, when we think of the future layout of the city models, we cannot take photos of the future of the city or the planning roads. Instead of taking pictures, the 3D shapes and material attributes of buildings and other objects will be prepared according to requirements of urban designers. Usually and traditionally, urban designers make planning proposals to present the future layouts of the town by drawing the site plans. There may be several site plans (digital maps) as planning alternatives. It is convenient for urban designers to present the proposed design alternative in 3D virtual space if the site plan can immediately be converted into virtual 3D city model. The integrated GIS and CG system of our work can automatically generate virtual 3D city model in real time so that the stakeholders can check and deliberate on design alternatives with each other in planning meetings.

Figure 13.10 shows the digital map of Meijo District in Nagoya city where an urban regeneration project is under consideration.

Figures 13.11–13.18 show virtual 3D city models automatically generated for local planning committee of Nagoya city. Figure 13.10 shows an aerial photo of the planning area for automatic generation: Meijo district in Nagoya city that is the fourth largest city in Japan. This area is almost central area of Nagoya city with two million populations and is located near Nagoya castle.

As shown in Figs. 13.10–13.11, this area is currently occupied by apartment houses for government officials and citizen run by municipal government. The area as a whole was selected as one of the area for city revitalization project by the Japanese government. Figure 13.11 is a generated virtual 3D city model for representing the current situation of the planning site, where three or four stories apartment buildings for government officials and citizen are located.

Figures 13.12–13.15 are the four visualized alternative plans A-D using our automatic generation system. Figure 13.12 shows alternative plan A proposed by an urban designer, which is planned as low density in the middle area for green spaces and urban facilities, and high density in both side areas for government officials and citizen. Figure 13.13 shows alternative plan B that is proposed by municipal government officials, in which parking spaces are secured in front of apartment

Fig. 13.10 Aerial photo of the target area for automatic generation: Meijo district

13 Automatic Generation of Virtual 3D City Models for Urban Planning

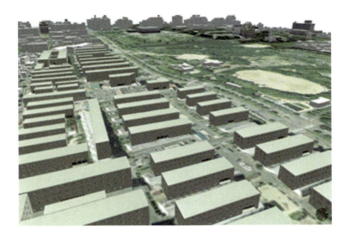

Fig. 13.11 Current 3D city model generated using our CG module: Apartment houses for government officials and citizen

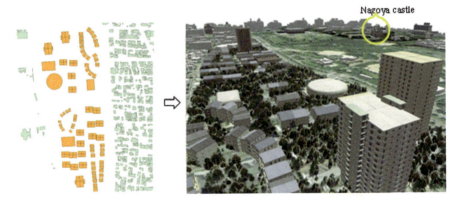

Fig. 13.12 Design alternative A: Low density in the middle area and high density in both side areas

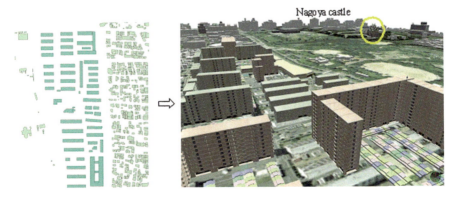

Fig. 13.13 Alternative plan B: Proposed by municipal government officials

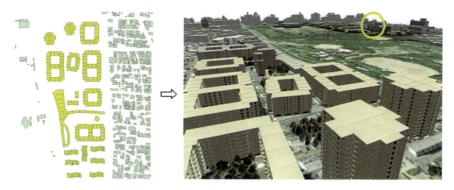

Fig. 13.14 Alternative plan C: Terrace houses sharing green in courtyard

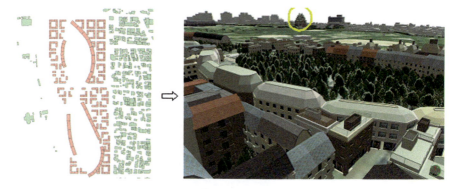

Fig. 13.15 Alternative plan D: Terrace houses laid out for viewing Nagoya castle

buildings. Figure 13.14 shows alternative plan C proposed as terrace houses sharing green in courtyard and buildings in the north standing tall for government officials and citizen. The courtyards are enclosed by terrace houses, providing a shared park-like space for residents. In the proposed alternative plan D as shown in Fig. 13.15, terrace houses sharing green spaces in large courtyard are laid out for viewing Nagoya castle. As shown in Figs. 13.16–13.18, viewpoints of alternative plan D for viewing Nagoya castle are reviewed in the planning committee.

Summary of Our Work

In this chapter, we present a GIS and CG integrated system that automatically generates 3D building models. In case of pitched roof building, we propose a new partitioning theme for generating 3D building model based on orthogonal polygons. In case of flat roof building, we propose to generate inner building polygon by straight skeleton computation and it is possible to deal with non-orthogonal

Fig. 13.16 Alternative plan D: Terrace houses sharing green in large courtyard

Fig. 13.17 Alternative plan D: From another angle increasing in height away from castle

polygons, meanwhile we propose a new scheme for creating a complicated shape of building models based on multiple bounded polygon. As a result, we successfully develop a new tool for generating 3D building model, which can be applied for generation of different types of building.

Fig. 13.18 Alternative plan D: Buildings are increasing in height away from castle

In urban planning practice, for all stakeholders including residents, citizens as well as planners and designers, a virtual 3D city model is quite effective in understanding what will be built, what image of the town will be changed in an urban area. In case of urban planning, urban designer will follow the conventional way of drawing site plans for presenting the future layout of a planning site. Depending on the building polygons drawn in the site plan, a virtual 3D city model can be generated automatically by using our tool in order to visualize the planning alternatives.

In this chapter, we do not analyse the effectiveness of application of our proposed virtual 3D city model in local the planning committee, which is remained as our further research work.

Acknowledgments We would like to thank the financial support of the Grants-in-Aid for Scientific Research (No. 22560537C), Japan Society of the Promotion of Science.

References

Aichholzer, O., Aurenhammer, F., Alberts, D., and Gärtner, B. (1995), "A novel type of skeleton for polygons", Journal of Universal Computer Science, 1 (12),752–761.

Aichholzer, O., and Aurenhammer, F. (1996), "Straight skeletons for general polygonal figures in the plane", Proc. 2nd Ann. Int. Conf. Computing and Combinatorics (COCOON '96), 117–126.

Aliaga, D. G., Rosen, P.A., and Bekins, D. (2007), "Style Grammars for interactive Visualization of Architecture", Visualization and Computer Graphics, IEEE Transactions on Vol.13, 786–797

13 Automatic Generation of Virtual 3D City Models for Urban Planning

Bekins, D., and Aliaga, D. G. (2005), "Build-by-number: rearranging the real world to visualize novel architectural spaces", Visualization, 2005. VIS 05. IEEE, 143 – 150.

Gruen, A. and Wang, X. (1998). "CC Modeler: A topology generator for 3D city models", ISPRS J. of Photogrammetry and Remote Sensing, 53, 286–295.

Gruen, A. et al., (2002). "Generation and visualization of 3D-city and facility models using CyberCity Modeler", MapAsia, 8, CD-ROM.

Jiang, N., Tan, P., and Cheong, L.-F. (2009). "Symmetric architecture modeling with a single image", ACM Transactions on Graphics - TOG, Vol. 28, No. 5

Laycock, R.G., and Day, A.M. (2003). "Automatically Generating Roof Models from Building Footprints", WSCG posters proceedings, 346–351

Müller, P., Wonka, P., Haegler, S., Ulmer, A. and Luc Van Gool, L. V. (2006). "Procedural modeling of buildings", ACM Transactions on Graphics 25, 3, 614–623.

Parish, Y. I. H. and Müller, P. (2001). "Procedural modeling of cities", Proceedings of ACM SIGGRAPH 2001, ACM Press, E. Fiume, Ed., New York, 301–308.

Sugihara, K. (2005). "Automatic Generation of 3D Building Model from Divided Building Polygon", ACM SIGGRAPH 2005, Posters Session, Geometry & Modeling, CD-ROM.

Sugihara, K. (2006). "Generalized Building Polygon Partitioning for Automatic Generation of 3D Building Models", ACM SIGGRAPH 2006, Posters Session Virtual & Augmented & Mixed Reality & Environments, CD-ROM.

Suveg, I. and Vosselman, G., (2002). "Automatic 3D Building Reconstruction", Proceedings of SPIE, 4661, 59–69.

Vanegas, C.A., Aliaga, D.G., and Beneš, B. (2010). "Building reconstruction using Manhattan-world grammars", Computer Vision and Pattern Recognition (CVPR), 2010 IEEE Conference, 358–365

Zlatanova, S. and Heuvel Van Den, F.A. (2002). "Knowledge-based automatic 3D line extraction from close range images", International Archives of Photogrammetry and Remote Sensing, 34, 233 – 238.

Chapter 14
An Urban Growth Control Planning Support System for the Beijing Metropolitan Area

Ying Long, Zhenjiang Shen, and Qizhi Mao

Introduction

This chapter is a case study of a planning support system (PSS) for urban growth control, in which the Beijing Metropolitan Area is used as an example to demonstrate the implementation of the planning support system. Generally, encroachment on open space and natural resources caused by urban sprawl has drawn worldwide attention and posed enormous challenges to sustainable human development. The possible negative impacts of urban sprawl include increased land consumption, infrastructure construction costs, commuting distance, traffic congestion, energy consumption, and air pollution (Burchell 1998; Anas and Rhee 2006). However, little attention has been paid to urban growth control planning (UGCP), considering various control factors and their indicators, or the possible application of PSS to these problems.

Currently, PSS is regarded as the latest form of computer-aided planning system (Klosterman and Richard 1997; Brail and Klosterman 2001; Stillwell 2002; Geertman and Stillwell 2004). It was initially proposed by Harris (1960). Several books have been edited in recent years regarding the most frequently discussed PSS (Brail and Klosterman 2001; Geertman and Stillwell 2002; Stillwell 2002; Brail 2008; Geertman and Stillwell 2009). The fundamental characteristics of PSS include its systematic, dynamic and interactive nature. It has been mainly applied in spatial plans (Kammeier 1999; Geneletti 2008), urban environment improvement plans (Edamura and Tsuchida 1999), industrial location choices (Piracha and

---

Y. Long (✉)
Beijing Institute of City Planning, School of Architecture, Tsinghua University, Beijing, China
e-mail: longying1980@gmail.com

Z. Shen
School of Environmental Design, Kanazawa University, Kanazawa, Japan

Q. Mao
School of Architecture, Tsinghua University, Beijing, China

Z. Shen, *Geospatial Techniques in Urban Planning*, Advances in Geographic
Information Science, DOI 10.1007/978-3-642-13559-0_14,
© Springer-Verlag Berlin Heidelberg 2012

Kammeier 2002), and land-use plans (Klosterman 1999). Typical PSSs currently used in urban planning are INDEX for land-use planning, *What If?* for land-use and urban planning, CommunityViz for community planning, CITYgreen for evaluating the ecological value of urban green space, GB-QUEST for future development prediction, NatureServe Vista for biodiversity protection, WEAP for water resource evaluation and planning, AEZWIN for land resource evaluation, RAMCO for coastal management, and BLM ePlanning for planning information sharing (Long 2007). However, there are no existing PSS packages related to urban growth control. Because the spatial distribution of control factors and the specific content of control indicators can all be temporally dynamic in UGCP, UGCP planning schemes need to be adjusted occasionally. Therefore, we construct the urban growth control planning support system (UGC-PSS), a planner-oriented system, so that it can adapt to these practical requirements by interacting with urban planners, re-calculate the planning scheme, and output the planning drawings with high efficiency.

To address UGCP issues, land-use suitability analysis for urban growth was first proposed by McHarg (1969). It has become a mature method in the field of land-use planning, used to identify appropriate regions for feasible urban land use (Diamond and Wright 1988; Carver 1991; Thill 1999; Sui 1993). To conduct land-use suitability analysis, overlay functions and multi-criteria evaluation (MCE) are commonly combined to create suitability maps for various land uses. The land-use suitability analysis approach, however, has its limitations in simultaneously taking into account numerous control factors (CFs) together with their multiple control indicators that are involved in UGCP. For this purpose, we developed a specific urban growth control planning support system, rather than employing the land-use suitability analysis directly, that supports the whole planning compilation process, including displaying control factors and automatically calculating and exporting the UGCP scheme.

In urban planning practice, urban growth control aiming to alleviate the negative effects of urban sprawl is needed in many urban areas in Europe, the USA, Japan, and China. In Europe and the USA, urban growth control toolkits can be classified into four types: neotraditional development, urban containment, compact city, and eco-city toolkits. Recently, urban containment policies, targeting improved development density and open space protection, have been extensively applied to control urban growth (Nelson and Duncan 1995). They usually include three forms: greenbelts, urban growth boundaries (UGBs), and urban service boundaries (USBs), all of which are designed to contain the future urban form within pre-defined boundaries in order to control urban sprawl and encourage "infill development". Urban planners have also advocated the adoption of the compact city model and other intensive forms (Echenigue 2005). In America, more than 100 metropolitan areas, together with their sub-regions, have adopted urban containment policies to control urban growth by 2004 (Nelson and Dawkins 2004). The main purposes of those policies were to protect open space and improve the efficiency of urban land development (Nelson and Dawkins 1999; Pendall and Martin 2002). In Japan, the "Planned Urban Area" policy is applied in land-use planning to regulate urban growth. During periods of rapid growth, urban growth has not been effectively controlled in Japan, resulting in over-scattered urban forms. Meanwhile, in China, construction-forbidden areas and UGBs need to be established to control urban sprawl according to the latest "Urban and Rural Planning Act" that took effect on January 1, 2008. In addition, green belt policies have also been implemented in some cities, such as Beijing and Shenzhen. Generally speaking, according to European and American experience, as well as practices in China, urban planners should view urban growth control as a critical element during the process of creating urban plans.

In Chinese urban planning and management practice, however, these borrowed urban containment policies, especially green belts and UGBs, that aim to determine rigid control boundaries for urban growth cannot control it effectively because China's over-abstracted control means that "development permitted" and "development not permitted" rules can only be applied in absolutely controlled areas, not in relatively controlled areas. In the relatively controlled areas, urban development may be permitted subject to some development requirements, such as maximum (protecting the perspective of old buildings) or minimum (preventing flood inundation) building heights, constrained land-use types, or constrained urban activities. Therefore, urban growth control can be divided into two classes, absolute control (forbidding all types of urban developments and activities), and relative control (allowing development meeting some designated requirements). To address this objective issue, we propose the concept of "urban growth control planning" and apply it to solve the comprehensive problem of urban growth control. In contrast to existing urban containment policies, UGCP requires not only the establishment of absolute boundaries restricting urban growth, but also specific control requirements based on various control aspects in regions where urban growth is relatively controlled.

In this chapter we discuss how to deal with absolute and relative control using the UGC-PSS and introduce the theoretical methodology, key techniques, and dataset requirements, as well as documenting one case of UGC-PSS implementation. To begin with, we discuss our approach to urban growth control planning in Sect. "Research Approach", and investigate how to establish UGCP planning using spatial control factors and their indicators in Sect. "Spatial Factors Related with Urban Growth Control". In Sect. "The System Framework of UGC-PSS", we analyze and discuss UGC-PSS implementation, and in Sect. "UGC-PSS in the Beijing Metropolitan Area" we report the application of UGC-PSS in UGCP in the Beijing Metropolitan Area. Finally, in Sect. "Conclusions and Discussion", we draw conclusions and offer further discussion about UGC-PSS.

Research Approach

In considering urban growth control, urban planners first want to know what control factors are available. For this, UGS-PSS should provide a well-structured menu of numerous control factors and corresponding control indictors, thus enabling planners to organize a dataset using the PSS to support their ideas in the planning process. UGC-PSS should take all possible control factors, together with their control indicators, into account to show their comprehensive and combined effects. Planners will continue to work out planning rules for urban growth control, and so UGC-PSS must be capable of not only calculating planning schemes using numerous control factors and their control indicators, but also of automatically exporting the calculated planning scheme. Moreover, UGC-PSS output is also expected to set conditions on other spatial plans for the built-up urban space layout at the various levels of the entire city, urban districts, or blocks in order for developers and urban planners to be able to understand the control conditions at their project sites.

To meet these system requirements, we choose the uniform analysis zone (UAZ) scheme, a vector dataset of irregular polygons, as the basic UGC-PSS data model to express control cells (in the following context, a UAZ can be regarded as a control cell). The UAZ scheme was first proposed by Klosterman (1999) and introduced into his WHAT IF? planning support system. In WHAT IF?, the UAZ is the fundamental analysis unit of modules, such as Land Suitability Evaluation and Land-Use-Demand Allocation. In the California urban future model (CUF) proposed by Landis (1994, 1995) for simulating urban development, the fundamental modeling unit is the developing land unit (DLU), which is similar to the UAZ in that they are both based on irregular polygons. Unlike WHAT IF? and CUF, the proposed UGC-PSS emphasizes natural resource protection and risk prevention in urban growth control, rather than analyzing and predicting urban growth.

In view of the large number of control factors located in a control cell, the GIS spatial analysis toolbox UNION can be employed to generate the UAZ dataset. UGC-PSS generates the UAZ layer based on different sets of control factors (see Fig. 14.1). The UAZ dataset denotes the spatial distribution of the control cells as the basic system input to UGC-PSS. The UAZ attributes, namely T, H, A, and U control indicators, can be calculated to generate planning rules for UGCP. The control cells, together with planning rules, are used to produce the UGCP planning scheme based on different spatial partitions.

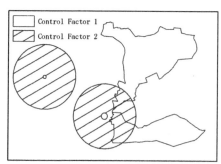
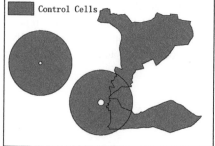

Fig. 14.1 Diagrams of control factors and corresponding control cells in a small part of the Beijing Metropolitan Area

In the next section, we discuss how to generate planning rules based on the UAZ dataset in UGC-PSS. Most of our control factors are prepared and edited in vector format. However, if a raster dataset is chosen for UGC-PSS, all the vector-based planning drawings should be converted into raster formats, which should be saved in the same grid size. This may reduce the data precision. In addition, compared with the regular rectangle raster, the irregular polygon-based UAZ dataset is more capable of recording numeric attributions for further calculations. However, one disadvantage of the UAZ data model is that it takes more computation time than a raster data model using map algebraic operators. Therefore, the arithmetic in UGC-PSS still requires further optimization.

Spatial Factors Related with Urban Growth Control

Spatial Factors

UGCP is based not only on the spatial distributions of factors that curb urban growth, but also on their control indicators determined by existing materials, including laws, codes, regulations, technical standards, research results, international treaties, and agreements, as well as already approved plans. The approved plans may consist of rules for geological hazard prevention, water resource protection, cultivated land protection, and ecological space protection, for example. UGCP can be regarded as "a plan of the plans" or a combination of the plans. Generally, UGCP's foundamental inputs are control factors and corresponding control indicators, as distinguished from land-use suitability analyses based on natural elements such as terrain, geological conditions, climatic conditions, and soil types.

On the basis of the processes and nature of the urban growth, it can be divided into urban developments and urban activities. Urban developments can be divided into the above-ground and underground types. Accordingly, the essence of UGCP in this chapter includes not only above ground urban development, but also underground development and the corresponding urban activities. This view of urban growth control is multi-dimensional in practice and includes controls on development type, building height, and city activities, as well as on underground development. The tasks involved in UGCP are to demarcate control zones (Z) as control indicators so as to distinguish between rigid absolute control areas and flexible relative control areas, and then carry out more detailed controls from a multi-dimensional perspective, utilizing the other four control indicators: land-use type control (T), construction height control (H), urban activities control (A), and underground development control (U) (see Long et al. 2006; He 2008).

The five control factor indicators, reflecting the various essential means of controlling an urban form, can be defined as the term of the negative form (NF) in this study in that every control indicator plays a role in modelling the urban form by controlling development types, building height range, urban activities, or underground development. In contrast to other studies on the urban form, a NF does not recommend types of urban forms to be developed, but rather defines the urban forms that are forbidden or restricted to control growth. In the process of urban growth, such forms should be controlled, so as to avoid encroaching on ecological resources or incurring other risks. The NF, as the final, multi-dimensional product of the UGCP, can be applied to guide the determination of urban forms in land-use planning, urban master planning, or detailed regulatory planning.

To conduct UGCP, it is first necessary to identify the spatial distribution of control factors (CFs). The spatial distribution of the CFs can be obtained from various local government departments. The CF control indicators, a primary basis of UGCP compilation, include the following main aspects:

Control zones (Z): divided into absolute control (AC) and relative control (RC) classes.

Absolute control (AC):

In this zone all types of urban construction, urban activities, and underground developments are forbidden. The absolute control zone, as the rigid control limit on urban growth (the other four indicators are irrelevant here), is identical to the currently existing plan drawings based on policies forbidding construction in certain areas in China.

Relative control (RC):

In this zone, there are strict constraints on urban construction, urban activities, and/ or underground developments. Special requirements for other control indicators (such as T, H, A, and U) need to be met to permit urban growth within this zone. Relative control is identical to the existing policy governing construction-limited areas in China. The other four indictors are not required to be simultaneously active for the construction-limited policy, which, in practice, considers only the boundaries of the limited construction areas.

Land-use type control (T): Forbidden urban land-use types. If there is no limitation, then the value of T is null.

Construction height control (H): Specifies the forbidden building height in meters. If there is no height limitation, the value of H is null.

Urban activities control (A): Refers to controlled urban activities. If there is no limitation on this indicator, the value of A is null.

Underground development control (U): Indicates whether underground development is controlled; if yes, $U = 1$, otherwise U is null.

Detailed descriptions of the control indicators are shown in Table 14.1 ("AC" stands for the absolute control zone, and "RC" for the relative control zone).

We define the detailed content of each control indicator of every control factor as shown in Table 14.1. For instance, the control elements of indicator T are "R", "C", and "M". The relationship between control elements and control indicators is

Table 14.1 Descriptions of control indicators for urban growth control

Control indicators	Z = "RC"				Z = "AC"
	T	H	A	U	
Description	Regulations for different types of land use	Regulations on building height	Regulations for different types of urban activities	Whether underground development is permitted	No construction will be permitted
Sample control elements	Residential (R)	>27	Control of facilities destruction (a)	1	
	Commercial (C)	<9	Control of pollutant emissions (b)		
	Industrial (M)	9–27	Control of resource exploitation (c)		
			Control of site occupation (d)		

illustrated in (14.1), where t_m, h_p, a_q, u_r are, respectively, the control elements of control indicators T, H, A, and U and m, p, q, and r are, respectively, the total numbers of control elements in the corresponding control indicator sets, K_T, K_H, KA, and K_U are the control element sets, i is the control factor ID, I is the total number of control factors, j is the control indicator ID, l_{ij} is the value of the control indicator j for control factor i, and L is the set of all control indicators for all control factors. From (14.1), we can see that the control indicator of a given control factor may include more than one control element. If $l_{ij} = \Phi$ (the empty set), control factor i is not controlled by control indicator j. On the other hand, if it includes all the control elements of the control indicator, namely $l_{ij} = K_j$, the control factor i is absolutely controlled by the control indicator j. Moreover, if all the control indicators are absolutely controlled, namely, $l_{iT} = K_T, l_{iH} = K_H, l_{iA} = K_A$ and $l_{iU} = K_U$, the Z indicator of control factor i is "AC", which stands for absolutely controlled.

$$K_T = \{t_1, ...t_m\}$$
$$K_H = \{h_1, ...h_p\}$$
$$K_A = \{a_1, ...a_q\}$$
$$K_U = \{u_1, ...u_r\}$$
$$L = \{l_{ij} | i = 1, 2, ..., I, j \in \{T, H, A, U\}\}$$
$$where \ l_{iT} \subseteq K_T, l_{iH} \subseteq K_H, l_{iA} \subseteq K_A, l_{iU} \subseteq K_U \qquad (14.1)$$

The control factor with all its control indicators corresponds to one NF. For example, Z denotes whether urban developments are allowed. If $Z = $ "AC", then urban developments are absolutely forbidden, and the NF should be a purely natural landscape. H and T play roles controlling the aboveground urban form. A controls urban activities within the urban form and is related to the social space rather than the physical space. U controls the underground development of the urban form. Equation (14.2) expresses the relationship between these indicators and NF, where F is all the possible urban growth patterns within the whole study area, UF is the urban form, and NF is the negative form. We divided all the possible urban growth patterns from aspects of T, H, A, and U within the whole study area into two parts, the negative form (NF) and the urban form (UF), based on T, H, A, and U.

$$F = \{UF \cup NF\}$$
$$UF = \{x | x \in F, x \notin NF\} \qquad (14.2)$$
$$NF = \{(T, H, A, U)\}$$

To address the issue of multiple overlapping control factors, we propose using the control cell to stand for the combination of control factors in the same geographical space. Since each control factor corresponds to one NF, every control cell also stands for one NF, which integrates the NFs of the control factors, and is the

14 An Urban Growth Control Planning Support System

basic unit of the UGCP analysis. The calculation method for the control cell using the control factors will be investigated in detail in the next section.

Calculating Planning Rules for Urban Growth Control

In view of the complexity of factor composition and the spatial distribution of control cells (being the basic UGCP analysis unit), a planning support system based on GIS has been developed to assist in compiling the UGCP, in which the NF of every control cell is calculated according to the UGCP scheme.

In this chapter, the term "planning scheme" is defined as the control cells, together with their planning rules, as generated from the control factors and their indicators. Equation (14.3) shows how to calculate the planning scheme using different control factors' indicators. In (14.3), UAZ stands for the generated control cell dataset using UNION, x for its ID, uaz^x for a control cell, $R(uaz^x)$ for the planning rule for control cell uaz^x, uaz_j^x for the planning rule j for control cell uaz^x, s_x for the control factor set contained in uaz^x, $\bigcap_{i \in s_x}^{spt}$ for the overlapping space of control factors s_x (the spatial intersection set), and f^j for the function to calculate the planning rule j.

$$UAZ = \{uaz^x | x = 1, 2, 3, ..., X\}$$

$$R(uaz^x) = (uaz_T^x, uaz_H^x, uaz_A^x, uaz_U^x) = l^{s_x} = \bigcap_{i \in s_x}^{spt} l_i \qquad (14.3)$$

$$uaz_j^x = \underset{i \in s_x, j \in \{T,H,A,U\}}{f^j} (l_{ij})$$

The planning rules can be calculated from the control factors contained in the control cell. However, methods of calculating the (f^j) of the planning rules may vary.

The System Framework of UGC-PSS

Urban Planners and UGC-PSS

UGC-PSS is important in the current planning practice requirements in Beijing. Given the spatial extent of particular developing sites, the system will be helpful in developing many urban and rural spatial plans to be implemented in Beijing, such as the respective coordinated planning of the eastern and western parts of Beijing, zoning of the central city, and master plans for nine new cities and 12 towns and the Badaling-Ming Tombs scenic area plan. Most of these spatial plans are at the master plan level.

For spatial planners with an urban planning background who are not very familiar with the regulations or standards related to urban growth control, UCG-PSS should be a comprehensive tool covering all the other fields, such as forests, environment protection, water resources, and earthquake prevention, adhering to all the *relevant* urban growth control regulations or standards, such as land-use control and building height control, thus making it possible for urban planners with better understanding to accept their urban growth control ideas and formulate better spatial plans meeting urban growth control requirements.

To present the results of the comprehensive urban growth control system in the negative form for spatial planners, all possible control factors for controlling the urban forms of land-use, building height, urban activities, and underground development in the spatial plan should be listed in the UGC-PSS.

To present the spatial distribution of the control cells and the planning rules for each cell, it is necessary to present the GIS dataset in terms of control cells and to print it out at different map scales. Since the combination of control factors and the massive number of control cells generated make it difficult to express the planning scheme in a single page, we partition the UAZ layer of the whole study area into multiple partitions according to the requirements of the result display. For instance, partitions can be generated by dividing the whole study area into several rows and columns, or according to standard Chinese map scales, such as 1:25,000, 1:10,000, 1:5,000, 1:2,000, and 1:500, for various purposes. Generally, to support the master plan, the UGCP scheme should be at the scale 1:25,000, the detailed plan at 1:2,000, and the architectural construction projects at 1:500. Thus, in checking development conditions, planners and developers can understand the objective conditions for urban growth control by reading the hardcopy of the scheme or querying the UGC-PSS.

This indicates that the main challenges in UGC-PSS implementation are the process for retrieving the spatial patterns of the urban growth control using UAZ and presenting the results for urban planners.

Development of UGC-PSS for Urban Planning and Design

The UGC-PSS system that we developed consists of six modules: the system configuration, spatial dataset display, comprehensive query, control cell generation, automatic planning scheme calculation, and automatic planning scheme output. The last three modules make up the core of the UGC-PSS.

14 An Urban Growth Control Planning Support System

The UGC-PSS operating procedure is as follows:

1. Users log in to the system through the common user interface, which has access control.
2. Users set system configurations, including defining the workspace for the input and output datasets, selecting the control factors to be involved in the planning scheme calculation, setting the scheme partitioning, and modifying the users' privilege levels, using the "System configuration module".
3. Users load the corresponding required datasets, including control factors and other base layers, and set the display parameters using the "Spatial dataset display module". The spatial distribution of control factors can be browsed by users along with basic feature classes, such as road networks, rivers, and residential areas, as spatial location references.
4. Using the "Control cell generation module", the control cell feature classes (planning rules are not calculated yet) can then be generated based on selected control factors and their indicators using the spatial analysis settings. The control factors are stored in the Geodatabase as feature classes, and the indicators for all the control factors are stored as an info table in the same Geodatabase.
5. The "Control cell generation module" can be employed to calculate planning rules for those control cells already generated by the "Control cell generation module". The results of the calculation can be regarded as the planning scheme, which can be accessed in two ways.

   - Users can conduct comprehensive queries of the planning scheme from aspects of the spatial distribution of control cells and their planning rules, and can analyze the construction conditions objectively within a given yet-to-be-developed area using the "Comprehensive query module". The module is developed especially for spatial planners to use when they query the urban growth control conditions for the designated planning area.
   - The planning scheme can be exported automatically using the "Planning scheme automatic output module" in the form of many digital figures, which can also be printed as one collection of figures for decision makers' creating the UGCP.

The fundamental structure and workflow of the UGC-PSS are as shown in Fig. 14.2.

Based on the diagram above, a UGC-PSS was developed using the ESRI ArcGIS Engine 9.0 by Visual Basic 6.0, while the automatic planning scheme calculation module was developed in Visual C++ for computational efficiency. We used the ESRI Personal Geodatabase based on Microsoft Access. As shown in Fig. 14.3, the main interface of the system is composed of seven parts: the title bar, menu bar, toolbar, map window, layer control window, thumbnail window, and query results output panel. It is possible to perform data browsing, queries, and other related functions through the main graphic user interface, while further system functions can be accessed through the menu bar.

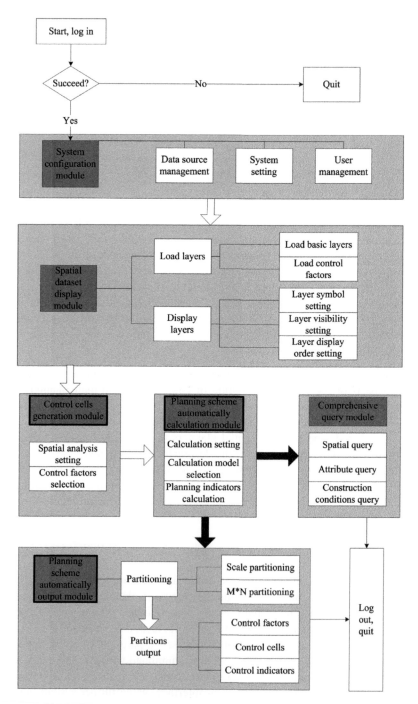

Fig. 14.2 Framework and flow diagram of the UGC-PSS (The modules with the thick frames are the three key modules of the UGC-PSS)

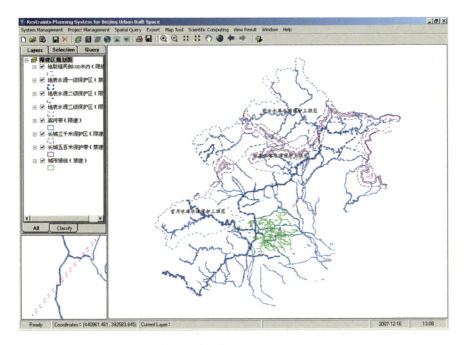

Fig. 14.3 Main UGC-PSS graphic user interface

UGC-PSS in the Beijing Metropolitan Area

Study Area: The Beijing Metropolitan Area

During the process of rapid development in Beijing, the tendency to disorderly urban expansion has not yet been effectively controlled. Within the Beijing Metropolitan Area, urban growth is restricted by many control factors. Urban growth control planning in Beijing has been conducted to solve the problems caused by urban growth in the mega-metropolis, and UGC-PSS played an essential role in planning support. The spatial extent of the planning region is the Beijing Metropolitan Area (see Fig. 14.4) with an area of 16,410 km^2. It lies in northern China, to the east of the Shanxi altiplano and south of the Inner Mongolia altiplano. The southeastern part of the study area is a plain about 150 km from the Bohai Sea. Mountains cover 10,072 km^2, 61% of the area.

Integrating All Control Factors and Indicators Using UGC-PSS

The basic datasets used in UGC-PSS include fundamental terrain datasets, administrative division boundaries, and control factors. The fundamental terrain datasets

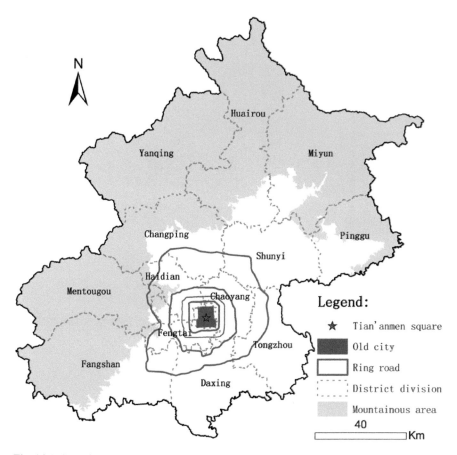

Fig. 14.4 Spatial structure of the Beijing Metropolitan Area

are topographic maps of different scales, aerial maps with 1 m resolution, digital elevation models (DEM), existing roads, existing wetlands, current land-use maps, and the urban master-planning scheme. There are administrative division boundaries at the district, town, and ward levels. The first two datasets, which are not involved in the calculations of the UGCP scheme, mainly aid in expressing the planning scheme spatially in terms of the base map and assist with spatial queries.

Control factors are among the most important datasets in the UGC-PSS. Within the study area, there are 60 control factors controlling urban growth, including water conservation, biodiversity conservation, agricultural land protection, flood control, geological disaster prevention, steep slope areas, and pollution source protection. Four of Beijing's control factors are displayed in Fig. 14.5. Within the spatial distribution of each control factor, control guidance remains the same, or must be subdivided into new control factors.

In the Beijing Metropolitan Area, the elements of each control indicator are, respectively, $K_T = \{\text{``R''},\text{``C''},\text{``M''}\}$, $K_H = [0, 1000)$, $K_A = \{\text{``a''},\text{``b''},\text{``c''},\text{``d''}\}$, and

14 An Urban Growth Control Planning Support System 299

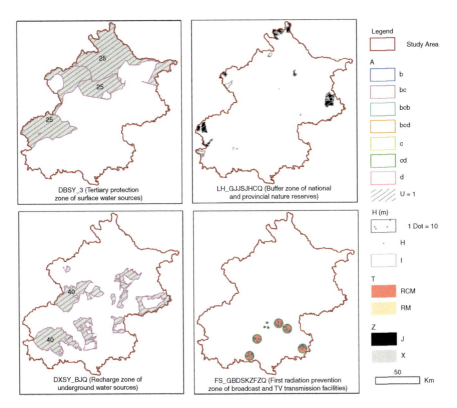

Fig. 14.5 Negative urban forms for four control factors (DXSY_3, LH_GJJSJHCQ, DXSY_BJQ, and FS_GBDSKZFZQ denote the control factors, and a, b, c, and d are control elements of the indicator A. R, C, and M are elements of the control indicator T. In the control indicator H, one dot stands for a maximum building height of 10 m)

$K_U = \{1\}$. The names and control indicators of all control factors are listed in Table 14.2, and the spatial distribution of every control factor is available from the Beijing Municipal Planning Commission (2007). For instance, one control factor in Table 14.2 stands for the secondary protection zone of surface water sources, which belongs to the relative control level, and urban activity with pollutant emissions is forbidden. Within the spatial extent of it, moreover, residential, commercial, and industrial land uses are not permitted. Another control factor denotes the first protection zone of surface water resources, where all urban activities and developments are absolutely forbidden.

All the planning rules are calculated using the same function, namely the union of sets, in this model. That is to say, the value of the planning rule equals the union of the corresponding control indicators of the factors contained in the control cell. In terms of planning practice, this means that the control cell obeys all the control rules of the factors within the control cell.

Table 14.2 Inventory of control factors and corresponding control indicators

Description	Z	H	U	A	T
Secondary protection zone of surface water sources	RC			b	R C M
Tertiary protection zone of surface water sources	RC			b c	
Area of over-exploited underground water	RC			c	R M
Recharge zone for underground water sources	RC		1	b c	
Protected zone of underground water sources	RC		1	b c	
Area with poor engineering geology	RC	>45			R C M
100 m to 500 m buffer zone around ground fissures	RC				RM
Secondary radiation prevention zone for broadcast and TV transmission facilities	RC			c d	
Primary radiation prevention zone for broadcast and TV transmission facilities	RC	>120		c d	R C M
Zone forbidding night soil treatment plants	RC				R C M
Zone forbidding waste compost plants	RC				R C M
Zone forbidding waste incineration plants	RC				R C M
High-risk flooding area	RC			d	
Existing forests in the second greenbelt	RC				R C
First greenbelt	RC	>18			
Secondary scenic spot protection area	RC	>9		b c d	
Tertiary scenic spot protection area	RC	>18		b c d	R M
Planned scenic spots	RC	>27		b c d	
Experimental zone of national and provincial nature reserves	RC			b c	
Forest parks	RC	>21		b c	
County-level nature reserves	RC			b c	R M
Other existing scenic spots	RC	>21		b c d	R CM
General ecological forests	RC			c d	RM
General cultivated lands	RC			c	RC
General farmlands	RC	>9	1	b c d	
River buffer zone	RC			a b c d	R C M
Reservoir buffer zone	RC			a b c d	
Piedmont ecological protection zone	RC			b c	R
Protection zone for south-to-north water diversion infrastructure	RC			d	
Secondary protection zone for oil/gas pipelines	RC			d	
3 km buffer zone for the Great Wall	RC	>9			
Protection area for underground cultural relics	RC	>15			
Historical culture protection area	RC	>15			
Secondary noise prevention zone around city roads	RC				R C
Secondary noise prevention zone around light railways	RC				R C
Tertiary noise prevention zone around Beijing Capital International Airport	RC				R M
Primary protection zone for surface water sources	AC				
Radiation prevention zone near 110 KV power stations	AC				
Radiation prevention zone near 220 KV power stations	AC				
Radiation prevention zone near 550 KV power stations	AC				
Flood diversion gates	AC				

(continued)

Table 14.2 (continued)

Description	Z	H	U	A	T
Flood detention area in central city	AC				
Green wedges	AC				
Planned green lands within the first greenbelt	AC				
Primary scenic spot protection area	AC				
Special grade protection area for scenic spots	AC				
Buffer zone around national and provincial nature reserves	AC				
Core zone for national and provincial nature reserves	AC				
Green spaces in central city	AC				
Key ecological forests	AC				
Basic farmlands	AC				
River-type wetlands	AC				
Pond and reservoir-type wetlands	AC				
Steep areas with slopes greater than 25°	AC				
Prevention zone around 110 KV transmission lines	AC				
Radiation prevention zone around 220 KV power transmission lines	AC				
Radiation prevention zone around 500 KV power transmission lines	AC				
Primary protection zone around oil/gas pipelines	AC				
500 m buffer zone around the Great Wall	AC				
Protected historical units	AC				

Presenting Planning Scheme Using UGC-PSS

In the Beijing urban growth control planning, UGC-PSS has played a supporting role in plan compilation from the beginning. Most importantly, UGC-PSS can perform the functions of calculating and outputting the planning scheme, tedious tasks in routine planning practice. Control cells are created based on control factors (Fig. 14.6), and 60 control factors generate 356,689 control cells using the UNION function. Based on the generated control cells, the UGCP scheme for Beijing was then calculated, which took 9.5 h.

The resulting planning scheme can be expressed and exported as digital pictures or paper drawings for decision makers, planners, or the public. The spatial distribution, calculated control cell attribution table, and base map, were utilized to express the planning scheme. Due to the complexity of urban growth conditions in the study area, we partitioned the whole study area into 213 partitions using a scale of 1:25,000. The basic partitioning method was as follows. First, a polygon dataset was created based on partitioning parameters to record the distribution of various partitions. Each polygon represented the spatial boundaries of a partition. Second, in accord with the boundaries, map control limits were determined by the boundaries of spatial datasets within the partition, and certain control cells within the partition were selected using spatial analysis functions. The attribute table of selected control cells within the partition was output as the planning rules table.

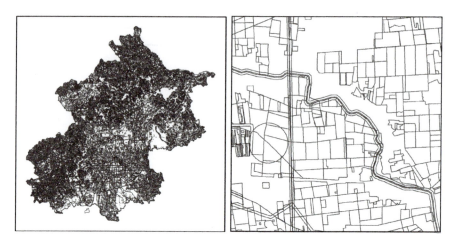

Fig. 14.6 Spatial distribution of control cells in the Beijing Metropolitan Area (*left*: the entire study area; *right*: the partial zoomed-in area)

Fig. 14.7 Spatial distribution of control factors of a partition (The main map is an enlarged view of the red polygon shown on the mesh in the upper right corner)

Third, auxiliary map elements were added to the map, such as the title, legend, compass, scale bar, and partitioning index map. Finally, the print configurations were set, so the partitions could be printed on paper or exported as JPEG digital files.

The planning maps exported by UGC-PSS are illustrated in Figs. 14.7 and 14.8. In Fig. 14.7, the control factors in the partition were rendered and displayed

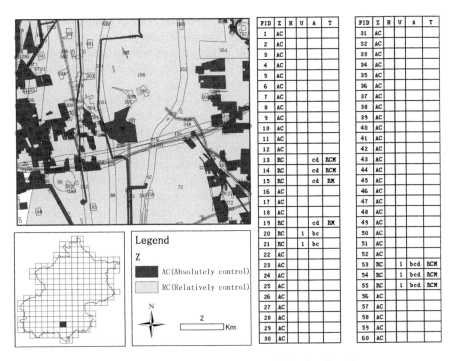

Fig. 14.8 Control cell distribution map and the first page of the planning rules

together with their legends. The control factor map can be used to gain an understanding of the basic urban growth control situation. In Fig. 14.8, the spatial distribution of control cells and the corresponding planning rules are shown, which can also be viewed as the NF information within the partition. If there are excess control cells that make it impossible to list all the planning rules on one page, UGC-PSS can export the planning rules information automatically in multiple pages. The application of UGC-PSS improved the efficiency with which the UGCP was compiled considerably, and reduced the time necessary to establish and express the planning scheme by over 90%.

Conclusions and Discussion

The UGC-PSS proposed in this chapter offers a platform for urban growth control analysis that can consider large numbers of control factors in a synthetic framework and use them to calculate and automatically output a planning scheme. For the Beijing Metropolitan Area, 60 control factors, together with their indicators, were included to reflect urban growth conditions in the study area. The PSS enables complex urban growth control problems to be solved in the mega-metropolitan area by using multi-disciplinary knowledge, such as flooding control, eco-zone protection, noise prevention, and disaster prevention. The UGC-PSS can integrate all these urban growth control-related factors in one platform, which has not previously been possible for a single urban planner with limited knowledge on various control factors. Therefore, the UGC-PSS can also be regarded as a knowledge base for urban growth control in planning practice.

Planners, decision makers, and citizens can benefit from the use of the UGC-PSS. First, for urban planners, the precision and efficiency of UGCP can be significantly improved by using the PSS to integrate all possible control factors and their indicators, which would be quite complicated using routine means. Furthermore, spatial planners can refer to the outcome of the UGCP stored in the UGC-PSS as a precondition for spatial plans. Second, decision makers in the municipal government or the planning department can query and check the objective construction constraints in their concerned areas using the UGC-PSS as a decision support system. Lastly, the UGC-PSS has the potential to let *citizens* participate in urban development by monitoring illegal development conflicting with the UGCP if the PSS is upgraded to be web-based. The application of the UGC-PSS in the Beijing Municipal Area has proven the benefits noted above.

In the UGC-PSS, the method of calculating the planning scheme using UAZs with several planning rules is not only suitable for research regarding urban growth control, but can also be applied to urban growth research on the condition that more factors that promote urban growth are also included, such as institutional factors and accessibility factors (a preliminary attempt has been conducted in Long et al. 2008). The prototype of UGC-PSS may also be applied to other geographic analysis and geocomputing fields where various spatial factors with multi-indicators are required.

Due to the limited availability of datasets, some control factors were not included in the empirical PSS implemented described herein, such as protected cultural & historical sites in the central city, environmental capacity, and resource carrying capacity. Therefore, the UGC-PSS needs to be improved in future to more objectively and comprehensively reflect the state of urban growth control in Beijing. Moreover, we also intend to upgrade the UGC-PSS to a web-based PSS to accommodate more stakeholders in public participation in the planning of urban growth control.

Acknowledgements We would like to thank the National Natural Science Foundation of China (No. 51078213) for the partial financial support.

References

Anas, A., & Rhee, H. (2006). When are urban growth boundaries not second-best policies to congestion tolls. *Journal of Urban Economics,* **61,** 263–286.
Beijing Municipal Planning Commission. (2007). *Ecologically Limited Land-use Planning in Beijing (2006–2020)* Unpublished report (in Chinese).

Brail, R. K. (2008) *Planning Support Systems for Cities and Regions* Lincoln Institute of Land Policy (Ed.).

Brail, R. K., & Klosterman, R. E. (2001). *Planning support systems: Integrating geographic information systems, models and visualization tools.* California: ESRI Press.

Burchell, R. (1998). *Cost of Sprawl Revisited.* New York: Rutgers.

Carver, S. J. (1991). Integrating multi-criteria evaluation with geographical information systems International. *Journal of Geographical Information Systems, 5,* 321–339.

Diamond, J. T., &Wright, J. R. (1988). Design of an integrated spatial information system for multiobjective land-use planning. *Environment and Planning B: Planning and Design, 15,* 205–214.

Echenigue, M. (2005). Forecasting the sustainability of alterative plans In Mike Jenks and Nicola Dempsey. *Future Forms and Design for Sustainable Cities.* London and New York: Elsevier.

Edamura, T., & Tsuchida, T. (1999). Planning support system for an urban environment improvement project. *Environment and Planning B-Planning & Design, 26,* 381–391.

Geertman, S., & Stillwell, J. (2002). *Planning Support Systems in Practice* (Eds.), Advances in Spatial Science.

Geertman, S. Stillwell, J. (2004). Planning support systems: An inventory of current practice. *Computers Environment and Urban Systems, 28,* 291–310.

Geertman, S. & Stillwell, J. (2009) *Planning Support Systems Best Practice and New Methods* (Eds.). Berlin: Springer.

Geneletti, D. (2008). Incorporating biodiversity assets in spatial planning: Methodological proposal and development of a planning support system. *Landscape and Urban Planning, 84,* 252–265.

Harris, B. (1960). Plan or projection: An examination of the use of models in planning. *Journal of the American Planning Association, 26,* 265–272.

He, Y. (2008). Planning in restrains of Beijing urban none built-up space. *Proceedings of 44th ISOCARP Congress,* Dalian, China, 19–23 Sept.

Kammeier, H. D. (1999). New tools for spatial analysis and planning as components of an incremental planning-support system. *Environment and Planning B-Planning & Design, 26,* 365–380.

Klosterman, R. E., & Richard, E. (1997). Planning support systems: A new perspective on computer-aided planning. *Journal of Planning Education and Research, 17,* 45–54.

Klosterman, R. E. (1999). The What if? Collaborative planning support system. *Environment and Planning B: Planning and Design, 26,* 393–408.

Landis, J. D. (1994). The California Urban Future model: A new generation of metropolitan simulation models. *Environment and Planning B: Planning and Design, 21,* 399–420.

Landis, J. D. (1995). Imaging land use futures: Applying the California Urban Future model. *Journal of American Planning Association, 61,* 438–457.

Long, Y., He, Y., Liu, X., & Du, L. (2006). Planning of the controlled-construction area in Beijing: Establishing urban expansion boundary. *City planning review, 30,* 20–26 (in Chinese).

Long, Y. (2007). *Planning support system: Theory and practices.* Beijing: Chemical engineering press, (in Chinese)

Long, Y., Shen, Z., Du, L., Mao, Q., & Gao, Z. (2008). BUDEM: An urban growth simulation model using CA for Beijing metropolitan area. *Proc. SPIE* Vol. **7143** 71431D

McHarg, I. L. (1969). *Design with Nature.* New York: Wiley.

Nelson, A. C., & Dawkins, C. J. (1999). *Unpublished survey of US Metropolitan Planning Organizations.*

Nelson, A. C., & Dawkins, C. J. (2004). *Urban containment in the United States: History, models and techniques for regional and metropolitan growth management Chicago.* IL, American Planning Association.

Nelson, A. C., & Duncan, J. B. (1995). *Growth management principles and practices Chicago.* Washington D.C.: Planners Press. IL, American Planning Association.

Pendall, R. J., Martin, J., & Fulton, W. (2002). *Holding the Line: Urban Containment in the United States Washington, D C.* (Brookings Institution Center on Urban and Metropolitan Policy).

Piracha, A. L., & Kammeier, H. D. (2002) Planning-support systems using an innovative blend of computer tools: An approach for use in decisions about industrial locations in Punjab, Pakistan. *International Development Planning Review, 24,* 203–221.

Stillwell, J. (2002). *Planning Support Systems in Practice.* (Ed.) *(Advances in Spatial Science).* Berlin: Springer.

Sui, D. Z. (1993). Integrating neural networks with GIS for spatial decision making. *Operational Geographer, 11,* 13–20.

Thill, J. C. (1999). Multicriteria Decision-making and Analysis: A Geographic Information Sciences Approach. New York: Ash gate.

Chapter 15
A Planning Support System for Retrieving Planning Alternatives of Historical Conservation Areas from Spatial Data Using GIS

Zhenjiang Shen, Mistuhiko Kawakami, Fangfang Lu, Lanchun Bian, Ying Long, Lin Gao, and Dingyou Zhou

Introduction

Urban conservation has drastically changed in theory, in practice, and even its basic definition, over the past 50 years. The definition of urban conservation is not limited to the scope of human society, which has continued to evolve since the 1931 Athens Charter. Historical towns and urban areas, as well as their natural landscapes and residential environments, are also included as subjects of urban conservation. In China, urban conservation is currently of interest to urban designers and city administrators, who are concerned with providing a historical urban identity as well as an authentic urban identity. The historical conservation areas in Beijing have become the most congested areas due to the lack of effective planning control and management, and these areas have gradually became warrens, which have been settled by numerous migrant laborers.

Local researchers and planners in the field of urban policy and urban sociology have expressed interest in conserving historical urban identity. The complexity of this field stems from both the theoretical and practical development of urban planning and management. In order to advance conservation planning in Beijing, a new method is necessary in order for discussions to take place between these different disciplines. In addition, a new ordinance issued by the Beijing Municipal

---

Z. Shen (✉) • M. Kawakami • F. Lu
School of Environmental Design, Kanazawa University, Kanazawa, Japan
e-mail: shenzhe@t.kanazawa-u.ac.jp

L. Bian • Y. Long
School of Architecture, Tsinghua University, Beijing, China

L. Gao
Architecture Design Institute in Beijing, Academy of Science of China, Beijing, China

D. Zhou
School of Architecture, Dongnan University, Nanjing, China

Z. Shen, *Geospatial Techniques in Urban Planning*, Advances in Geographic
Information Science, DOI 10.1007/978-3-642-13559-0_15,
© Springer-Verlag Berlin Heidelberg 2012

Government in 2003 described a new planning framework in historical conservation areas with restoration planning through public participation. In this chapter, the relevant concepts reflecting the contribution of the present research, which are comprehensive planning approaches including new supplementary ordinances of the bylaw system, a solution for complicated relationships regarding real estate titles, conservation standards, and restoration planning in urban planning and design, are discussed.

This chapter focused on policy reorientation of the restoration plan in the historical conservation areas in Beijing, China, and explores how to effectively support participatory planning from a viewpoint of promoting the implementation of conservation planning. The geographic information system (GIS), the decision-making model, and the visualization tool are recognized as the three core modules used for planning support, because historical conservation planning includes the spatial analysis of related datasets, retrieving planning measures based on planning policies and guidelines, and the visualization of various planning alternatives. Accordingly, this research will benefit from the application of the planning support system (PSS).

In recent years, the PSS has attracted extensive attention as the latest form of computer-aided planning system (Klosterman 1997; Brail and Klosterman 2001; Geertman and Stillwell 2004). The concept of the PSS was initially proposed by Harris (1960). Several books for the PSS have been edited in recent years (Brail and Klosterman 2001; Geertman and Stillwell 2003; Brail 2008; Geertman and Stillwell 2009). The PPS has been mainly applied in spatial plans (Kammeier 1999; Geneletti 2008), urban environment improvement plans (Edamura and Tsuchida 1999), industrial location choices (Kammeier 1999), and land use plans (Klosterman 1999). In the integrated conservation and development programs, participatory approaches have been used to engage local people in protected area management and conservation actions. Wang et al. (2007) proposed a decision support model for estimating the historical buildings conservation budget. In addition, Shen and Kawakami (2010) developed an online visualization tool for Internet-based local townscape design in Japan, and citizens can access this tool for planning participation. However, there are few PSSs for historical conservation planning from a comprehensive viewpoint of conservation planning policy and related design guideline.

Here, the concept of a multidisciplinary approach for finding an appropriate solution of enforceable restoration planning is proposed in this chapter in the form of the PSS, which includes several tools. In order to investigate how to carry out restoration planning from the viewpoint of promoting public participation that has been stipulated in a new ordinance of the Beijing Municipal Government, the conservation planning policies and guidelines for plan implementation can be transcribed as scenarios consistent with the existing urban bylaw system in Beijing. For this task, GIS is used to setup a database of historical conservation areas with attribute information, including statistical survey data and data on architectural style, preservation state, real estate information, and planning control indicators. Scenarios of planning implementation are configured according to urban bylaws so that conservation plans have legal basis and become practicable. Moreover, the survey data of each courtyard house restored in the GIS database and the conservation standards established by urban planners should be employed to run a reasoning process using GIS customized program in order that the planning measures for each courtyard house can be stored as attributes of GIS features in a database. Furthermore, VR data are created based on the CAD and GIS database for arguments as to how to gain consensus in the restoration planning and design of courtyard houses between stakeholders.

Relative Planning Regulations of Urban Conservation and Property Ownership in Beijing

As metropolises, some capital cites establish huge social political engines and have the power to enact complex bylaw systems based on the national legal system. On the other hand, various studies have investigated urban conservation systems of historical urban districts and traditional houses throughout the world. In the context of these bylaw systems, the importance of the conservation of physical characteristics is emphasized, along with cultural, social, and economic aspects (Townshend and Pendlebury 1999; Ipekoglu 2006; Ahmed and Salah 2001; Jim 2005). Ericson (2006) evaluated the strengths and weaknesses of the participatory approach used in an applied research program conducted in three Ejido communities in the Calakmul Biosphere Reserve on the Yucatan, Peninsula of Mexico. Table 15.1 shows the local bylaws of Beijing regarding urban conservation that can be applied to cultural properties and historical conservation areas. Control regulations, construction supervision, and building management are all addressed in the bylaws in Beijing.

However, starting from 1949, preservation planning was not implemented smoothly due to confusion with respect to real estate management. From 1949 to 1980, private properties were gradually made into public properties as a result of procurement under the publicly owned housing policy under the Chinese socialism regime. All land parcels in urban areas were nationalized, and the titles of buildings were signed over to the government. Since the 1980s, policy reforms in China have stated that government-owned buildings should be returned to their original proprietors. Thus, confusion regarding the original proprietors has emerged because of missing information and documents of previous title.

On the other hand, local administrative provisions regarding conservation and renovation affairs of historical conservation areas (enacted, November 18, 2003) illustrate Beijing's policy reorientation of the restoration plan in historical conservation areas and consider how to perform restoration planning while promoting public participation. Thus, local government should cultivate the sense of identity of stakeholders in conservation planning so that stakeholders can take the initiative in solving planning problems in the conservation process. However, confusion in

Table 15.1 Local bylaws in Beijing regarding historical conservation areas

Relative regulations	Types of local bylaws
Local administrative provision regarding the preservation of cultural properties in Beijing (enacted June 23, 1987, revised October 16, 1997)	Regulation of the Congress of People's Deputies in Beijing
Local administrative provision regarding preservation areas of cultural properties and planning controls in Beijing (enacted November 13, 1987)	Bylaw of the Beijing Municipal Government
Local administrative provision regarding the construction supervision of cultural properties (enacted November 13, 1987)	Bylaw of the Beijing Municipal Government
Strategies regarding the refinement and management of traditional building in cultural property areas in Beijing (enacted July 31, 1989)	Bylaw of the Beijing Municipal Government
Verdict regarding planning action of the 25 historical conservation areas in Beijing (enacted February 1, 2002)	Bylaw of the Beijing Municipal Government
Local administrative provision regarding conservation and refinement affairs of historical conservation areas (enacted, November 18, 2003)	Bylaw of the Beijing Municipal Government
Ordnance regarding the historic culture city of Beijing (preliminary regulation) (enacted, March 1, 2004)	Bylaw of the Beijing Municipal Government
Local administrative provision regarding the purchase of courtyard houses by companies and legal persons in historical conservation areas of Beijing (April 1, 2004)	Bylaw of the Beijing Municipal Government

real estate management caused by the transformation of property ownership in urban areas influenced the primary role of residents who are confused regarding their property building rights.

In addition, there are too many renters who are immigrant laborers from regions outside the Beijing metropolitan area. Their existence has resulted in the overpopulation of historical cultural areas. As such, their relocation to other residential areas is indispensability for the realization of urban conservation in the inner city in Beijing. Therefore, the enablement requirements for conservation plan implementation are ownership-based conservation planning and decrease in population through the movement of immigrant families.

Conservation Plan Based on the New Ordinance of Conservation Standards of Beijing

Due to government-owned buildings being returned to their original proprietors, the conservation plan should be changed to an ownership-based plan. Residents of traditional buildings follow the instructions of conservators in the conservation process. Thus, the conservation efforts of residents require financial assistance from the government. In other words, a subsidy system should be established in Beijing.

15 A Planning Support System for Retrieving Planning Alternatives   311

Nevertheless, if the population of immigrant families in traditional houses increases and the living conditions in the conservation areas are not sufficient, then residents will not be able to implement ambitious plans to continue living in historical conservation areas. The present situation makes it difficult for residents to make a living in the conservation areas of Beijing relative to in other residential areas of Beijing. Thus, the implementation of a conservation plan will require a great deal of effort, even though the subsidy system that supports residents in conserving traditional houses can be introduced in historical conservation areas.

As a practical matter, if the Beijing authorities carry out historical conservation as described in the new ordinance while taking public involvement into account, the residents of conservation areas will be responsible for implementing the conservation plan. The scenario for ownership-based conservation should take planning implementation into account. In this scenario, conservation standards designed by urban planners and designers are used as preconditions in the decision process.

Moreover, the relocation of residents in historical areas is also a precondition of implementation of the conservation plan. Therefore, the process illustrated in Fig. 15.1 shows who should first be relocated from the historical areas. Next, the process is to determine planning alternatives for traditional houses and who should be the reasonable party for the restoration of traditional buildings and the landscaping of conservation areas. Herein, the term "landscaping" means that, in the

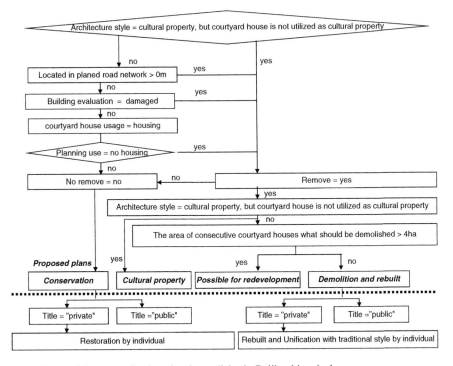

Fig. 15.1 Decision tree reflecting planning policies in Beijing historical areas

conservation areas, existing traditional style buildings should be restored and new buildings should be constructed in accordance with the traditional style.

First, the bifurcation at the beginning of Fig. 15.1 is related to whether the house is a culturally valuable property and is used as a cultural property. If this is not the case, and if the house is not situated along a planned road, the building is in good condition, the courtyard house is used as a private residence, and the planned use is also as a private residence, then the residents can continue living in the traditional buildings because there is no need to demolish their houses. In such cases, if the courtyard house is privately owned, then the courtyard will be supported through the subsidy system. However, if the courtyard house is publicly owned, then the government is responsible for the restoration or reconstruction of the courtyard house.

Second, if the residents should vacate the traditional courtyard houses, there are several cases to be considered. If the courtyard house is designated as a cultural property, then ownership should be transferred to the public, even though the house is privately owned. If the courtyard is not a highly valued cultural property, and if the house is situated along a planned road, if the house is not in good condition and is currently being used as a private residence, whereas the planned use is not as a private residence, then the residents cannot continue to live in the house because the house must be demolished. In this case, the courtyard house should be rebuilt and will not be an original traditional building continuously. If the total area of adjacent courtyard houses in neighbor to be dismantling is larger than 4 ha, then the site will be defined as a redevelopment area because of a related regulation in the local bylaws of Beijing (Enacted, 18.11.2003, as shown in Table 15.1). Otherwise, the buildings will be rebuilt in the conservation areas. If the buildings are privately owned, then the owners will be asked to maintain the traditional architectural style upon reconstruction.

Planning Support Using GIS to Retrieve Planning Alternatives

For this chapter, the study area was chosen from among the 25 historical conservation areas in Beijing. Figure 15.2 shows the location of the study area, which is a typical area including government and municipality-designated cultural properties, selected traditional buildings, and new buildings that should be renewed and built in the traditional style. This area also faces planning issues such as complicated real estate ownership and high population density, as mentioned above. Table 15.2 briefly summarizes this area.

In this area, approximately one third of the courtyard houses have been returned to the original owners. Figure 15.3 shows the ownership distribution of the historical cultural area. The actual planning is implemented by the house inhabitants of the courtyard houses, who may be individuals, the government, or companies. Figure 15.4 shows recent photographs of the buildings in the study area.

15 A Planning Support System for Retrieving Planning Alternatives 313

Fig. 15.2 Location of the case study area

Table 15.2 Overview of the study area

Historical conservation project area	70.38 ha
Length from east to west	1,400 m
Total floor area of all buildings	499,480 m^2
Number of traditional courtyard houses	1,120
Residents (households)	24,030 (9970)

As described above, Fig. 15.1 presents a decision tree based on local policies regarding real estate management, urban planning and design, and conservation standards designated in the local bylaw system, which is a comprehensive planning approach. The GIS is used to facilitate the decision process described in this decision tree, which is shown as Fig. 15.5. In Beijing, survey data and planning drawing were edited in Excel format and CAD format, which are converted into a GIS database as dbf files and shape files. A rule-based tool is developed as an if-then customized planning support system using VBA of ARCOBJECT. In this if-then decision tool, the planning alternatives, namely, the planning measures of each courtyard house, can be deduced based on spatial features and their GIS data attributes.

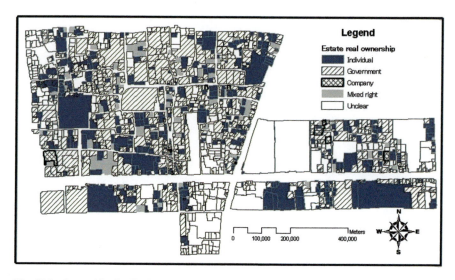

Fig. 15.3 Ownership distribution map of the study area

Fig. 15.4 Extension or reconstruction of buildings occupied by immigrant families

15 A Planning Support System for Retrieving Planning Alternatives 315

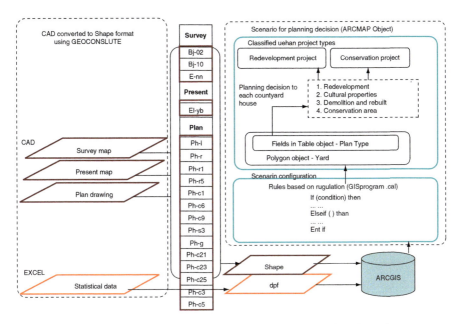

Fig. 15.5 System framework of the proposed tool for generating planning alternatives

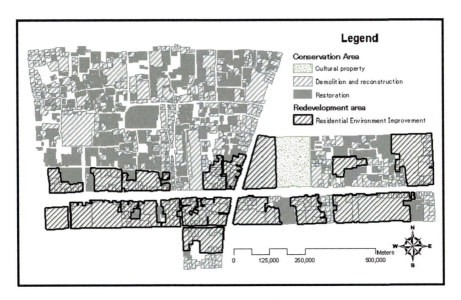

Fig. 15.6 Results of planning measures

Figure 15.6 shows the results of planning alternatives of courtyard houses generated by the proposed tool, which includes the four alternatives: residential environment improvement in redevelopment areas, cultural property, demolition and reconstruction and restoration in conservation areas. The local government in

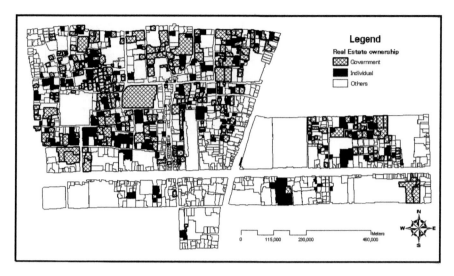

Fig. 15.7 Distribution of courtyard houses identified as restoration targets

China supports cultural property preservation through public finance. Moreover, the government-owned courtyard houses should be restored or rebuilt by the government, which will not be included among the targets of a possible subsidy system. The privately owned courtyard houses in conservation areas require financial support from the local government for restoration or reconstruction. All of the buildings in redevelopment areas will be demolished and rebuilt in traditional architectural style, but no financial support will be provided through public finance.

Figure 15.7 shows the courtyard houses identified as restoration targets retrieved from the entire area. In the figure, the black polygons are privately owned courtyard houses, and the hatched pattern polygons are government-owned courtyard houses. The other courtyard houses are landscaping targets that should be rebuilt in the traditional architectural style.

Alternatives of Housing Environment Improvement for Restoration Targets Using VR

For openly providing planning information on the conservation area over the Internet, we develop a visualization tool with which to visualize design alternatives using VRML based on planning standards issued by local planning authorities for the public to determine possible design alternatives for a courtyard house. In order to obtain a consensus with respect to housing environment improvement, the number of houses that should be removed and planning budgets should be considered among stakeholders. However, all possible alternatives can be visualized through different building forms. Thus, we suggest a design tool that can simulate

15 A Planning Support System for Retrieving Planning Alternatives

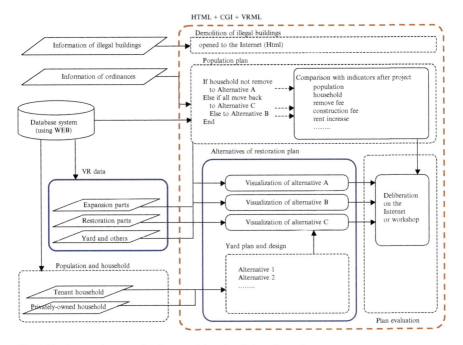

Fig. 15.8 System framework of our tool for visualizing alternatives

these planning alternatives and their possible building forms using VRML for discussion and evaluation, which can also be provided openly to the public over the Internet. Users include residents, experts, and administration officers.

In practice, after the restoration targets are determined using the GIS tool, the planning alternatives for each restored courtyard house should be considered on a case-by-case basis. For the possible future public participation, we examine how to use the proposed visualization tool to represent the alternatives.

Figure 15.8 shows the system framework of this visualization tool, which is developed as a web application for stakeholders to access using the Internet. This system is expected to be helpful for stakeholders to obtain a better understanding of the objectives of the housing environment improvement in the conservation area according to a planning implementation process in "Verdict regarding planning action of the 25 historical conservation areas in Beijing (Enacted 1.2.2002)", which includes the following steps:

1. Demolition of illegal buildings,
2. Population planning,
3. Alternative restoration plans, and
4. Plan evaluation.

As shown in Fig. 15.5, the VR data set for representation alternatives is prepared in a web database and can be selected by users to represent design alternatives with planning information as well. In the first planning process, namely, the demolition of illegal buildings, the illegal buildings that should be demolished are made known to the public. For population planning, the system is developed as a tool for presenting alternative and relative indicators based on the decisions of the stakeholders. The alternative restoration plans can be visualized using VRML to share a virtual world and related planning information between stakeholders. Finally, an evaluation of the plan is prepared in order to collect the opinions of stakeholders.

Figure 15.9 shows the interface of the population plan. The web contents in the left-hand frame show the planning process and are linked to the right-hand frames, which show the alternatives in a planning drawing and a visualized VR world.

A typical courtyard house has been prepared for the purpose of educating the stakeholders. First, buildings in the courtyard house without a construction permit should be demolished: the positions of these illegal buildings are indicated on the homepage along with information on the floor area and the number of residents. The residents of the illegal houses are renters. When the housing environment improvement project is carried out, the renters must decide whether to move or to pay increased rental fees.

As a result of overpopulation, population adjustment is necessary. Population planning must be conducted according to the housing standard of "residential floor area per person" enacted in the ordinance in historical preservation areas, which is at least $10\,m^2$ per person. In order to explain the objective of the housing environment improvement project, the visualization tool should coordinate different population planning needs with different alternative building forms. Finally, stakeholders should attempt to make decisions on planning alternatives through plan evaluation. Figures 15.9 and 15.10 show the three alternatives prepared using the system, and Table 15.3 shows some of the indicators for each alternative.

As with the described planning process, the fundamental component of the proposed tool is for representing the alternative plans using VRML to propose a population plan to address the high population density of the preservation area. The primary goal of the proposed tool is to allow stakeholders to understand the situation and the choices available to them.

Conclusions

The economic reforms of the past few decades have physically transformed China's cities. Private ownership of real estate has gradually increased after the decline of the socialist ownership policy that began in 1949, and the transformation of property ownership in the urban area has influenced the system of urban development. This chapter has focused on the policy reorientation of the restoration plan in the historical conservation areas in Beijing, China, which considers how to carry out restoration planning from the viewpoint of promoting public participation, which has been stipulated in a new ordinance by the Beijing municipal government in 2003.

15 A Planning Support System for Retrieving Planning Alternatives 319

Fig. 15.9 Representing the population plan using the visualization tool: (**a**) Alternative plan A. (**b**) Alternative plan B. (**c**) Alternative plan C

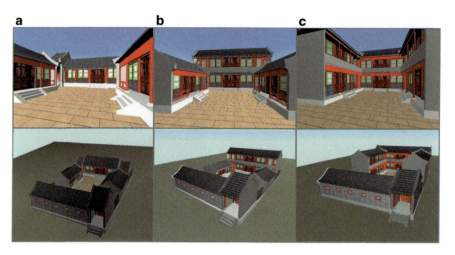

Fig. 15.10 Alternative plans (**a**) A-Restoration. (**b**) B-Expansion I. (**c**) C-Expansion II

Table 15.3 Alternatives plans

	Alternative A	Alternative B	Alternative C
Scenario	Restoration	Expansion I	Expansion II
Expansion of houses	None	One building	Three buildings
Leave the house intact			
Privately owned house	4	4	7
Tenant house	0	2	3
Remove house			
Privately owned house	3	3	0
Tenant house	0	1	0
Residential floor space (m^2)	337.3	475.5	922.1

For the restoration of historical conservation areas in accordance with the policy of public participation, a subsidy grant system for restoration that is supported by public expenditure is necessary. Therefore, a comprehensive approach is proposed in order to find a solution for a complex relation among real estate ownership, conservation standards, and restoration planning in the conservation area.

As a result of this research, the possibility of introduce a decision tool and a visualization tool to support the conservation plan is clarified. A decision tool using GIS is used to integrate urban policy with respect to real estate management and urban design and planning, through which the conservation plan can be changed to a practicable ownership-based plan.

In the proposed system, we used ARCGIS, VR to support the conservation plan in Beijing. We succeeded in data integration using GIS and data conversion software, experts in local urban planning can explore planning policies using the GIS tool based on if-then rules reflecting local planning ordinances. Residents can learn about restoration plan alternatives for courtyard houses through VR simulation.

15 A Planning Support System for Retrieving Planning Alternatives

Acknowledgements This project is supported by the Matsushita International Foundation (Project Reference No.01-524). We also would like to thank Z. Jing, F. Zheng and P. Ru for their collaboration and great contribution.

References

Ahmed M, Salah O (2001) Authenticity and the Sense of Place in Urban Design, *Journal of Urban Design*, 6(1), 73–86. [DOI: 10.1080/13574800120032914]

Brail, R. K (Ed.). (2008) Planning Support Systems for Cities and Regions. (Lincoln Institute of Land Policy, Cambridge, MA)

Brail, R. K., & Klosterman, R. E (Eds.) (2001) Planning support systems: Integrating geographic information systems, models and visualization tools. (ESRI Press, Redlands, CA)

Edamura, T., & Tsuchida, T. (1999) Planning support system for an urban environment improvement project. *Environment and Planning B-Planning & Design*, 26, 381–391.[DOI:10.1068/b260381]

Ericson J A (2006) A participatory approach to conservation in the Calakmul Biosphere Reserve, Campeche, Mexico, *Landscape and Urban Planning*, 74(3–4), 242–266. [DOI:10.1016/j.landurbplan.2004.09.006]

Geertman, S., & Stillwell, J. (2004) Planning support systems: an inventory of current practice. *Computers, Environment and Urban Systems*, 28, 291–310.[DOI:10.1016/S0198-9715(03)00024-3]

Geertman, S., & Stillwell, J (Eds.) (2003) Planning Support Systems in Practice, Advances in Spatial Science. (Springer, Berlin)

Geertman, S., & Stillwell, J (Eds.). (2009) Planning Support Systems Best Practice and New Methods. (Springer, Berlin)

Geneletti, D. (2008) Incorporating biodiversity assets in spatial planning: Methodological proposal and development of a planning support system. *Landscape and Urban Planning*, 84, 252–265. [DOI:10.1016/j.landurbplan.2007.08.005]

Harris, B. (1960) Plan or Projection: An Examination of the Use of Models in Planning. *Journal of the American Institute of Planners*, 26, 265–272.[DOI:10.1080/01944366008978425]

Ipekoglu B (2006) An architectural evaluation method for conservation of traditional dwellings, *Building and Environment*, 41(3), 386–394. [DOI:10.1016/j.buildenv.2005.02.009]

Jim C Y (2005) Outstanding remnants of nature in compact cities: patterns and preservation of heritage trees in Guangzhou city (China), *Geoforum*, 36(3), 371–385. [doi:10.1016/j.geoforum.2004.06.004]

Kammeier, H. D. (1999) New tools for spatial analysis and planning as components of an incremental planning-support system. *Environment and Planning B: Planning & Design*, 26, 365–380.[DOI:10.1068/b260365]

Klosterman, R. E. (1997). Planning support systems: A new perspective on computer-aided planning. *Journal of Planning Education and Research*, 17, 45–54.[DOI: 10.1177/0739456X9701700105]

Klosterman, R. E. (1999) The What if? Collaborative planning support system. *Environment and Planning B: Planning and Design*, 26, 393–408.[DOI:10.1068/b260393]

Shen Z, Kawakami M (2010) An online visualization tool for Internet-based local townscape design, Computers, Environment and Urban Systems, 34(2), 104–116. [DOI:10.1016/j.compenvurbsys.2009.09.002]

Townshend T, Pendlebury J (1999) Public participation and the conservation of historic areas: case-studies from the North-east of England. *Journal of Urban Design*, 4(3), 313–331. [DOI: 10.1080/13574809908724453]

Wang H-J, Chiou C-W, Juan Y-K (2007) Decision support model based on case-based reasoning approach for estimating the restoration budget of historical buildings, Expert Systems With Applications, 35(4), 1601–1610 [DOI:10.1016/j.eswa.2007.08.095]

Chapter 16
Visualization of the District Ecological Network Plan at Urban Partitions for Public Involvement

Zhenjiang Shen, Mitsuhiko Kawakami, and Satoshi Yamashita

Introduction

This chapter explores methods of supporting ecological network planning at the district level (biotope) of local cities by obtaining the necessary information for environmental learning programs through public involvement. Recently, environmental learning activities among communities in Japan have become a useful way to improve residents' awareness of the ecosystem around residential areas. However, ecological network database sources for the planning of such eco-city concepts are limited at the urban or regional levels of municipalities. Environmental learning programs are also a way for local governments to gather useful information on biotopes in urban districts. In order to gain consensus in the decision-making process of an ecological network plan, particularly in districts located on the edges of urban and suburban areas, some local governments collect information provided by residents via on-line investigation of possible biotopes around their residences.

In previous reports (Ahern 1991; Cook 2002; Barbosa et al. 2007) regarding ecological networks, metrics for connectivity and circuitry for imagery classification or network analysis have been employed, most of which are conducted at the regional and urban levels. The concept of traditional open spaces typically reflects a response to land-use controls and property boundary locations. Cook (2002) showed that notable improvements in ecological value can be achieved by implementing a planning strategy for open space systems that embraces the concept of ecological networks. As the core of nature conservation and biodiversity planning (Jones-Walters 2007), ecological networks (Jongman 1995) are based on certain core areas, corridor zones, buffer zones and, if needed, nature rehabilitation areas. Opdam et al. (2006) proposed the ecological network concept as a suitable basis for inserting biodiversity conservation into sustainable landscape development.

Z. Shen (✉) • M. Kawakami • S. Yamashita
School of Environmental Design, Kanazawa University, Kanazawa, Japan
e-mail: shenzhe@t.kanazawa-u.ac.jp

Z. Shen, *Geospatial Techniques in Urban Planning*, Advances in Geographic
Information Science, DOI 10.1007/978-3-642-13559-0_16,
© Springer-Verlag Berlin Heidelberg 2012

Furthermore, Urban and Keitt (2001) presented a method of measuring ecological networks using the number of linkages, the number of nodes, and the link suitability factor of network analysis (Maruli and Mallarachi 2005; Cook 2002; Hagget et al. 1977). Ahern (1991) presented a strategy for planning the conception, implementation, and management of an extensive open space system for rural landscapes based on spatial configuration and connectivity. These open spaces (Thompson 2002) are ideally comprised of public and privately owned lands aggregated into a network. Li and Wang et al. (2005) argued that urban parks, forestry, agriculture, and water should be planned and designed in an integrated manner with the objective of achieving the "eco-city," which would interconnect and integrate urban greenspaces. Thus, ecological networks can facilitate stakeholder decision-making in achieving the goals of natural conservation, feasible biodiversity, and land-use control. For planning support, Vuilleumier and Dautz (2002) presented a flexible tool using a GIS dispersal model to analyze landscape dispersal. This tool allows the simulation of different policy decision scenarios. However, research reports regarding ecological network planning at the district level are insufficient.

In planning conducted in Japan, the Geographical Survey Institute (GSI) and a number of private companies ordinarily conduct surveys and make digital maps, which include vector data sources of natural parks, sanctuary areas, and natural and environmental preservation territories and are all edited at the regional or urban level. It is possible to generate a base map of ecological networks at the regional or urban level from existing data sources. However, because all of these reports consider urban and regional planning, the necessary information at the district level is missing from existing data sources.

The Organization for Landscape and Urban Greenery Technology Development of Japan (GLUGTDJ 2000) has published a new guideline regarding its ecological network plan (ENP), which is composed of the urban ecological network plan (UENP), the district ecological network plan (DENP), and the regional ecological network plan (RENP) for green design, as listed in Table 16.1. The district ecological network plan is targeted at green design in residential areas. The plan also considers biotope areas, which are spaces reserved for the activities of small animals in the natural environments of urban districts so as to enable the coexistence of humans and animals. For the purpose of reserving natural land cover in urban districts, we discuss how to use the Geographic Information System (GIS) to retrieve the district ecological network (DEN) from aerial photography. This GIS data will prove helpful in district ecological network planning.

There are several existing reports on the distribution of green spaces that make use of remote sensing data that can be used in the retrieval of ecological network data through information technology. Up-to-date, high-resolution (below 10 m) satellite images have proven to be an excellent tool (Pascual-Hortal et al. 2007) for the identification of sources of existing natural habitats, strategic points, existing biotopes, and empty spaces. These images are important instruments for planning the management of tree corridors and forests in the suburban and urban landscape. Forest patches and corridors in the suburban landscape can be viewed as stepping-stones between urban forests and forests in the adjacent woodlands. All of these

Table 16.1 Concepts of the district ecological network

Type	Space areas	Statement
Core patch	Most important central area in a DEN	Growth and development of green spaces for animal and plant life in wooded areas with good conditions
Base patch	Takes advantage of the natural environment in urban parks	Growth and development of green spaces for animal and plant life in regenerative natural areas of urban parks
Corridor	Wooded corridor	Reservation of green spaces between houses and roads that support the movement of animal and plant life in land and air
	River corridor	Reservation of natural land-use spaces of river and green spaces near rivers that support the movement of animal and plant life in water, land, and air
	Noncontiguous corridor	Green spaces that support the movement through air of animal life

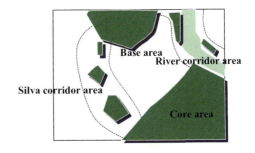

Fig. 16.1 Spatial image of a district ecological network

studies have been conducted at the urban or regional planning level. However, it is also necessary to consider the urban district level.

In this chapter, we discuss how to generate a network map and how to develop a tool for online field surveys of animal and plant information through public involvement. According to the DENP criteria, we attempt to automatically generate the ecological base map at the district level (Fig. 16.1). We also attempt to gather the necessary information from residents' observations of their own daily lives.

System for Supporting DENP Through Public Involvement

As mentioned above, in most municipalities there have been no field surveys to collect animal and plant information at the urban district level. In Japan, environmental learning programs are carried out for students in elementary and high schools or for related non-profit organizations. The local government makes necessary information concerning animal and plant life accessible through the Internet to gain a consensus for the DENP. In this chapter, we present suggestions for gathering district-level ecological network information in two parts. The first part is a GIS tool that enables ecological planners to generate a base map of DEN according to some of the criteria of DENP. The second part is a web system for public participation that allows local residents in a district to enter their observation information into the system. Ecological planners can make ecological network plans after gathering the residents' information through the public participation on the Internet.

First, we discuss how to generate a base map of ecological networks at the district level through aerial photography. The participants can provide animal and plant life information through on-field environmental learning to support DENP decision-making. It is possible to retrieve natural land cover data through imagery classification. However, with aerial photography at the district level, the shapes of buildings and other chaotic data will be reflected, thereby negatively impacting the classification accuracy. How, then, should these data be retrieved from aerial photography of areas having buildings?

In order to resolve this problem, GIS is used to generate a base map of a DEN. A dataset including a local urban planning database and aerial photographs is prepared for analysis. As a convenient method of data acquisition, aerial photography is widely used in Japan.

Here, we describe the proposed procedure primarily with respect to the automatic generation of an ecological network base map at the district level using GIS. Knowledge-based rules are discussed according to the concept of district ecological networks, as shown in Table 16.1. Generally, two rules are necessary. First, the land cover of an ecological network should be limited to natural cover, which includes water, wooded areas, and grassy areas. Second, core patches, base patches, and corridors, as defined in Table 16.1, must have their constant area indexes set according to the DEN criteria. However, a discussion of DENs, including biological types and paths of animal movements remains for field surveys during the period of environmental learning. In this process, we use the land cover types and the area indexes as the rules of a knowledge base for generating the DEN.

After generating the base map, we present a pragmatic suggestion that the relevant plant and animal information be gathered by local residents, which requires a web-based public participation system. After the residents have observed animal and plant behavior in their district, they can use their personal computers to access a server that will post their observations. In order to realize public participation via the Internet, we have developed a prototype system.

Retrieving Natural Land-Use Data in Urban Districts from Aerial Photographs Using GIS

We suggest using GIS to generate a base map that includes central core patches, base patches, and corridors in a DEN. In order to achieve this goal, a GIS data set that includes a local urban planning database, a digital map of the local city, and aerial photographs should be prepared.

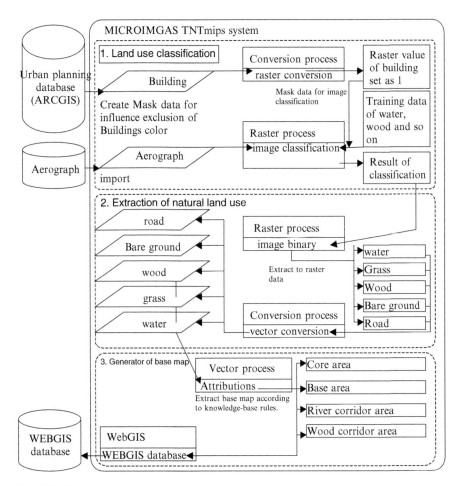

Fig. 16.2 Approach to generating an ecological network base map

As shown in Fig. 16.2, there are three steps involved in the generation of a base map for a district ecological network: (1) land cover classification, (2) extraction of natural land cover from the classification results, and (3) generation of a base map. The GIS system for imagery classification is also the product of microimages referred to as TNTmips.

Aerial Photography for Imagery Classification

There have been several reports on the regional scale, such as Maruli and Mallarachi's study (2005), which use 1:25,000-scale land-use maps. However, this method cannot be used at the urban district level. When retrieving green

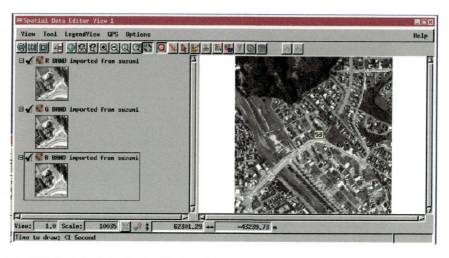

Fig. 16.3 RGB band for the classification of images extracted from an aerograph (Suzumi-cho, Kanazawa City)

space data at this level, chaotic data will influence the accuracy of the imagery classification because the roofs of buildings are included in the images.

A residential area that includes rivers, wooded areas, houses, and roads is appropriate for the analysis of a DEN located at the edge of an urban or suburban area, as shown in Fig. 16.3. In this aerial photograph, the image resolution is 0.88 pixels/m, its actual size is 398 m × 569 m, and its pixel size is 350 × 500. For retrieving an ecological network, natural land-use types should be analyzed according to the knowledge-based rules of the district ecological networks shown in Table 16.1. The core patch is a wooded area, and the base patch is a reservation area of urban parks. Corridors are areas that include rivers, wooded areas, and grassy areas. Therefore, wooded areas, water, and grassy areas should be classified in the process of image classification. Other land-use types, which include buildings, roads, and bare ground, can be excluded.

A high-resolution satellite image is well suited for discriminating existing natural habitats, strategic points, existing biotopes, and empty spaces. Clever et al. (2008) contested the accuracy of pixel-based and object-based classification methods for mapping the wildland–urban interface with high-spatial-resolution aerial photography. In Clever's study, the pixel-based classification assumes that the individual pixels on each image are independent and that the pixels are treated in the classification algorithm without considering any spatial association with neighboring pixels. The object-based classification approach provides a higher overall accuracy than the pixel-based approach. However, for the tree/shrub class, similar degrees of accuracy are achieved by both approaches. Therefore, we explored pixel-based classification using the algorithm set, namely Maximum Likelihood Classifier (MLC).

Creation of Mask Data and a Training Dataset

Retrieving natural land cover such as wooded areas, grassy areas, and water from aerial photography is comparatively easier than retrieving artificial land cover such as buildings. In fact, roofs are constructed using various materials and many be any number of colors. Therefore, minimizing the impact of roofs is important. In Kanazawa City, based on aerial photography, building features are edited as poly-lines in the urban planning GIS database. The building poly-lines can be used to create mask data that exclude the building shapes during the image classification process.

The mask data set is shown in Fig. 16.4. The conversion process from building vector data to mask data is the same as the process from vector data to raster data. The raster value of the masked area is set to 1, and the value of the area outside the mask area is set to 0. Since the vector data source of the buildings in the study area is created based on aerial photographs, the mask data are well overlain, as shown in Fig. 16.5.

Retrieving Natural Land Cover from Image Classification Results

Figure 16.6 shows an edited training data set that includes different types of natural land cover, such as grassy areas, wooded areas, water, and roads. Imagery classification is implemented using the maximum likelihood method with the training data set and the mask data set (Fig. 16.7). As shown in Fig. 16.8, the overall accuracy is up to 91.77%. The highest accuracy is obtained for water, whereas the lowest accuracy is obtained for bare ground. Among natural land cover types,

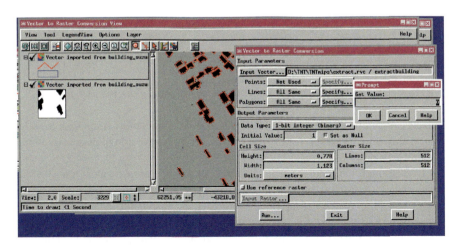

Fig. 16.4 Mask data created by building polygons

Fig. 16.5 Mask data overlain with an aerograph

Fig. 16.6 Training data set of water, wooded areas, grassy areas, roads, and bare ground

misclassification between wooded and grassy areas occurs more often than for other types. However, the results of the image classification have a high degree of overall accuracy.

Even though the land cover is classified as wooded areas, water, grassy areas, roads, and bare ground, only the natural land cover is useful for generating a base map. According to the definitions of core patches, base patches, and the different types of corridors in the GLUGDJ guidelines, the knowledge-based rules of area index and land covers can be used to generate a base map for a district ecological

16 Visualization of the District Ecological Network Plan at Urban Partitions 331

Fig. 16.7 Results of image classification (pixel-based)

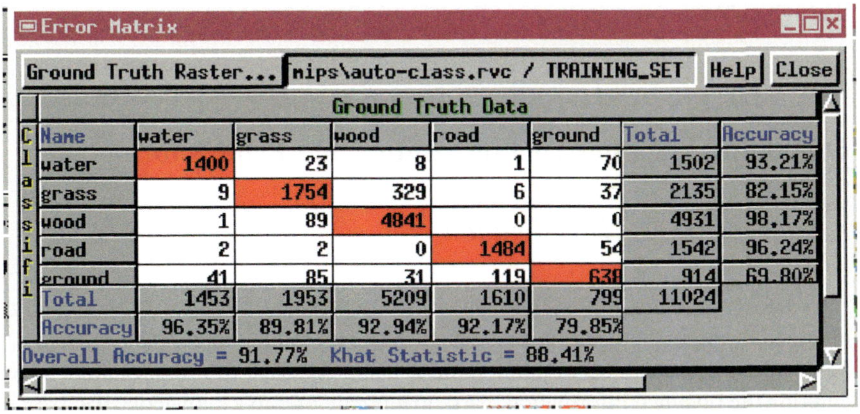

Fig. 16.8 Table of error matrix and degree of accuracy

network using GIS. Thus, the standards of DEN are basically established by the area indicator, which are difficult to retrieve directly from raster datasets.

A conversion process from classified raster to vector is necessary. A color binary function of the raster process is thus conducted in order to separate the different land cover types, which can be changed to vector data. In the color binary process, the natural land cover can be divided into three raster objects: wooded areas, water, and grassy areas. Each of these objects can be further converted to a vector object as a polygon dataset, so that the area index of each polygon can be calculated for the implementation of the knowledge-based rule (Fig. 16.9).

Consequently, natural land cover is retrieved for the DENP decision-making process. However, the network of animal activity is also necessary for DENP. In the

Fig. 16.9 Extraction of natural land use from the classification results

next section, we develop an online tool for the collection of animal activity data through public involvement.

Online Investigation Tool for Animal and Plant Information

A Tool for Online Investigation

Information on plant and animal activities in a district ecological network can be obtained through photographs and text descriptions. Both photographs and text can be integrated as attributes of point features using WebGIS over the Internet. When entering the obtained information, participants can refer to a knowledge database of plants and animals.

As shown in Fig. 16.10, this online investigation tool consists of three components, namely, a WebGIS database system, a database system for storing pictures and comments input by users, and a VR database. In WebGIS (Fig. 16.11), the retrieved DEN map can be provided as a background. Users can enter their information as attribute data of point features by creating point features in a text format file from client sites on the Internet. The attribute data can be file names of photographs of plants or animals and user comments. In the developed tool, we use JSHAPE and its CMD script object to construct links to photographs. The point features can be published on the web pages using JSHAPE, and the attribute data of these points can be published using the CMD script command.

The database for information provided by users can be entered using the Common Gateway Interface (CGI) through the Perl script. This tool provides an interface for residents to upload photographs of plants and animals. Point features

16 Visualization of the District Ecological Network Plan at Urban Partitions 333

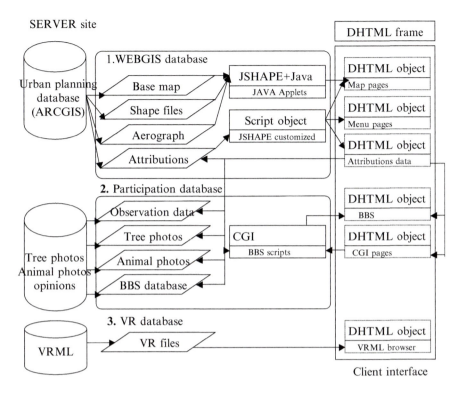

Fig. 16.10 WebGIS for public participation

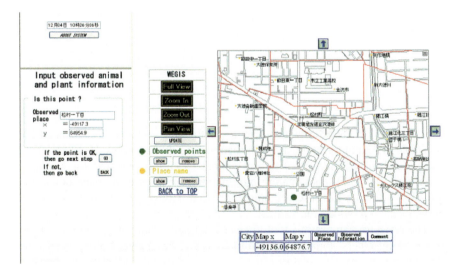

Fig. 16.11 WebGIS interface of public participation (example)

Fig. 16.12 List of plants and animals

can be created in WebGIS, which links to photographs and text through an attribution table of the point feature layer. Furthermore, as shown in Figs. 16.12 and 16.13, categorized information of plant and animal species is prepared for reference. Otherwise, this point coordination information can be used to create points in a VRML world with the observed animal or plant photographs. In addition, a BBS system can be provided for residents to discuss the district ecological network.

The residents can use this tool to enter information and check the information provided by others. Finally, the information can be analyzed along with other datasets on the ArcGIS platform using a local personal computer, as shown in Fig. 16.14.

16 Visualization of the District Ecological Network Plan at Urban Partitions 335

Fig. 16.13 Input of observed data

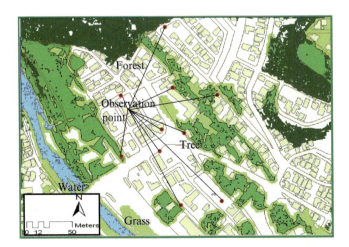

Fig. 16.14 Investigation information and retrieved natural land covers (ArcGIS)

Implementation of Online Investigation

In order to establish an ecological network for a district, the network of animal activities should be clarified. There were 15 responses during the 2-month open access period. These responses cover the locations of the activities of animals, such as homeless cats and dogs and the tracks of bears moving from forests to parks near residential homes in the winter. Thus, interactions between humans and animals would be the main information collected through the Internet. However, if we attempt collecting information on animals and plants in urban districts using the Internet, we must conclude that the ecological network of different species is difficult to clarify through public participation. This is because residents pay more attention to substantial harm due to animals in daily life and how to prevent animals from invading residential areas. Although the ecological network is composed of connected patches and corridors of natural land cover, most of the information collected concerned bear tracks and the activities of homeless cats and dogs. In conclusion, the investigation of plant and animal species should be conducted through environmental learning programs under expert supervision because information provided by non-experts will always concentrate on the spaces outside natural land cover.

We have discussed how to retrieve natural land cover from aerial photography and how to develop a tool for collecting user information on plants and animal activities provided by experts and non-experts. However, we have not yet discussed the problem of how to make a DENP decision.

Visualization of DENP for Public Involvement

According to the DENP guidelines, an ecological network at the district level is defined by land-use type and an area index. However, the network of animal activities is not sufficient to build the ecological network at the level of urban districts, where there are not only boundaries but also coexisting spaces between human and animals. In addition, the tracks of dangerous animals and their impacts on each site are also important and should be considered when building an ecological network. In order to make a DENP decision, we have visualized the DEN according to the criteria of GLUGDL for further deliberation between stakeholders. We hereinafter discuss how to visualize the ecological network based on the GLUGDJ guidelines of DENP for public involvement.

With respect to the existing metric standards of ecological networks, Pascual-Hortal et al. (2007) provided guidelines for the appropriate selection of connectivity metrics at the regional level. Fransco et al. (2003) assess the impacts of agroforestry network planning outputs on the perception of landscape in terms of scenic beauty. Zhang and Wang (2006) described a planned ecological network as a graph consisting of patches and corridors in which nodes were defined as greenway patches no smaller than 6 ha. Fifty greenway patches were selected as the nodes for Xiamen Island. Bierwagen (2005) developed a metric to estimate ecological connectivity for cities and regions. These projects presented a regional scale for retrieving the ecological network. On a smaller scale, Barbosa et al. (2007) measured the distance along a transport network to the available public green spaces for households, as compared with the distribution of private garden spaces. Pirnat (2000) suggested that new corridors of trees established along motorways should connect with the fragmented remains of the ecological infrastructure. Generally speaking, connectivity and circuitry is measured using the number of linkages and nodes, as well as link suitability factors (Urban and Keitt 2001; Cook 2002; Hagget et al. 1977; Lee et al. 1999). As mentioned above, the network of animal activities is not sufficient for DENP. Interactions between humans and animals should also be taken into account. For deliberation on these issues through public involvement, we suggest a visualized DENP for stakeholders.

Generating DENP Based on the GLUGDJ Guidelines

After the polygon features of wooded areas, water, and grassy areas are created through the process described in Sect. "Retrieving Natural Land-Use Data in Urban Districts from Aerial Photographs Using GIS", an area index of each polygon can be calculated and saved as its attribute. In Table 16.2, we provide knowledge-based rules for generating a base map of a district ecological network according to GLUGDJ guidelines. The results of the polygon features extracted by the knowledge-based rules are shown in Fig. 16.9. Concretely, the area index and land-use types of each polygon feature can be employed in the retrieval process. First, because the core patch should be a wooded area over $5000\ \mathrm{m}^2$, the wooded vector data is selected to create the core patch. Second, because the base patch should be a reserved wooded area over $5000\ \mathrm{m}^2$ within urban parks, a wooded area over $5000\ \mathrm{m}^2$ can be extracted as a base patch. Finally, in contrast to the core patch and the base patch, the corridors are wooded and

Table 16.2 Criteria for base rules for a district ecological network base map

EN	Knowledge-based rule
Core patch	Wooded area maintained in good condition over $5000\ \mathrm{m}^2$
	Largest wooded area in an EN district
Base patch	Regeneration of green space in urban parks
	Woodland over $5000\ \mathrm{m}^2$ within an EN district
Wooded corridor	Wooded area in a district over 300–$5000\ \mathrm{m}^2$, including nearby grasslands over $500\ \mathrm{m}^2$
River corridor	River in a district including nearby wooded (300–$5000\ \mathrm{m}^2$) and grassy (over $500\ \mathrm{m}^2$) areas. A pond over $1000\ \mathrm{m}^2$
Noncontiguous corridor	Natural land use is noncontiguous
	Wooded area over 300–$5000\ \mathrm{m}^2$
	Grassy area over $500\ \mathrm{m}^2$

river areas that have 300–5000 m^2 of nearby grassy and wooded areas. The extracted results for a generated base map are shown in Fig. 16.15. However, the generated base map is only the beginning of ecological planning at the urban district level.

Visualization of DENP

In local GIS, ArcScence is used to launch an ecological network image in three dimensions as shown in Fig. 16.16. This tool is helpful in representing the

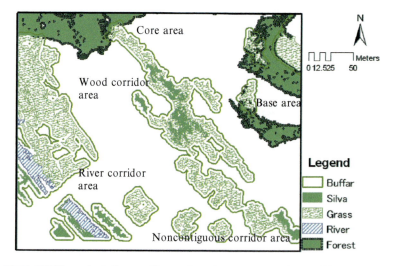

Fig. 16.15 DENP map including corridors and patches generated using ArcGIS

Fig. 16.16 Visualized map of district ecological network in ARCGIS, which can also present the point layer with the observed information

16 Visualization of the District Ecological Network Plan at Urban Partitions 339

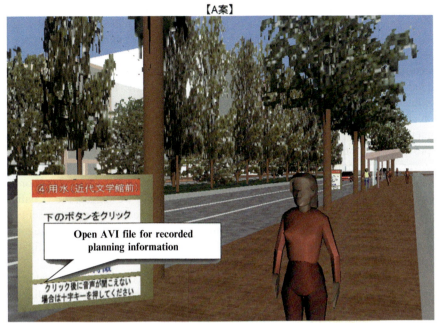

Fig. 16.17 VR world for DENP using a VRML that recorded visualized information to represent the ecological information

boundaries of the ecological network. However, the tool used with local GIS can combine the collected residents' opinions submitted via Webgis, which can be referred by planners for the presentation and analysis of DEN. For DENP decision-making through public involvement, a visualized DEN is helpful for providing DEN information to the public.

For additional details concerning the DEN, the integration of observed information concerning landscape and buildings in a VRML world is helpful, but the boundaries of human and animal activities are difficult to identify. As shown in Fig. 16.17, the observed data integrated in the VRML world are AVI format videos or sound files, which are helpful for representing biological activities.

Conclusion

We have succeeded in generating a base map of an ecological network at the district level. Public participation is necessary during the planning process in order to arrive at a consensus on a district ecological network plan. We have also suggested the development of an online investigation tool.

As a result, image classification that allows the extraction of natural land cover from aerial photographs and excludes the influence of building colors is possible. According to the GLUGDJ guidelines, the area index of natural land-use types is available as a knowledge-based rule for generating a base map. Although an ecological network is composed of connected patches and corridors of natural land cover, most online investigation information at the urban district level is related to bear tracks and the activities of homeless cats and dogs. Consequently, the investigation of animal and plant species should be conducted through environmental learning programs under expert supervision because the information provided by non-expert users is concentrated on spaces outside the natural land cover. Furthermore, the network of animal activities is not sufficient for the DENP, and interactions between human and animals should be taken into account. For the deliberation of these issues through public involvement, we suggest the visualization of the DENP for stakeholders. Local GIS systems can be used to present the boundaries of the DEN. The VRML can be used to allow access to integrated information, such as texts, videos, and sounds, so as to present investigation information over the Internet.

At the same time, urban greenspaces in cities also occur as semi-natural areas, managed parks and gardens associated with roads and incidental locations (Jim and Chen 2003). In the present paper, the retrieval process using aerial photography cannot reorganize DEN from the viewpoint of amenity-recreation venues while considering the design context between blocks, roads, and buildings. Otherwise, even though existing urban parks are considered in this retrieval process, new parks set up within this area are not included. In order to accomplish this task, an approach that integrates existing urban greenspaces and planned urban greenspaces is necessary.

Acknowledgment This project is supported by Kanazawa city government (Project Reference No. 1178).

References

Ahern, Jack (1991), Planning for an extensive open space system: linking landscape structure and function, Landscape and Urban Planning, Vol 21, No. 1–2, 1991, pp. 131–145

Barbosa, O., Tratalos, Jamie A., Armsworth, Paul R., Davies, Richard G., Fuller, Richard A., Johnson, P., Gaston, Kevin J. (2007), Who benefits from access to green space? A case study from Sheffield, UK, Landscape and Urban Planning, Volume 83, Issues 2–3, 19, 187–195

Bierwagen, B G, 2005, "Predicting ecological connectivity in urbanizing landscapes" Environment and Planning B: Planning and Design 32(5) 763–776

Cleve, C., Kelly, M., Kearns, Faith R., Moritz, M. (2008), Classification of the wildland–urban interface: A comparison of pixel- and object-based classifications using high-resolution aerial photography, Computers, Environment and Urban Systems, Volume 32, Issue 4, 317–326

Cook, Edward A. (2002), Landscape structure indices for assessing urban ecological networks, Landscape and Urban Planning, Volume 58, Issues 2–4, 15, 269–280

Franco, D., Franco, D., Mannino I., and Zanetto G. (2003), The impact of agroforestry networks on scenic beauty estimation: The role of a landscape ecological network on a socio-cultural process, Landscape and Urban Planning, Volume 62, Issue 3, 119–138

GLUGTDJ, (2000), A guideline of urban plan for harmonious coexistence of nature and humans, urban ecological network, Gyosei, Tokyo

Hagget, P., Cliff, A.D., Fry, A. (1977), Location Analysis in Human Geography, the 2nd Edition. Wiley, New York, 454

Jongman, Rob H. G. (1995), Nature conservation planning in Europe: developing ecological networks, Landscape and Urban Planning, Volume 32, Issue 3, 169–183

Jones-Walters, L. (2007), Pan-European Ecological Networks, Journal for Nature Conservation, Volume 15, Issue 4, 262–264

Lee, S.E., Morioka, T. and Fujita, T. (1999), Evaluation of biotope connectivity and planning of ecological network in urban landscape, Environment system research reports, Vol.27, pp. 285–292

Li, F., Wang, R. S., Paulussen, J., Liu, X. S. (2005), Comprehensive concept planning of urban greening based on ecological principles: a case study in Beijing, China, Landscape and Urban Planning, 72, 325–336

Jim, C. Y., Chen, S. S. (2003), Comprehensive greenspace planning based on landscape ecology principles in compact Nanjing city, China, Landscape and Urban Planning, 65, 95–116

Marulli, J., and Mallarach, Josep M. (2005), A GIS methodology for assessing ecological connectivity: application to the Barcelona Metropolitan Area, Landscape and Urban Planning, Volume 71, Issues 2–4, 243–262

Opdam, P., Steingr€over, E., and Rooij, van S. (2006), Ecological networks: A spatial concept for multi-actor planning of sustainable landscapes, Landscape and Urban Planning, Volume 75, Issues 3–4, 322–332

Pascual-Hortal, L. and Saura, S. (2007), Impact of spatial scale on the identification of critical habitat patches for the maintenance of landscape connectivity, Landscape and Urban Planning, Volume 83, Issues 2–3, 176–186

Pirnat, J. (2000), Conservation and management of forest patches and corridors in suburban landscapes, Landscape and Urban Planning, Volume 52, Issues 2–3, 135–143

Thompson, C. W. (2002), Urban open space in the 21st century, Landscape and Urban Planning, Volume 60, Issue 2, 59–72

Urban, D., & Keitt, T. (2001). Landscape connectivity: A graph theoretic perspective. Ecology, 82, 1205–1218

Vuilleumier, S., and Prélaz-Droux, R. (2002), Map of ecological networks for landscape planning, Landscape and Urban Planning, Volume 58, Issues 2–4, 157–170

Zhang, L. and Wang, H. (2006), Planning an ecological network of Xiamen Island (China) using landscape metrics and network analysis, Landscape and Urban Planning, Volume 78, Issue 4, 449–456

Chapter 17
Simulating Land-Use Patterns and Building Types after Land Readjustment at the Urban District Level Using the CAUFN Tool

Zhenjiang Shen, Mitsuhiko Kawakami, Takaaki Kushita, and Ippei Kawamura

Introduction

This chapter focuses on visualizing the build-up process after the implementation of a land readjustment project in Japan. A prototype simulation tool, called CAUFN (cellular automata for urban form of neighborhoods), is developed on the ArcGIS platform for the simulation of land-use patterns and building types after land readjustment. This tool can be operated by planners to represent the build-up process at the urban district level to reveal the future forms of urban neighborhoods.

Accelerated by the widespread use of the geographic information system (GIS), some basic laws were issued in order to improve the use of geospatial data sources in Japan. One such law, referred to as the Basic Law of Promoting Geospatial Information, which was intended to exploit the advantages of geospatial information and accelerate the use of the e-statistic web database, was published in 2007. Geospatial and statistical data can be freely downloaded from several online databases in Japan and are expected to be used for planning support. In addition, public participation in urban planning has become a hot topic among planners and researchers (Sugisaki et al. 2003; Pløger 2001). In order to better reflect public opinion, residents are motivated to take part in the urban design of their own cities. Therefore, visualization of urban planning has been recognized as a powerful tool for representing the vivid future of urban areas to the public, helping the public to more easily participate in urban planning (Shen and Kawakami 2010). Moreover, urban simulation based on geospatial information and the Geodatabase has been proven to be useful as a method for predicting the impacts of policy instruments on urban development, because such a method can simulate different scenarios according to urban planning and design (Stevens and Dragicevic 2007; Shen et al. 2009). As mentioned earlier, in order to make planning alternatives more

---

Z. Shen (✉) • M. Kawakami • T. Kushita • I. Kawamura
School of Environmental Design, Kanazawa University, Kanazawa, Japan
e-mail: shenzhe@t.kanazawa-u.ac.jp

Z. Shen, *Geospatial Techniques in Urban Planning*, Advances in Geographic
Information Science, DOI 10.1007/978-3-642-13559-0_17,
© Springer-Verlag Berlin Heidelberg 2012

reasonable and more practical, we attempt to simulate the impacts of planning policy on urban development. Therefore, we visualize the build-up process of an urban development area on the district level based on planning conditions, such as land-use zoning and road network plan.

With respect to existing research and theory, many researchers have focused on developing simulation tools for simulating urban or regional growth by CA model. Most simulation tools, such as SLEUTH (http://www.ncgia.ucsb.edu/) and DINAMICA (http://www.csr.ufmg.br/dinamica/), are developed for simulating land-use patterns on the urban or regional level. In Japan, a number of research projects in the field of urban simulation have also been conducted based on the CA theory. For instance, Watanabe et al. (2000) and Takizawa et al. (1998) conducted research on entire cities in which CA was the primary component of the simulation tools developed on the GIS platform and used to conduct urban growth simulation through considering the urban land unit as adjacent regular cells. Most of these CA models use the capabilities of standard GIS and program the models in general programming languages or macro-languages. Therefore, in most of these case studies, programming knowledge is needed in order for modelers to implement the models, which makes it difficult for non-expert users to understand and apply such models (Santé et al. 2010).

However, parcels and blocks in urban districts are irregular polygons. Thus, another problem emerges in that it is difficult to deal with actual land-use simulation through adjacent irregular cells in the CA models. In order to solve this problem, we (Shen et al. 2007) proposed a model to handle irregular urban blocks and parcels for simulating land-use patterns after the land-use adjustment project. iCity (Stevens and Dragicevic 2007) is a tool that is similar to CAUFN, through which users can easily set parameters, select the initial GIS data containing the land-use parcels, and then observe the simulation of each developable land use, such as residential, commercial, and park uses. Some simulation tools, such as OBEUS (Benenson and Torrens 2004) and CAGE (Blecic et al. 2003), can also deal with simulations with irregular parcels in urban district. However, these tools are difficult for planners to operate. If planners plan to use OBEUS and CAGE in simulations, they must program the CA model themselves.

In this chapter, we focus on how to develop a simulation tool on the ArcGIS platform and how to build related simulation models. In Sect. "Research Approach", we introduce the research approach. Simulation models and system frameworks will be discussed in Sects. "Simulation Models for Visualizing the Build-up Process Through GIS" and "System Framework for Developing the CAUFN Tool Using ArcGIS", respectively. In Sect. "System Test and Additional Applications", a case study involving Kanazawa City will be conducted in order to verify the applicability of this simulation tool. After the simulation is validated using the case study area, the present simulation tool and a number of existing tools are compared in Sect. "Conclusion" in order to verify the applicability of the proposed tool for a planning support system. Finally, conclusions are presented.

Research Approach

The present research is an attempt to develop a clear and easily operated platform for the planner to investigate the impacts of planning conditions on the designated urban project area through simulating different scenarios. The tool introduced in this chapter can be used to represent the future images of the urban area through simulation of different land-use types and building types, and the building types will be divided into different real estate property types during the simulation.

Therefore, the following simulation models are integrated into the CAUFN simulation tool: simulation of land-use types by cellular automata (CA), simulation of building types by neural network (NN), and real estate management by portfolio theory. These simulation models have been validated by Shen et al. (2007) and Kawamuwa et al. (2008), respectively. In addition, ArcToolBox, which is an extension of the ArcGIS platform of ESRI, will be used for development of the CAUFN tool. We use some components, such as a model and scripts of ArcToolbox to develop the CAUFN simulation tool. The simulation model is programmed using the Python script, which is combined in the ArcToolbox of ArcGIS Desktop. In addition, Python is an object-oriented programming computer language that can be freely customized by users. Moreover, the "Model Editor" option in ArcToolbox is used to organize the simulation scripts during the implementation of all simulation processes.

With respect to tool validation, the entire simulation process will be carried out on the ArcGIS platform. Thus, an example area sourced from the land readjustment projects in Kanazawa is selected for the case study. During the verification of the simulation tool, we investigate the simulated results with respect to land-use patterns and building use types in the case study area.

Simulation Models for Visualizing the Build-up Process Through GIS

Figure 17.1 shows the simulation flow regarding how to simulate the build-up process based on irregular parcels. The essential concepts of the proposed simulation model are discussed in this section.

Concept of Land-Use Simulation

The parcel will be the spatial unit for land-use types, building types and asset management. Thus, the simulation results for land-use types can be taken as a basis for the simulation of building types. The definitions of building types and real estate management included in this research are defined in Fig. 17.1.

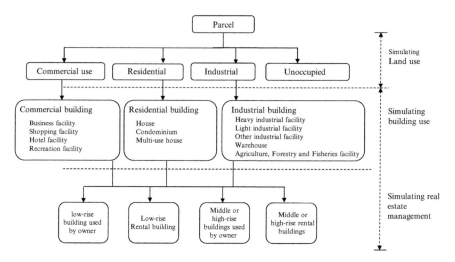

Fig. 17.1 Simulation flow of real estate management

As reported in a previous study on the CA model (Shen et al. 2007), land-use types in urban areas usually include residential, commercial, industrial, and unoccupied land use. As shown in Fig. 17.1, this division of land-use types is used to represent land-use patterns in urban areas during the process of urban development. In addition, public land use has been further divided into four types: parks, hospitals, schools, and other public land uses. These types of public land use generally tend to be stable in their locations during the process of urban development. Thus, the simulation does not include these types of parcels.

Accordingly, the attributes of parcels are variables for the simulation that should be predefined in the simulation model. As the following paragraph shows, it is necessary to define the neighborhoods and transition rules for the CA model when considering the impacts of neighborhoods or the entire district on land-use state.

Concept of the Neighborhood

In this chapter, the neighborhood is divided into two different types: neighbor parcels and neighbor blocks. The common idea of the neighbor in the CA model is similar to the concept of the Von Neumann neighborhood, in which the adjacent parcels are defined as neighbors. The land-use formation of adjacent parcels affects the parcels. According to this principle, neighbor parcels are defined as parcels that share edges, as shown in Fig. 17.2. However, from the broader viewpoint of neighbor conception, neighbor blocks in this research can be considered in the manner shown in Fig. 17.3. Similar to the principle of neighbor parcels, neighbor blocks are the blocks that share edges. Thus, the concept of the neighborhood in this chapter has been constructed using hierarchical and multiple definitions.

17 Simulating Land-Use Patterns and Building Types after Land Readjustment

Fig. 17.2 Neighborhoods of parcels

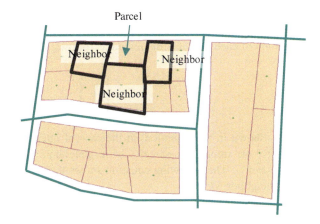

Fig. 17.3 Neighborhoods of blocks

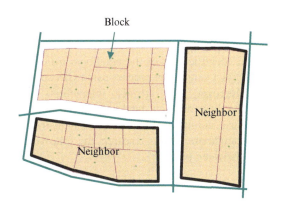

Calculate the Land-Use State of Parcels

This system performs land-use simulation by calculating the development potential of each parcel. The development potential of each parcel is defined as parcel state x for different land-use types. The parcel state will be affected by the surrounding land use, the geometric characteristics, such as the area and perimeter of the parcel, and the planning conditions (including land-use zoning and front road, namely the road in front of the site). The following equation, which is based on the equation reported by Shen et al. (2007), represents the mathematic principle of state calculation:

$$x_i^{(m)}(t+1) = \{x_i^{(m)}(t) + \Delta x_i^{(m)}(t) + \Delta x_I^{(m)}(t) \\ + Metab_i^{(m)}(t) + GloTransp_i^{(m)}(t)\} \cdot \left(\frac{1}{3}\sum_{j=1}^{3} P_j^{(m)}\right)^q \qquad (17.1)$$

Equation 17.1 is designed to calculate the states of land use by CA simulation and represents the state transformation within parcels. The neighborhood impact for the state of each parcel is expressed as follows:

$$\Delta x_i^{(m)}(t) = G^{(m)}\left\{\frac{1}{N_i}\sum_{k=1}^{N_i(t)} x_k^{(m)}(t) - x_i^{(m)}(t)\right\} \qquad (17.2)$$

As shown in this equation, in the case of land-use type m, the neighbor state, namely, the impact on parcel i of its neighbor Ni can be adjusted through the parameter G. The following equation shows the state impact on block I associated with neighbor NI:

$$\Delta x_I^{(m)}(t) = G^{(m)}\left\{\frac{1}{N_I}\sum_{l=1}^{N_I(t)} x_l^{(m)}(t) - x_I^{(m)}(t)\right\} \qquad (17.3)$$

Even though these equations are designed for the calculation of the neighbor state, identification of the neighbors from parcels and blocks must be conducted through a simulation process using GIS software.

Moreover, there are also interactions between different land-use types. We denote these different land-use types as m, $m+$, and $m-$, which denote residential land use, commercial land use, and industrial land use, respectively, within parcels and neighbors. In this research, the following equation represents the state transition between parcels:

$$Metab_i^{(m)}(t) = \frac{D^{(m)} \cdot x_i^{(m)}(t) \cdot x_i^{(m+)}(t)}{1 + H \cdot x_i^{(m+)}(t)} - \frac{S^{(m)} \cdot x_i^{(m)}(t) \cdot x_i^{(m-)}(t)}{1 + H \cdot x_i^{(m)}(t)} \qquad (17.4)$$

where Metab denotes the interactions between different land-use types within parcel i at step t, where D and H are parameters. The following equation is designed for the state of GloTransp that reflects the interaction between parcel i and all other parcels for which the land-use types are the same as for parcel i, and GT and H are parameters:

$$GloTransp_i^{(m)}(t) = \frac{GT^{(m)} \cdot x_i^{(m)}(t) \cdot \frac{X^{(m)}(t)}{N_{all}}}{1 + H \cdot \frac{X^{(m)}(t)}{N_{all}}} \qquad (17.5)$$

Using (17.4) and (17.5), the global impact of the entire area on the parcel state can be taken into account.

Kamide et al. (1998) and Shen and Ishimaru (2000) conducted studies on the build-up process in the land readjustment project area by CA models and analyzed the impact of planning conditions on land-use formation of parcels. In order to reflect the policy impact of urban planning, namely, planning conditions, on the formation process of land use in urban areas, this research takes planning conditions as simulation parameters for the calculation of the state value. These planning conditions cover the concepts of land-use zoning, geometric characteristics, and position. The following equation introduces the state influence of planning conditions, such as land-use zoning, roads, and corner position on state P for land use m:

$$\sum_{j=1}^{3} P_j^{(m)} = P_{1u}^{(m)} + P_{2r}^{(m)} + P_{3c}^{(m)} \tag{17.6}$$

Moreover, the following equation is used to reflect how the geometric characteristics, q, (Maniruzzaman et al. 1994) affect the state of a parcel having area a and perimeter p:

$$q = c^2 = \left(2\frac{\sqrt{\pi a}}{p}\right)^2 \tag{17.7}$$

Both of these equations reflect the local parameters of planning conditions and the influence of geometric characteristics on the land-use state.

Transition Rules

The CA transition rules used in this research are taken from the methodology proposed by Wu and Takizawa based on the state of regular cells (Wu 1996; Takizawa et al. 1998). However, on the urban district level, the state of irregular polygons must be determined for each parcel. During the simulation, three state values for each parcel, which represent the different potentials of land-use formations for different land-use types, will be calculated using the equations described above. In addition, there will be two threshold values embedded in each parcel for determining the land use change, namely, a maximum threshold and a minimum threshold regarding the sum of all land-use types. The maximum threshold is set to determine the land-use change if the state of a parcel is larger than the maximum threshold, and the minimum threshold is set to determine the land-use type turn to unoccupied if the state of a parcel is smaller than the minimum threshold. The transition rules of land-use change in each parcel are described in detail as follows:

Rule 1: Parcel to be developed and assigned to one type of land use

Land-use types of each parcel are divided into commercial use, residential use, industrial use, and unoccupied use. If the sum of states of all these land-use types within a parcel exceeds the maximum threshold, then the land-use type of this parcel will change in the next simulation step. The new land-use type will be set as the land-use type, the state value of which is the largest among the three land-use types: commercial, residential, and industrial.

Rule 2: Return to unoccupied

If the sum of states within a parcel of all these land-use types is smaller than the minimum threshold, the land-use type of this parcel will become unoccupied.

Rule 3: Maintain land use

If the sum states within a parcel of all of these land-use types is between the maximum and minimum thresholds, the land-use type of this parcel will remain stable in the next simulation step.

The implementation of this transition rule is based on the if-then rule on the ArcGIS platform. The attributes of adjacent polygons are retrieved from a map in order to calculate the neighbor state. On the other hand, the state values of each parcel with respect to the planning conditions of, for example, geometrical characteristics, land-use zoning, and front road can be obtained from the attributes of this parcel. Thus, the calculation of relevant state values can be conducted through a Geodatabase.

Concept of Building-type Simulation

In the present study, neural network (NN) analysis is used to simulate the formation of different building types. The process of simulating building types by NN analysis should be conducted after land-use simulation by the CA model. Neural network analysis is a learning process that is based on planning conditions, such as geometrical characteristics, land-use zoning, and front road of land parcels. Within this process, the building types that do not match simulated land-use types can be automatically ruled out based on the simulation results for land-use types. The parameter settings of geometric and planning conditions should be retrieved from sample urban areas through a NN learning process and predefined in a Geodatabase for simulation. Since this chapter focuses on the development of the CAUFN simulation tool, the details of the NN learning process for obtaining geometric and planning parameters are not discussed in this chapter. Instead, we directly apply the parcel geometric and planning parameters to the simulation of building types on these parcels.

$$y_k = \varphi_0 \left(\alpha_k + \sum_k w_{hk} \varphi_k \left(\alpha_k + \sum_i w_{th} x_i \right) \right) \qquad (17.8)$$

where x is the input data (planning conditions and geometric characteristics of each parcel), y is the output data (residential, commercial, or industrial building types), w is the weight of an input parameter, α is a constant, and φ is a logistic function.

In addition, the prediction process by NN is expressed as (17.8). Although the learning algorithms for retrieving parameters are not supported by the ArcGIS platform, it is possible to integrate ArcGIS VBA with statistical analysis tool R (COM) to develop a learning tool. This tool can be implemented in order to obtain all of the necessary parameters by accessing the attributes of spatial features in the NN learning process. The retrieved parameters can be output to a Geodatabase, which is connected to the CAUFN tool for predicting building types (Table 17.1). Therefore, it is possible to simulate building types on parcels after the implementation of CA simulation on land-use types. Moreover, if we predefine the rules for determining the building types that do not match the simulated land-use types, the computing times during the simulation can be reduced. Therefore, we showed the relationship details between simulated land-use types and building types for the prediction process in Table 17.2.

Table 17.1 Relationships between land use and building type

Land-use type of parcels	Building types on parcels	Geometric parameters	Planning parameters
Commercial use	Business facility	Area	Front road
	Shopping facility	Perimeter	Land-use zoning
	Hotel facility	Shape	Floor area ratio
	Recreation facility	Corner	Building coverage ratio
Residential use	House		
	Condominium		
	Multi-use house		
Industrial use	Heavy industrial facility		
	Light industrial facility		
	Other industrial facility		
	Warehouse		
	Agriculture, forestry, and fisheries facilities		

Table 17.2 Prediction of building type using a neural network

Simulation results of land-use types of parcels by CA	Predicting building types		
Simulated commercial land-use type by CA	Commercial building types		
Simulated residential land-use type by CA		Residential building types	
Simulated industrial land-use type by CA			Industrial building types

Concept of Real Estate Management Simulation

In the following, we introduce the principle of simulation of real estate management using an existing simulation model. This model was proposed by Takahiro and Takeshi (1992) and Kawamuwa et al. (2008), where the portfolio theory (mean-variance approach) and the Bayesian probability theory were used to analyze household decision-making for the management of their own properties. This model (Kawamuwa et al. 2008) allows us to simulate the process of land owners making decision regarding their methods of real estate management. Based on the portfolio theory, land owners decide to sell or rent their properties through the comparison of risk and returns. The rules of land owners in making decisions on real estate management are presented through the following equations:

For sale:

$$\pi_i = (aP_h - C_f) \cdot f \cdot n/L \tag{17.9}$$

For lease:

$$\pi_i = \left(a \sum_{t=0}^{t=T} R_h - C_f\right) \cdot f \cdot n/L + P_L(1 + \rho)^T \tag{17.10}$$

where π represents for the annual returns of construction investment per unit land area. Equation 17.9 is the expected profit obtained by selling a property, where π_1 ($i = 1$) represents the expected returns on investment for a single house, π_3 ($i = 3$) represents the expected returns on investment for an apartment. Equation 17.10 is the expected rental returns, where π_2 ($i = 2$) represents the expected returns on a rental single house, and π_4 ($i = 4$) represents the expected returns on all rental house units in an apartment.

In these two equations, a is the ratio of available areas for sale or lease in a building, n is a random number representing the story number of a building, by which the building coverage ratio and the floor area ratio cannot exceed the planning criteria, C_f is the construction cost per unit floor area within a building, and f is a random number that represents the building covered area governed by the building coverage ratio. In (17.10), R_h is the profit of leasing per unit floor area during t years, L is the land area, and ρ is the depreciation rate of real estate investment. In (17.9), P_h is the exclusive selling price per unit of floor area, and, in Equation 17.10, P_L is the price of the building per unit area of land, and T is the entire period of investment returns by tenants. In these equations, we assume that the room area is the same for all floors of each building.

Equations 17.9 and 17.10 give the expected returns of construction investment. The developer or land owner will make a decision on how to manage their real estate properties according to the profit and transaction costs between different managing forms. Moreover, when the developer or land owner selects a management method of real estate properties according to the expected returns calculated

17 Simulating Land-Use Patterns and Building Types after Land Readjustment

using (17.9) and (17.10), the risk R of the investment will also be considered (Johansson 2002). In this simulation, the Bayesian probability of investment returns $P(\pi_i/R)$ will be calculated by considering the investment risk R, which is expressed by (17.1). However, we use a random value for the consideration of the risk without discussion of how to estimate the risk. In this equation, $P(\pi_i)$ can be calculated by expected returns π_i, in the form of a utility model. The probabilities $P(R/\pi_i)$ and $P(R)$ for decision making under uncertainty can be also taken into account. However, in the present simulation, the uncertainty is reflected through random factors. In this situation, $P(R/\pi_i)$ should be estimated as a value smaller than $P(R)$, and $P(R)$ is generated by a random value smaller than 1.

$$P(\pi_i|R) = \frac{P(R|\pi_i) \times P(\pi_i)}{P(R)} \tag{17.11}$$

$$P(\pi_i) = \frac{e^{\pi i}}{\sum_{i=1}^{i=j} e^{\pi i}} \tag{17.12}$$

The concept of simulating real estate management by agents is stated above, and we next integrate this concept into the CAUFN tool. Actually, this can be easily accomplished by defining the attributes of spatial features as the variables used in (17.9) through (17.12). Thus, the calculation function based on the ArcGIS platform connecting a Geodatabase is sufficient for achieving the simulation of decision-making for real estate management.

System Framework for Developing the CAUFN Tool Using ArcGIS

System Structure

In this Section, ArcToolbox and the Geodatabase of ArcGIS are used to perform the simulation. Figure 17.4 shows the main system structure of the CAUFN tool. The simulation implementation is divided into two parts, namely, spatial processing and database processing. For spatial processing, a function for retrieving necessary variables reflecting the neighbor state from attributes of adjacent polygons is necessary. For database processing, the functions of retrieving all necessary variables from the records in a Geodatabase, conducting necessary calculation for state calculation and rule-based reasoning, and storing the simulation results in a system database are necessary. Here, the Geodatabase is a database of ESRI, which is contained in ArcGIS and managed by DBMS under the Windows System.

In ArcToolbox, Python is used as the program language for editing program scripts. The basic features of CAUFN are as follows. First, since Geoprocessing functions of ArcGIS can be integrated in Python scripts of ArcToolBox, if ArcGIS library has been imported, the attributes of adjacent polygons can be retrieved through Python scripts. Second, the retrieved attributes of spatial features, such as polygons and their adjacent polygons in ArcMap, will be saved to a Geodatabase by Python script. As such, a calculation function can be implemented by Python script for retrieving all necessary variables from the records in a Geodatabase, conducting the necessary calculation, rule-based reasoning, and storing the results as predefined procedures. Accordingly, the CAUFN simulation tool can realize the spatial processing and database processing using the ArcGIS platform. In addition, the use of Python allows Com object technology of Windows in ArcGIS Geoprocessing and Geodatabase (Access database) to be accessed directly and simply.

Furthermore, as one component of ArcToolBox, "Model" can combine all of the scripts in "Script" as a new unified ArcTool and conduct simulation according to the running orders of the scripts. Finally, the simulation results are saved automatically by "Model" to a Geodatabase, which, in the present case, is a Microsoft Access database, through ODBC.

In the following, we describe the CAUFN components in detail. Compared with Fig. 17.4, Fig. 17.5 shows all the components of the CAUFN tool. One of the components, which is referred to as VBA, is a learning tool for the estimation of parameters for building type simulation. This tool is developed using ArcObject, which incorporates a neural network tool of R (COM) with the ArcObject VBA program. As shown in this figure, a component caller ArcScene is used to present the simulation results in the 3D world. However, neither the VBA components nor the ArcScene tool are incorporated into the "Model."

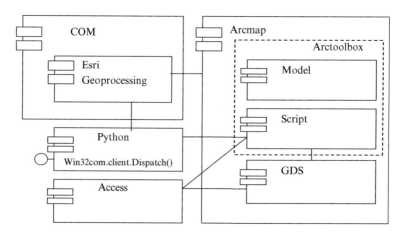

Fig. 17.4 System structure

17 Simulating Land-Use Patterns and Building Types after Land Readjustment   355

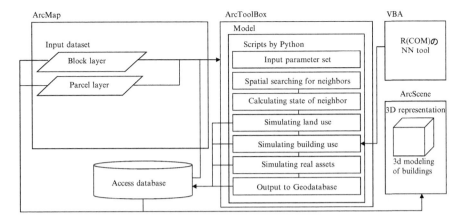

Fig. 17.5 Simulation model

Table 17.3 Input data for simulation

Field name	Type	Content
Land use	Integer	Land use of parcel
Front road	Double	Type of front road
Zoning	Integer	Land-use zoning
Corner	Double	Parcel position
AmountR	Double	State of Residential use
AmountC	Double	State of commercial use
AmountI	Double	State of industrial use
Area	Double	Area of parcel

Development of the CAUFN Tool

As discussed in the previous section, ArcToolBox and the Geodatabase are required in order to implement simulation of the urban build-up process. Before the simulation procedures, the spatial features and attributes of land parcels and blocks must be set up. The attribute information of a planned parcel should be prepared as shown in Table 17.3. Within the simulation of urban build-up process, the parameter settings for scenario configuration should first be prepared as tables and stored in a Geodatabase for both spatial processing and database processing.

Scenario Configuration

In this research, the simulation focuses on the impacts from planning conditions such as land-use zonings on the urban build-up process. We visualize urban patterns through different scenario simulations. During the simulation, different scenarios are set according to different planning conditions. Therefore, the scenario configuration will be carried out based on land-use zonings and other planning conditions. Different parameters for land-use simulation, building type simulation, and real estate management simulation will be set and uploaded as part components of scenarios. We simply predefined the settings for different scenarios of land use, building types, and real estate management according to different land-use zonings. Thus, scenario configuration can be easily achieved by choosing a parameter setting that matches the planned land-use zoning in the simulation target area.

For scenario configuration, all of the parameter settings are predefined and saved as tables in a Geodatabase. The only task users must perform is to choose one table from the database and setup the parameter settings for simulation. Thus, the user can choose different parameter settings easily by adjusting the table names, which are labeled according to different land-use zonings in the database. In this chapter, since we focus on CAUFN development, detailed information about parameter settings will not be discussed further. With respect to the variables of the parameter settings for scenario configuration, the parameters for land-use simulation are explained in (17.1) through (17.7), and building type simulations are explained in (17.8). In addition, parameters for real estate management are explained in (17.9) and (17.10). Within the database, all of the tables for parameter settings are prepared in a Geodatabase and are based on different land-use zonings.

Retrieving Attributes of Neighbors from Spatial Features (Spatial Processing)

Traditionally, in CA models, interactions between cells and their neighbors are commonly considered to determine the state of land parcels and blocks affected by neighbors. As a method of identifying the cell neighbors, the pre-set attributes of parcels and blocks in GIS database can be used to define the cell neighbors, such as the works of Benensonand Torrens (2004) and Blecic et al. (2003). As discussed in previous sections, the neighbors of the parcel layer are defined as parcels that share edges. Thus, the spatial search function of ArcMap can be used for neighbor identification.

For the calculation of the neighbor state, land-use simulation using CA requires a function to search adjacent polygons for retrieving necessary attributes from adjacent polygons. Since the simulation in this chapter will only be conducted at the urban district level, it is unnecessary to deal with the system database by ArcSDE. However, the personal Geodatabase is compatible with the proposed system. Thus, Microsoft Access is used. Consequently, the function that saves the neighbor state along with its parcel ID in a Microsoft Access database for further calculation is necessary. Thus, it is necessary to connect the Microsoft Access database in the simulation process. The block ID of the block layer is also necessary in order to identify neighbors between the layers of blocks and parcels. It is possible to calculate the sum of the state values of all parcels within a block. Thus, the block ID of the block layer can be used to save the parcel state in the blocks' fields for calculating the state impacts from adjacent blocks by retrieving the state of neighbor blocks by the block ID.

Concretely, a single parcel will first be automatically selected for further calculation of state impacts from its neighbors. Then, "Search Space" (Boundary_Touches) will be implemented in order to identify the parcels that share edges with the selected parcel. After the neighborhood identify process, "summary statistic" (Summary Statistics) will calculate the state values of the selected parcels and save the average value of the neighbor state in a database. Thus, through the above processes, the state impacts of neighbors on each land parcel can be calculated, and the results can be saved in a Microsoft Access "table."

In addition, this process has advantages in that it allows users who have experience with ArcGIS to conduct simulations simply by selecting scenario settings and layers of parcels and blocks within an urban district. This is quite different from the simulation tools developed by Benenson and Torrens (2004) and Blecic et al. (2004), in which users must write programs for transition rules and neighborhood identification.

Calculation Function of Geodatabase Processing (Database Processing)

The simulation of land-use type, build type, and real estate management can be conducted by database processing.

With respect to land-use simulation, Microsoft Access is used as a Geodatabase in this research, where the state of the land parcel is determined by geometric characteristics, planning conditions, and neighborhood state in the spatial processing of each simulation loop. Thus, the geometric characteristics and planning conditions of a parcel should be stored as initial conditions through the attribute table of each parcel by a predefined dataset in a Geodatabase. During land-use simulation, the subsequent state of a parcel can be calculated through database processing by considering the influences from the fields of planning conditions, the current parcel state, and its neighbor state in a Geodatabase. The information of these different aspects of a parcel is stored in the same record of the Geodatabase, which is linked to the spatial feature in ArcView. Hence, the calculation process of the land-use state can be carried out using the field values saved in the Geodatabase after the process of retrieving the states of neighbors from spatial features.

With respect to the state calculation mentioned above, a calculation module is needed. In this research, a series of script modules (Table 17.4) is provided for this purpose. These modules can input values of different fields within the same record restored in a Geodatabase for calculation of the subsequent state of a parcel after its initial setting, and the results are output in a Geodatabase table. All of these processes are repeated until the end of the simulation. Within this process, we assume that one loop of the simulation represents one time period in real time.

After calculating the parcel state by Python script, it is necessary to determine land-use types by transition rule. In this research, the land-use types are also determined by Python script, which can add an attribute field to a table of the database and visualize the simulation results on an ArcGIS map. The land-use state will be updated to the attribute table after calculating the subsequent state of each parcel from the initial step, and so simulation can be carried out. Finally, the land-use type is output to the database. The above processes are shown in Fig. 17.6.

Table 17.4 Series of Python scripts

Script name	Content	Processing
Scenario (scenario configuration)		
Parameter_setup	Scenario set	Scenario
Simulation (CA simulation in ArcGIS)		
Neighbor_land Processing	Calculate the average state value of neighboring land	Geo
Mean_block	Calculate the average state value of the city area	
Mean_neighbor_block	Calculate the average state value of the areas near the city district	
Calculate1	Calculate the impact of the neighborhood	
Calculate2	Calculate the amount of land inside the affected state	Simulation
Calculate3	Calculate the amount of the entire area of the affected state	
Calculate4	Calculate the amount of land affected by characteristics	
Calculate5	Calculate the amount of the next state, land-use decisions	
Calculate6	Calculate building use and real estate choice	Output
Output_total_table	Create output table	
Delete_table	Delete unnecessary files in the process of calculating the output	

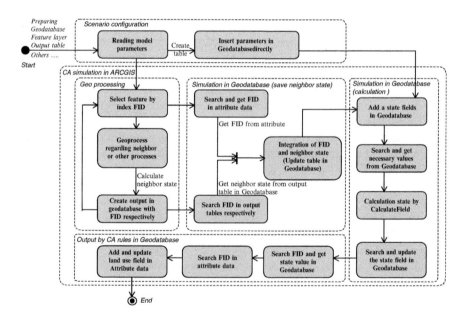

Fig. 17.6 CA simulation for land-use simulation

Computing Building Types and Real Estate Management (Database Processing)

Although simulation models for building types and real estate management are different from simulation models for land-use types, both the simulation processes of building types and real estate management can be performed through the same type of database processing without any spatial processing. The simulation of building types can be carried out based on parameter settings regarding the geometric characteristics, the planning conditions, and so on, which can be retrieved from sample areas by a neural network learning process. Monte Carlo simulation based on a Bayesian probability process can be used for the simulation of real estate properties.

In this chapter, since we focus on system development, the simulation algorithms are not discussed. Accordingly, even though the simulation models of land-use types, building types, and real estate properties are different, the simulation processes can be implemented by spatial processing and database processing, as mentioned above.

Interface of the CAUFN Simulation Tool

The ArcToolBox Model can perform all of the spatial processing and database processing described in the scripts, and can visualize the flow of the simulation process, which allows the simulation operation to be performed simply. As shown in Fig. 17.7, simulations can be implemented through the simulation flow, which integrates all of the scripts into the Model. Moreover, user can use this Model

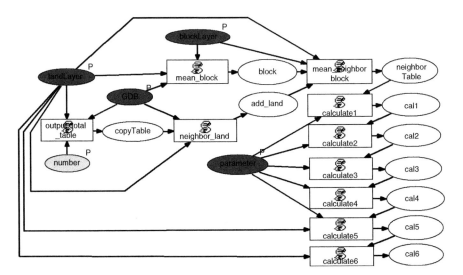

Fig. 17.7 Simulation model using Model of ArcToolBox

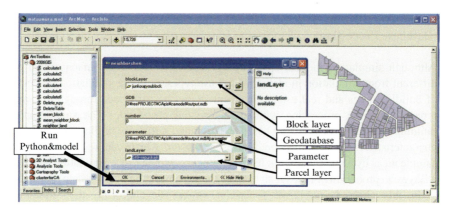

Fig. 17.8 Screen capture of GIS

through a user interface, as shown in Fig. 17.8. Users must open the Geodatabase, parcel layer, and block layer, and choose a parameter set from a table of a database as the scenario configuration.

In summary, the simulation of land use, building types, and real estate properties in urban districts can be automatically performed by the CAUFN tool, in which Python scripts can deal with the simulation models and can automatically save the simulation results to an Access database.

System Test and Additional Applications

As described in previous reports (Shen et al. 2009, Kawamura et al. 2008), the simulation model described in this chapter has been already tested. In this section, we simply consider the analysis of the simulation accuracy and confirm that the scripts embedded in the simulation tool can accurately reflect the simulation model.

Case Study

In order to test the simulation system, a case study is conducted in Matsumura, Kanazawa City. Matsumura is a semi-industrial zoning area in Kanazawa City, where the land-use adjustment project was implemented from 1969 to 1993. This area contains a main street and a sub-main street, as shown as Fig. 17.9. The simulation is carried out in this area, and the build-up process and the simulated build-up process are shown in Fig. 17.10. In addition, the actual land use and simulated land use in the study area are also shown in order to allow comparison of simulation results and real data (Fig. 17.11). In addition, the actual building types and simulated results for building types in the study area are shown in Fig. 17.12.

17 Simulating Land-Use Patterns and Building Types after Land Readjustment

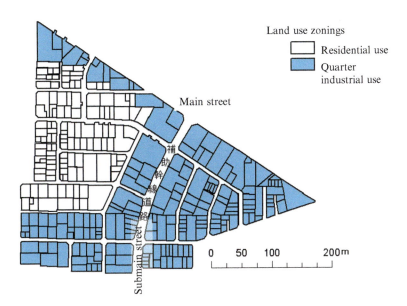

Fig. 17.9 Matsumura area, Kanazawa City

The map of Matsumura shown in Fig. 17.9 is used as a base map for the visualization of the simulation results. Accordingly, we compared the simulated results for land-use and building types with the original datasets for land-use and building types obtained from the basic urban planning survey of Kanazawa City. As shown in Fig. 17.10, with respect to land-use simulation, the simulation accuracy evaluated by the ratio of different types of land use in each step is more than 80%. Moreover, we analyzed the spatial patterns of real land use and simulated land use in the case study area. As shown in Fig. 17.11, the simulated spatial pattern is also very close to that in real urban area. Then, the same procedure was carried out for the evaluation of simulation accuracy of building types. A building type simulation was conducted after the land-use simulation, as shown in Fig. 17.12, and the simulation accuracy was confirmed to be 70%. However, no survey data supporting the accuracy evaluation of real estate management simulation was obtained because of data source limitations. Therefore, as shown in Table 17.5, the simulation accuracy for real estate properties remains unclear, and this part of the system test will be considered in a future study.

The simulation accuracy for land-use types is higher than for building types because the simulation order of land use is a priority. The land-use simulation model has a substantial impact on the simulation accuracy of building types and real estate properties. Based on the results of system test obtained here, the system can be said to perform well with the simulation model, and the simulation output exhibits sufficient accuracy.

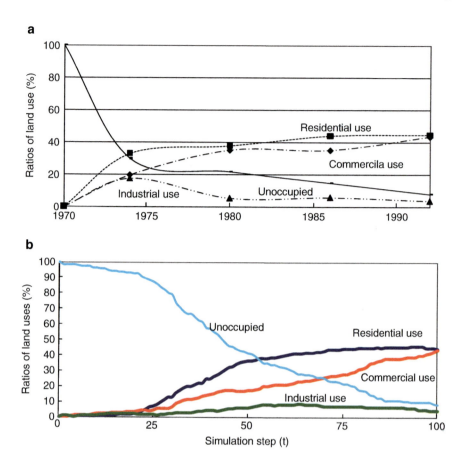

Fig. 17.10 Comparison of the build-up process. (**a**) Build-up process in real society (1969–93, 25 years). (**b**) Simulated build-up process (1969–93 years, 25 years)

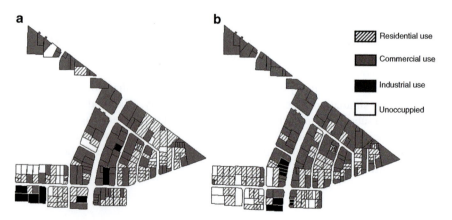

Fig. 17.11 Comparison of actual land use and simulated land use. (**a**) Land-use pattern in real society (1993). (**b**) Simulated land-use pattern

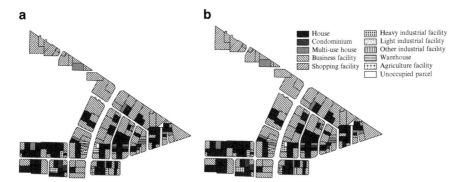

Fig. 17.12 Comparison of actual building types and simulated building types. (**a**) Building use pattern in real society (1993). (**b**) Simulated building use pattern

Table 17.5 Simulation results of asset management

	Low-rise buildings		Low-rise rental buildings		Middle- or high-rise buildings		Middle- or high-rise rental buildings		Total amount	
Commercial use	19	22%	17	20%	30	35%	19	22%	85	100%
Residential use	23	28%	19	23%	19	23%	21	26%	82	100%
Industrial use	1	13%	3	38%	4	50%	0	0%	8	100%
Unoccupied	–		–						13	
Total	43	23%	39	21%	53	28%	40	21%	188	100%

Further Application of the CAUFN Tool

In this system, the simulation results for land use and buildings types can be output in the attributes of spatial features in the format of ArcGIS data. The land parcels and buildings in ArcMap are edited as polygon, thus the buildings can be displayed in a 3D form on ArcScene, as shown in Fig. 17.13. Therefore, it is possible to visualize the simulated results in a 3D form if the information of building height is obtained from the simulated results stored in the Geodatabase tables.

The number of building stories should be estimated by floor area ratio and building coverage ratio in order to present the building height in 3D form. Meanwhile, the building forms can be automatically edited by the buffer function in order to better control the building coverage ratio. Thus, the building forms can be accessed according to their height attribute values recorded in the Geodatabase and can be presented in the 3D world on ArcScene. As shown in Fig. 17.13, the proposed tool will enable more realistic representation using 3D forms, as compared to a 2D map, which can be referenced for future urban development.

Next, we will discuss how these 3D VRML buildings can be used for planning support in planning practice. From the viewpoint of planning and design, the 3D VRML models shown in Fig. 17.14 can be used to review whether the buildings match planning and design guidelines before construction. These types of 3D

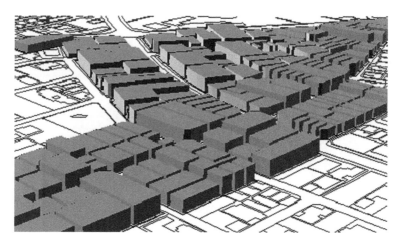

Fig. 17.13 Three-dimensional building form based on simulated building stories

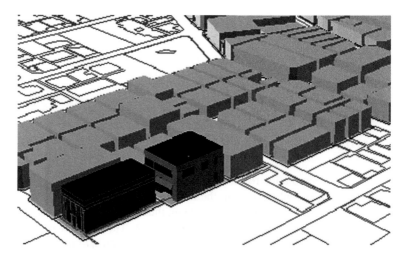

Fig. 17.14 VRML objects replacing building wire frames

images can also be used to check the townscapes of the urban area, and thus more vivid and realistic building models should be considered. Fortunately, 3D models created by VRML and GoogleSketchUP can be inserted directly into ArcScene. Therefore, the build-up process of the urban development area can be gradually represented by a VRML dataset after one building construction is permitted or constructed. In addition, the 3D database can be updated according to the actual building construction. This would be very helpful to stakeholders who learn planning regulations and design guidelines from the simulated urban landscape of their living space and the constructed buildings. Furthermore, the update can be used to provide planning information to the public and efficiently improve public

17 Simulating Land-Use Patterns and Building Types after Land Readjustment 365

participation. Using the Style editing function of ArcMap, these VRML 3D datasets can be incorporated into the database as a symbol. However, for registering VRML data for representing buildings on ArcScene, as shown in Fig. 17.14, only manually operation is available at this stage.

Comparison with Some of the Existing Tools

In Table 17.6, we compare the characteristics of the proposed system with a number of existing systems. Even though all these tools can basically handle irregular polygons, there are a number of different points in the dataset, the database, the neighbor relationship, and so on. First, the dataset for simulation in the present study is edited from the ArcGIS platform, whereas OBEUS is developed based on MapInfo. In the case of the CAGE system, the user must create a dataset using the CAGE tool. Second, Access is employed and OBEUS uses Microsoft's SQL server, whereas CAGE does not use any special database. Third, for defining the neighborhood in the present project, the Python script of ArcToolbox can be customized to handle neighbors automatically. Whereas both OBEUS and CAGE use C++ language for searching neighbor polygons and users must define neighbors by programming, which makes it difficult for ordinary PC users to define neighbors. In CAUFN, this is achieved using only two layers, namely, the parcel layer and the block layer. In both the parcel layer and the block layer, neighbors can be retrieved

Table 17.6 Comparison of simulation tools

	Research tool	Obeus	Cage	Tool using repast
Data	ArcGIS	MapInfo	Editing tools	ArcGIS
Database	Access	SQL	No	Access or others
Neighbor definition	Spatial processing (shared edge)	Spatial processing attributes	Spatial processing attributes	Spatial processing attributes and features that share edges
Relations between neighbor	Within a layer	Between layers	Between layers	Layer and between layers
How to define the neighborhood	Programming does not require input from the user	Programming and attributes specified by the user	Programming and attributes specified by the user	Programming
Engine	ArcToolbox, Python	OBEUS, Borland C++	CAGE, C++	Python
Mathematical model	Prepared in the tool	Developed by the user	Developed by the user	Developed by the user
Simulation time	Once, 5 min (180 polygons)	Approximately 10 s per tick	Approximately 10 s per tick	NA
Difficulty	Easy	Difficult	Difficult	Difficult

by spatial processing. In OBEUS/CAGE, neighbor relations between the multiple layers and spatial features can be defined by programming through attribute tables.

Consequently, the CAUFN tool conquered the limit of users of OBEUS/CAGE that can only be used by system developers, so that ordinary users, namely planners can use the CAUFN tool to simulate the build-up process at the urban district level. However, in the case of CAUFN, the system searches neighbors by python script so that the simulation loop for searching will consume a relatively long computing time.

Conclusion

In this chapter, we introduced the CAUFN tool based on a dataset of irregular polygons, which takes into account to the build-up process simulation regarding land use, building types, and real estate properties at the urban district level. If users have experience in operating ArcMap, a table can be selected using a parameter set as a scenario configuration, and the layers of blocks and parcels can be chosen in order to perform the simulation simply. In the proposed simulation system, the CA approach, the NN theory, and portfolio theory are integrated for the simulation of land use, building types, and real estate properties in a build-up process. Here, CA simulation is the initial step, and simulated land use will provide the basis of further building type simulation, thus the error issue of CA simulation result has substantial influence on the accuracy of building type simulation. Accordingly, the spatial processing and database processing of ArcGIS are used to handle the basic concept of the cell state, the neighborhood, the transition rule, and the prediction of building use and real estate properties. ArcScene can be used to provide a future image of a city in 3D form and it becomes possible for users to share a future image of their own city in planning practice.

Furthermore, we tested the CAUFN tool in a case study area and analyzed the accuracy of the CAUFN tool. The CAUFN simulation tool was found to be capable of visualizing high-precision simulation results. First, the proportion of land use and building types agrees well with the real dataset. Second, the simulated spatial patterns of land use and build types represent well the real spatial patterns in the case study area. However, since the land-use simulation is conducted before those for buildings and real estate properties, the simulation results for land use are crucial and greatly influence subsequent simulations.

Compared with other systems, such as CAGE (Blecic et al. 2003) and OBEUS (Benenson and Torrens 2004), which are designed for system developers, the CAUFN tool is easy for ordinary users to operate because the functions of defining neighbors and simulation have been prescribed in the CAUFN tool in this study. Thus, users do not need to write programs to define neighborhoods and their transition rules. This system has limitations, such as high computing time, and so reducing the simulation time is a future challenge. In addition, validation of the simulation results for simulating real estate management should be conducted. Moreover, a multi-agent system (MAS) would be helpful for the integration of build-up processes for simulating land-use types, building types, and real estate properties. Thus, a multi-agent system should be introduced to the ArcGIS platform, and Repast, a simulation tool of integrating agent-based model with GIS will be a useful tool for further research.

Acknowledgments This research have received supports from research grants for Scientific Research C (Proposal No. 19,560,613).

References

Blecic, I., Cecchini, A., Rizzi, P., Tronfio, Giuseppe A. (2003). Playing with Automata. An Innovative Perspective for Gaming Simulation, 5 C-3, CUPUM'03, Sendai, Japan, May 2003.
Benenson, I., Torrens, P.M. (2004): Geosimulation Automata-based modeling of urban phenomena. Wiley, England, 2004.
Johansson, H. (2002), Investment appraisal using quantitative risk analysis, Journal of Hazardous Materials, Volume 93, 77–91.
Kawamuwa, I., Shen, Z., Kawakami, M., Mochitsuki, S. (2008), Planning support system for Land Use Formation on Urban Partitions Using MAS, Proceedings of the 31th symposium on computer technology of information, systems and applications, 43–48.
Kamide, K., Kawakami, M., Kidani, H. (1998). Urbanization in Land Readjustment Project Area-Case of Land Owner's Project in Kanazawa City -, City planning review Vol. 33, 145–150. (In Japanese)
Maniruzzaman, K. M., Asami, Y., Okabe A, (1994), Land use and the geometry of parcels in Setagaya Ward, Tokyo, Theory and Application of GIS, Vol. 2 pp. 83–90.
Pløger, J. (2001), Public participation and the art of governance, Environment and Planning B: Planning and Design 28(2) 219–241.
Santé, I., Garćıa, Andrés M., Miranda, D., Crecente, R. (2010), Cellular automata models for the simulation of real-world urban processes: A review and analysis. Landscape and Urban Planning, Volume 96, Issue 2, 108–122.
Shen, Z., Kawakami, M., Kawamura, I. (2009), Geosimulation model using geographic automata for simulating land-use patterns in urban partitions, Environment and Planning B: Planning and Design 36(5) 802–823.
Shen, Z., Ishimaru, N. (2000): Planning influence factors of land use after land readjustment project in danbara area redevelopment project, Hiroshima city, Journal of Archit. Plann. Environ.Engng., Vol. 536, 191–198. (In Japanese).
Shen, Z., Kawakami, M., Kawamura, I., Kato, K. (2007). Study on development and application of micro simulation model for lot usage transition using ca, Journal of architecture and planning Vol. 620, 249–256.
Shen, Z., Kawakami, M. (2010), An online visualization tool for Internet-based local townscape design, Computers, Environment and Urban Systems, Volume 34, Issue 2, 104–116.
Stevens D., Dragicevic S. (2007), A GIS-based irregular cellular automata model of land-use change, Environ. Plann. B: Plann. Design 34, 708–724.
Sugisaki, K., Koizumi, H., Okata, J. (2003), Consideration on effects of public outreach program on plan making process by citizen participation, Journal of the City Planning Institute of Japan, Vol. 38, 140–145.
Takahiro Tashiro, Takeshi Ohteki (1992): Decision-making on construction investment based on the Portfolio Theory – estimated the ratio of land use in urban area –, City planning review Vol. 27, 757–762.

Takizawa, A., Kawamura, H., Akinori Tani (1998). Cities as cellular automata (Part 1), Applicability of ca and formation of urban land-use patterns. Journal of Archit.Plann.Environ. Engng., Vol. 506, 203–209. (In Japanese)

Watanabe, K., Ohgai, A., Igarashi, M. (2000). Cellular automata modelling for estimating historical change of urban area, Journal of Archit. Plann. Environ. Engng., Vol. 533, 105–112.(In Japanese)

Wu, F. (1996). A linguistic cellular automata simulation approach for sustainable land development in a fast growing region, Computers, Environment and Urban Systems, Volume 20, Issue 6, 367–387.

Chapter 18
Integration of MAS and GIS Using Netlogo

Zhenjiang Shen, Xiaobai A. Yao, Mitsuhiko Kawakami, Ping Chen, and Masahito Koujin

Introduction

In the past years, Japanese cities have widely practiced inner-city regeneration plan by implementing diverse planning policies. However, it is proven difficult to evaluate the impact of these policies properly as a city is a complex system with integrated socioeconomic components. The remarkable phenomenon of downtown decline in local Japanese cities may be partly contributed by locating large-scale shopping malls in urban fringe where there is lack of urban planning control. Considering dynamic interactions in an urban system, many researchers advocate for modeling urban sprawl systems from spatial and temporal perspectives via simulation.

Many existing urban simulations have been achieved with the use of Cellular Automata (CA) based simulation, which involves cellular decomposition of the space; neighbourhood relations among cells, and modeling dynamics based on local evolution rules. The more recent agent-based approach provides another promising alternative whose strength lies in its ability to simulate dynamics based on actions of autonomous agents in the simulated space. Vancheri et al. (2008a, b) integrate Cellular Automata approach with the idea of multi agent systems to describe dynamics of the urban system as decision processes of agents. Many good efforts of integrated approach to urban simulation for urban planning have been reported in the literature. For example, CityDev, a web-based system for city planning teaching (Sembolonia et al. 2004), uses an interactive MAS model of a city in grid system. Timmeramans group in the Netherlands developed MASQUE (Ma et al. 2004)

Z. Shen (✉) • M. Kawakami • P. Chen,

• M. Koujin
School of Environmental Design, Kanazawa University, Kanazawa, Japan
e-mail: shenzhe@t.kanazawa-u.ac.jp

X.A. Yao
Department of Geography, University of Georgia, Athens, GA, USA

Z. Shen, *Geospatial Techniques in Urban Planning*, Advances in Geographic
Information Science, DOI 10.1007/978-3-642-13559-0_18,
© Springer-Verlag Berlin Heidelberg 2012

which applies the approach in the context of retail. In another prior study of MAS for retail locations, Arentze et al. (2000) proposes a knowledge-based system to support the analysis of problems and formulation of actions in the field of retail location planning. Wu (1996) embeds micro-economic models of community and developer behaviour within a CA framework. His study simulates natural land use zoning under free market and incremental development control regime. The study takes into account land use zoning and development control policies that regulate land use intensity to the socially optimal level.

Crooks et al. (2008) pointed out that there are seven challenges of multi agent modeling. In this chapter, we want to present our solutions in response to some critical challenges in the integration of GIS and MAS for interactive household agents' shopping behaviours in multi-agent simulation. Firstly, the challenge is how to generate a virtual space that is on equal terms with a real city. Secondly, the challenge is how to generate agents in the virtual space that can really represent those in the real city. In our former reports of research along this line (Shen et al. 2004; Chen et al. 2008), simulation models have been presented and implemented in a virtual space. This study aims to further validate and test our simulation model with a real city.

Research Approach

As illustrated in Fig. 18.1, the basic idea of the designed simulator is to locate new shops and then find market shares of the suburban large-scale shops and those of the small downtown family shops. This simulator, Shopsim-MAS (Shen et al. 2011), is a simulation model specially designed and implemented in the Netlogo environment. Netlogo is a free simulation platform and is downloadable from the Internet.

To simulate the formation of market shares, the household agents' shopping behaviors, the distribution of households, locations of shop agents, and planning regulations with respect to locations are critical information. In the following sections, we will show how real world data of such information are fed into the simulation system.

Kanazawa city is selected for the case study. Our model will simulate the shopping behavior of residents in Kanazawa city. The following are the major types of space and agents information which are collected and processed in the geographic information system (GIS):

- Census survey 1985 for creating and locating household agents.
- Commercial survey 1985 for locating large-scale and small-scale shop agents and their shopping market share.
- Planning information provided by local government as new shop agents location regulation.

The GIS data are input into Shopsim-MAS and saved as the native Netlogo file format, named patch data. After the above preparation of real city data, we will load

18 Integration of MAS and GIS Using Netlogo 371

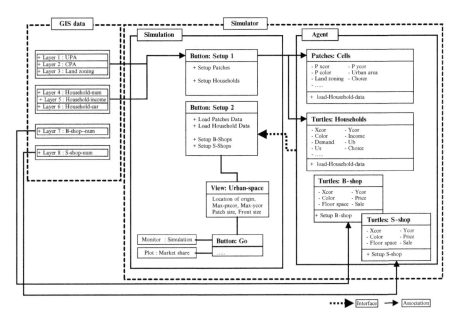

Fig. 18.1 GIS data input and Shopsim-MAS simulator (UPA: urban planning area; UCA: urban control area)

the patch (space) and agents datasets into Shopsim-MAS and apply our developed Shopsim-MAS model in Netlogo to simulate shopping behaviors and decision making dynamics.

Simulation Model

The simulation component of the system models households' shop choice behaviour. To let the simulation focus on the impact of B-shops, we make following assumptioms to simplify the situation:

1. The distribution of goods among shops is homogeneous, i.e. the household can buy the same goods at all the shops.
2. Each household has a constant demand for goods. When the total demands of all household agents are satisfied, the simulation process will end.
3. In each simulation iteration, a household wants to buy one unit of demand.
4. A household only considers shops within a threshold travel distance γ.
5. The shop with the highest utility will be chosen by the household. When available shops are under equal conditions in terms of utility, the household chooses one from them randomly.
6. Shopping choices of all household agents will reach a cooperative equlibrium for a better commerical environment.

Utility Function Without Consideration of Interactions Among Agents

Random Utility Model (RUM) has been widely made to account for the heterogeneity among variables' influences. The random parameter logit (RPL) approach (Lijesen 2006), the latent class logit (LCL) approach (Boxall and Adamowicz 2002) and the mixed logit (ML) model (e.g. Frew 1990) are three popular variations of the RUM methods and were proven to be able to forecast equally well (Provencher et al. 2004). The ML model allows the coefficients of observed variables to vary randomly for different people. Considering the heterogeneity of household agents in our study, we adopt the framework of the ML model and modified it to design the decision rules of household agents, as expressed in (18.1). While accounting for variations among individual preferences, the modified ML utility model is to ensure consistent behaviour of households in the same income group. Equation (18.1) defines the utility function of household i of income group g shopping at shop j.

$$U_{ijg} = \sum_{n=1}^{n} \mu\beta_{ign}X_{ijn} + \varepsilon_{ijg}, \quad \varepsilon_{ijg} = a - b(\text{In}(-\text{In}(\theta))), \ \theta \in [-\theta_g, +\theta_g] \quad (18.1)$$

X_{ij} is a vector of observable explanatory variables describing attributes of household i and the shop j. Examples of these variables include travel cost, urban amenity, price of goods, floor spaces of shop j, and others. The subscript n refers to the dimension of the vector (number of variables). The symbol $\mu\beta_{ig}$ is a vector of respective coefficients to the variables. The coefficient vector has two components, the vector of average coefficients for the gth income group and the vector of random values reflecting individual deviation within the group. The element ε_{ijg} in (18.1) represents unobserved random contribution to the utility. It is used to compensate for the inherent uncertainty of shopping behaviors. This random element follows Gumble distribution and can be generated using a random number θ_g following uniform distribution. The pre-defined range of θ_g represents the maximal magnitude of possible internal differences within the income group g. The parameters a, b in (18.1) are set as 0.5 and 2 in this study. After obtaining the utility measures from household i to every shop alternative, the probability that i shopping at shop j can be calculated from (18.2).

$$P_{ij} = \exp(V_{ij}) \Big/ \sum_{k=1}^{J} \exp(V_{ik}); \quad j, k \in J \quad (18.2)$$

where J is the collection of all shops and V_{ij} is the utility value calculated from (18.1).

18 Integration of MAS and GIS Using Netlogo

Interactions Among Agents

The interactions among agents are not considered in (18.2). This section models an additional component for the previously discussed utility to take the interactions into account. This new component is termed interaction utility. We model the combined effects of two types of interactive influences, the peer impact among neighboring household agents and information delivery from shop agents to household agents.

In this study, the peer impact refers to the influence a household receives from the shop choices of neighbors. Here the neighbors of a household are defined as those who are in the nine-cell neighborhood area around the home cell. The utility of a t type shop (say, S-shop) can be promoted by the peer impact from neighbors who go shopping at the same type of shops (any S-shop). It is represented by I_{ijg} in (18.3).

The information delivery type of interaction considers information of shops (such as prices, types of goods, shopping environment, etc.) being delivered from the shop agents and spread among the households. The spread of such information may attract shoppers who were previously patrons of other shops. The utility of a t type shop (say, S-shop) can be promoted by spreading information to households who are currently patrons of a different type of shop (e.g. any B-shop), thereby these households may be potentially attracted to the t type of shops. It is represented by D_{ijg} in (18.3). Collectively the interaction utility is INT_{ijg} in (18.3). It stands for the interaction utility of household i in income group g shopping at shop j is denoted as INT_{ijg}. In the equation, subscript t refers to the type of shop that j shop belongs to. In this study, there are obviously only two types of shops, namely the S type and the B type.

$$INT_{ijg} = I_{itg} + D_{itg}, \quad t \in \{S, B\}, \quad j \text{ is type } t \text{ shop} \tag{18.3}$$

The peer influence I_{itg} and information delivery D_{itg} are calibrated with (18.4):

$$I_{ijg} = k_g(N_{itg} - N_{iog})/N_{ig}$$
$$D_{ijg} = d_g(N_{iogd} - N_{itgd})/N_{igd} \tag{18.4}$$

where N_{itg} is the number of i's neighbors who are in gth income group and shop at t type of shop, and N_{iog} is the number of those who shop at the other type of shop. N_{ig} is the total number of i's neighbours in gth income group. The equation for information delivery has similar notations with the additional subscript d which is the distance between household i and shop j. A notation with subscript d means the respective number is counted within the search area of radius d around shop j. The parameter k_g is a scaling factor reflecting household agents' subjective reaction to such influences, which is a constant. In short, the impact of the number of any type of shops in a neighbourhood contributes to the interaction utility in two opposite ways through I and D respectively and thus makes the total interaction utility changing in a wave form.

After considering interactions among agents, the utility function defined in (18.1) should be modified as (18.5).

$$U_{ijg} = \sum_{n=1}^{m} \mu\beta_{ign}X_{ijn} + \mu INT_{ijg} + \varepsilon_{ijg}$$

$$\text{where } \varepsilon_{ijg} = a - b(\ln(-\ln(\theta))), \theta \in [-\theta_g, +\theta_g] \tag{18.5}$$

Modeling Transportation Mode in the Shop-Choice Model

Shop-choice decisions are heavily influenced by the transportation mode of the planned shopping trip. For this reason, transportation policy instruments can weigh in to influence the market shares of different shops. A traveler's choice of transportation mode and route are determined by many factors (Ortúzar and Willumsen 2001) with costs in time and distance being the most accountable factors (Outram and Thompson 1978). In this study, we use generalized cost as defined in (18.8) to incorporate time, distance, and monetary cost to model the choice of transportation mode and route. In the equation, $TM\ Cost_m$ refers to the generalized cost for transportation mode m. The notation D is the travel distance, C_m is the unit monetary travel cost for mode m, s_m is the average speed associated with the travel mode, and T_m is parking fee. The notation h is a weight used as a scaling factor, which follows normal distribution with mean of 1 and standard deviation of 0.5. Equation (18.6) defines the utility function for travel mode choice. The random coefficient $\mu\beta_{igm}$ in (18.6) is aimed to account for variations among household agents, in which μ is generated following normal distribution with mean of 1 and standard deviation of 0.5 and β_{igm} has different values based on the gth income group and different travel mode m. There is also a random element γ_{img} to account for variations due to other unobserved factors. It follows Gumble distribution generated as independent and identically-distributed random element. In the equation,

$$U_m = V_m + \gamma_{igm} = \mu\beta_{igm}X_{ijm} + \gamma_{igm} \tag{18.6}$$

$$X_{ijm} = TM\ Cost_m = DC_m + T_m + \frac{hD}{S_m} \tag{18.7}$$

$$P_m = \exp(V_m) \left/ \sum_{m=1}^{M} \exp(V_m) \right. \tag{18.8}$$

The process of shop and route choices is summarized in Fig. 18.2. In the figure, the travel mode choice probabilities (P_1, P_2, P_3 and P_4) and the shopping probabilities (P_b and P_s) are simulated by the mixed logit (ML) model (e.g. Frew 1990) based on utility values.

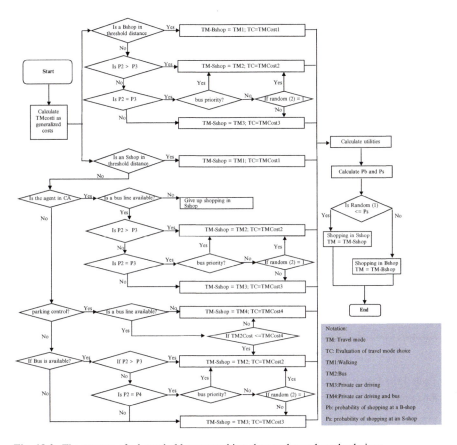

Fig. 18.2 The process of a household agent making shop and travel-mode choices

Figure 18.3 shows the interface of the simulation tool using hypothetical data.

Household Agent Attitude to Urban Policy: Interaction Between Agents

A Cooperative Equilibrium for Shop-Choice Simulation

Competitive equilibrium and cooperative equilibrium are two major alternatives to simulate the process of reaching the system's equilibrium in an MAS (Namatame et al. 1998). This study takes the cooperative equilibrium approach. We make the assumption that household agents do not compete with each other but instead they make shopping decisions cooperatively with consideration of the entire commercial and local policy environment.

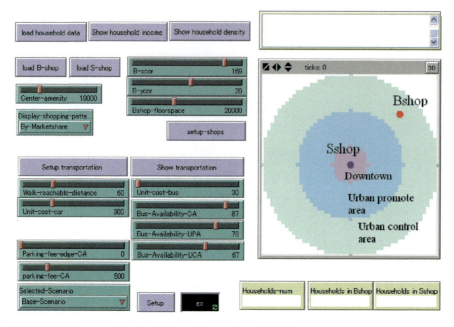

Fig. 18.3 A demonstrative simulation without realworld GIS dataset

Table 18.1 Payoff matrix

Strategy of household agents	Positive to policy (market share of S-shop is larger) (p_s)	Negative to policy (market share of B-shop is larger) ($1 - p_s$)
1. Shopping in an S-shop	$U_s + \beta_s$	U_s
2. Shopping in a B-shop	U_b	$U_b + \beta_b$

For this, household agents' attitudes to planning policy are considered in simulation model. The agents have two strategic choices in making decision of where to go shopping: an S shop or a B shop. In (18.9), w_i is agents' attitude to ratio of market share Ps of S-shop that is respective with the planning regulation of large-scale B Shop.

$$\beta_s = w_i(1 - P_s)$$
$$\beta_b = w_i(1 - P_B)$$
(18.9)

If an agent chooses a shopping strategy of favoring S-shop (supporting urban policy), the β_s will be used as an additional component of the utility function to determine the agent's final shopping choice. However, if the agent is negative to the urban policy and choices a strategy of favoring B-shop, the additional component β_b will be used in the utility function. The payoff matrix in Table 18.1 shows how the actual utility value is computed with the consideration of policy attitude.

Once the utility values are computed, shopping decisions will be made accordingly. For instance, when $U_s > U_b$, the agent will choose the S-shop. This instance is expressed mathematically in (18.10).

$$U_s > U_b$$
$$U_s = p_s(U_s + \beta_i) + U_s(1 - p_s) > U_b = U_b p_s + (U_b + \beta_i)(1 + p_s)$$
$$p_s > k = (U_s + \beta_s - U_b)/(U_s + \beta_s + U_b + \beta_b - U_s - U_b)$$
$$= (U_s + \beta_s - U_b)/(\beta_s + \beta_b) > 0$$

(18.10)

Model Test for Policy Attitude

In the above section, we describe the household agents' shopping behaviours while considering heterogeneous distribution of the agents as well as their interactions and policy attitude in the simulation process. In our former work (Shen et al. 2006, 2011; Chen et al. 2004), we have examined the parameters' sensitive of bus cost, parking fee, bus availability, good price, floor space and interaction between household agents on in shop-choice model. Therefore, in this section, we will focus on the model testing for the parameter of policy attitudes which will be at work in the dynamic process to reach market equilibrium.

In our case study, a global parameter (wi) – attitude-to-policy is set for all household agents to reach a cooperative equilibrium (Namatame et al. 1998), which is employed to modify the utility values individually as discussed in Sect. "A Cooperative Equilibrium for Shop-Choice Simulation". When too many household agents choose B-shops, the parameter and associated new utility component will make the utility of S-shop grow larger in the simulation process. Thus S-shop choices will improve and vice versa. This way the market share will gradually reach a balance between the market shares.

Figure 18.4 shows how the market shares fluctuate dynamically and reach the final equilibrium (stability). It also shows that different equilibrium of the market shares can be reached under different policy attitude. As shown in Fig. 18.4, we can understand how all the household agents reach a cooperative equilibrium of market shares by attitude-to-policy parameter. When more people shopping at S-shops, users can set the parameter as 5, thus the market share of S-shop became lager. When the parameter is set as -5, thus the market share became the result on the contrary.

In summary, both the neighbor impact and the attitude-to-policy are parameters that can change the utility of S-shop according to the utility of B-shop, thus if the parameters are set too large market shares will change in a wave form periodically. In this chapter, we keep these parameters under thresholds as shown in Fig. 18.2, which have been tested in our former work (Shen et al. 2011). In the next section, we like to apply this simulation model to a real world city – the Kanazawa city.

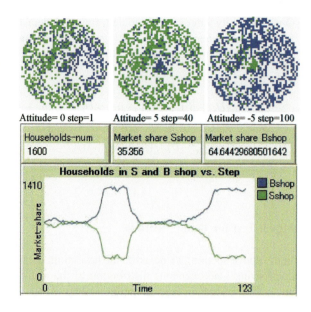

Fig. 18.4 Interactions between household agents in virtual space. Household agents in B-shop and S-shop

A Reality-Based Case Study

GIS Dataset

In order to make reality-based virtual conditions in the simulation space, household distribution and shop locations are generated from survey database. However, because the survey data are provided in aggregated forms, household distribution and shop locations cannot be precisely represented.

Figure 18.5 shows the flow of processes to convert from real city survey data to data ready for use in our developed simulation system on the Netlogo platform. The GIS dataset are from national surveys in Japan. Most of them are in vector format. The statistical units in these survey data are primarily of two types, namely mesh grids and survey zones. Mesh grid is a kind of polygons of adjoining equal-sized squares, which is always employed for the management of Digital National Information of Japan including commercial survey, industrial survey and land use survey. The size of the squares is either 1 km or 500 m wide. The survey zones are usually irregular polygons, which are statistical units of national census surveys.

Table 18.2 compares differences in data between our previously experimented virtual city simulation and this real city simulation. As dynamic interactions among agents in the urban is largely facilitated and constrained by spatial distributions of them, it is absolutely necessary to translate the spatial and attribute information in the vector GIS database into properties of patches (or cells) in the simulation space. However, the available household data are in aggregated format in the census

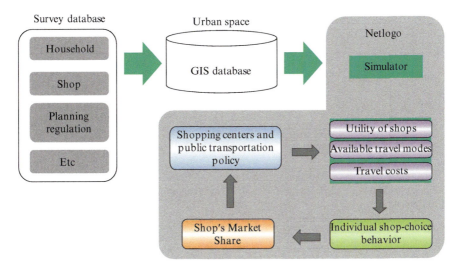

Fig. 18.5 GIS dataset and simulator

survey and other data sources, the household agents actually cannot be created directly from any database of national or regional survey.

Another tricky issue is about the spatial resolution. The commercial survey data stored in 500-m mesh is too big for a shop agent. In addition, a survey zone as statistical spatial unit of census survey is also too large for one household agent. If household agents are generated based on 500-m size mesh, the household will appear concentrated on a point matrix of the mesh centers. To deal with this problem, we partition each 500 m mesh to 50 m grid cells. The resolution of 50 m is chosen as it is a rational size for one household if living in a detached house. Numbers of households are evenly divided among all cells within a mesh. However, this may cause non-integer numbers to be assigned. The rule is that a cell assigned with a decimal number smaller than 1 (such as 0.9) is regarded empty. Thus the household numbers allocated at the 50-m meshes can be slightly smaller than the actual number of the real city. As shown in Table 18.2, su the simulated household number is 134,786 which is 4.16% smaller than the actual survey data of 141,097 households. It is noted that in the converted raster GIS data, 94.5% of all 50 m grid cells have more than one household in each cell. Figure 18.6 shows the household distribution in vector format based-on 1985-census survey of Kanazawa city vis-à-vis the converted raster GIS data of 50-m resolution. Through this process, the vector-based GIS data can be inputted into Netlogo as raster data and be saved as patch data in Netlogo. For integration of GIS dataset in the platform of Netlogo, the simulation space is properly georeferenced according to the converted raster dataset.

With respect to generating shop agents in the simulation space, we first investigate number of shops in each mesh according to national commercial survey. As the commercial survey data are collected for each 500-m mesh, we re-distribute the

Table 18.2 Comparison of data for the virtual city and the real city

		Spatial data of the virtual city	Spatial data of the real city
Source	Shape	Grid cells	Vector data
	Data type	50 m grids	Commercial survey 500 m mesh
Data			Census survey Irregular polygon
			Land use zones
	Area (m^2)	17,625,000	17,625,000
	Space size	466*746 cells	466*746 cells (converted to raster)
	Resolution	50 m	50 m (converted to raster)
Data prepared for simulation	Household	134,786 or a proportion of it	141,097 (census survey)
	Number of B-shops	20	20 (commercial survey)
	Number of S-shops	6,500	6,500 (commercial survey)

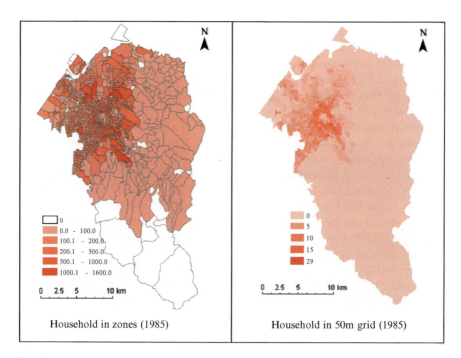

Fig. 18.6 Data conversion from vector data to raster data

numbers of shops from 500-m to 50-m mesh to be in consistency with the dataset of household. Thus the locations are somewhat different from the real shop locations. As a result, these shop agents are distributed like a spatial matrix in the simulation

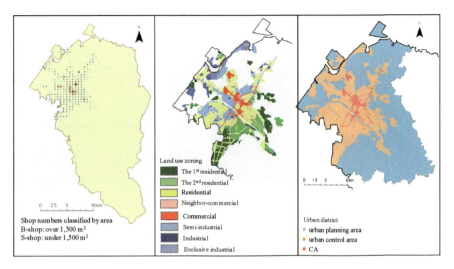

Fig. 18.7 Shop distribution and planning information that should be conversed to 50*50 m grid data

space as shown in Fig. 18.7a. In the converted raster data, the study made sure that the numbers of B-shops and S-shops are kept the same as that in the original data in commercial survey, and the locations of shops are on the centers of 50 m grid cells. In the preparation of simulation data, planning information are also fed into the simulation space of 50-m spatial resolution. The planning information such as planning areas and land use zones are from Kanazawa local government as shown in Fig. 18.7b, c.

To help in the sophisticated data preparation processes, a tool for inputting GIS dataset into Netlogo is developed. The interface of the tool is shown in Fig. 18.8. This tool employs GIS extension of Netlogo that can import and export GIS dataset. The GIS datasets in the map shown in Fig. 18.7 include land use zonings, urban planning areas (downtown, urban promote area (UPA) and urban control area (UCA)), household numbers, income type distribution and shop numbers. All the information of GIS dataset are exported as a patch file using the conversion tool and then be imported to Shopsim-MAS tool directly. Additionally, household numbers can also be imported to Shopsim-MAS as household agents instead of patches. Similarly, the file containing existing shop numbers and locations are imported to Shopsim-MAS directly as existing shop agents. In this interface, a new shop agent can also be located in simulation space by users directly (Fig. 18.9).

As shown in Fig. 18.10, the household agents, small shops (S-shop) and large-scale shops (B-shop) are created successfully in Netlogo. With respect to the locations of S-shops, there are S-shop agents in downtown and other areas. As the impact of B-shops to S-shops located in downtown is the focus of our simulation study, the market shares of the S-shops outside the downtown should be separated so that the impact of B-shops can be measured correctly. At the same time, the impact of new B-shops is another point of research interest, thus the market share

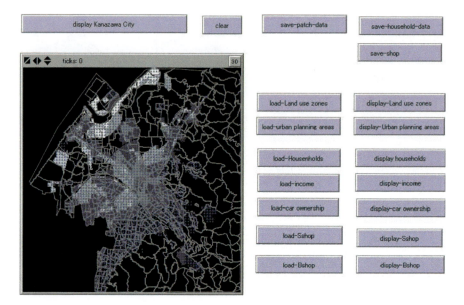

Fig. 18.8 Data conversion tool (for input raster dataset)

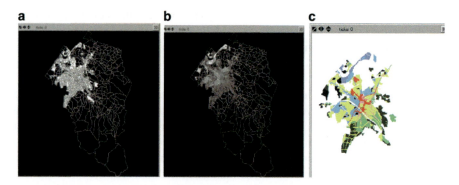

Fig. 18.9 Urban planning information inputted in Shopsim-MAS using Netlogo platform: (**a**) UCA and UPA; (**b**) land use zoning; (**c**) combined Netlogo map

of existing B-shops should also be separated when calculating market shares after simulation.

The Simulation Process

After loading the household, agents, land use planning, and shop data by procedures as discussed in the previous section, a user may interactively define other simulation

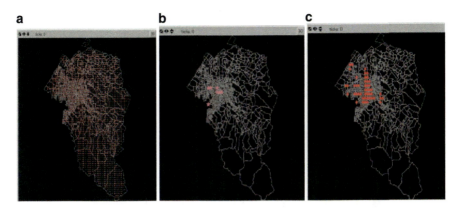

Fig. 18.10 Shop agents and household agents inputted in Shopsim-MAS using Netlogo platform: (**a**) generated household agents; (**b**) generated B-shops; (**c**) generated S-shops

parameters and launch the simulation of household agents' shopping behaviours. In this chapter, we focus on the impact of planning regulations on shopping behaviours and consequently shop's market shares.

To account for interactions of household agents in simulation, the neighbour impact and information delivery are considered in simulation of Kanazawa city. There was not such a policy against B-shops in 1985 when development projects of large-scale shopping centers were boosted energetically in Kanazawa. Therefore, the attitude-to-policy parameter is set as -1 to reflect household agents' preference for B-shops in 1985. All other parameters employed in this simulation are shown in Fig. 18.11.

Simulation with the scale of a real city is very computationally expensive. For 141,097 household agents which is the case for Kanazawa city, the simulation time for one iteration is 1½ h. A simulation usually needs 30 iterations for 1 month with each iteration representing 1 day's shopping. To control the computation time, we try to reduce population proportionally in each cell. Table 18.3 shows the result of such reduction of population size. When the proportion is 1/3, the simulating time for one iteration is around 19 min. If we choose 1/10, one iteration only takes 2 min to complete.

To investigate the impact of the choice of proportions on the population configuration in different land use types, a sensitivity analysis is conducted. We compared the relative shares of households in different land use zones in the simulated urban space by different choices of proportion of population included in the simulation. As shown in Fig. 18.12, the land use zones in comparison include commercial use zone, UPA and UCA in the simulated urban space. The majority of population is always found in the UPA and the relative share becomes pretty stable when the proportion of included household decreases up to 1/5. When the proportion of household numbers further decreases to 1/10 (10%), the shares of households in different urban areas start to change significantly. As a result, we make a conclusion

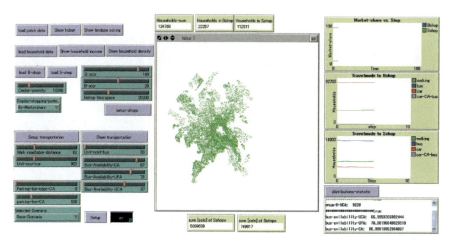

Fig. 18.11 The simulation tools in the Shopsim-MAS

Table 18.3 Simulation time and household numbers

Proportion of household numbers (virtual space/real space and its %)	Total household numbers in simulation space	Time/one step
1/1 (100%)	134,786	100′
1/2 (50%)	56,146	42′
1/3 (33.3%)	24,244	19′
1/5 (20%)	13,914	12′
1/10 (10%)	2,318	2′

Note: if the numbers of household in one cell is under 1 after division, the cells are accounted for no household

that the applicable proportions of household numbers for simulation is no less than 1/5, or 20% of household numbers of the real census data.

To further investigate the impact of above-mentioned choice on simulation result, we also examined output of the choices of transportation modes. It was found that most households living in commercial use zones are in preference to walking, most households living in the UPA have preferences for bus or car riding, while those households living on UCA are more likely to choose car riding for shopping trips. As shown in Fig. 18.13, two shopping behavior survey data and the corresponding simulated results are plotted together. In all cases of 1/1, 1/3 and 1/5 household proportions used in the simulations, the transportation mode share of walking is higher in the simulated results than those of both surveys. The mode share of bus riding is lower than those of real survey data. In the case of simulated based on 1/1 (100%) household data, the discrepancies between simulated results and real data are likely due to inaccuracies of the Shopsim-MAS simulation model. In the case of simulations based on 1/3 and 1/5 of households, we can see similar trends of discrepancies whereas all discrepancies increased in magnitude.

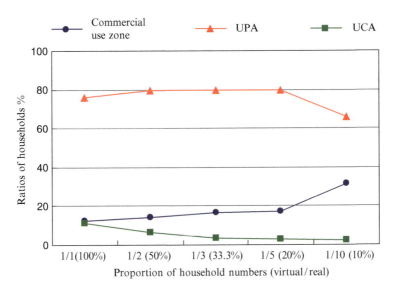

Fig. 18.12 Comparison of ratios of household numbers on land use zones

Finally, we chose the proportion 1/1 of households for simulation and the simulated market share results are compared against real survey data. Firstly, the household agents located in CA, UPA and UCA are compared with the census survey. As shown in Table 18.4, the simulated household numbers in CA and UPA are almost the same as those in the real city, while the simulated household number of UCA is somewhat smaller than that of the real city. It means that some households located in UCA are lost in the data conversion process. This is also the reason why the ratio of household agents choosing TM1 (walking) in simulation is higher than the ratio of walking in real city as shown in Fig. 18.13. With respect to market share, 30 ticks (iterations) of simulation were conducted to monitor the dynamic process of reaching the equilibrium of market shares. The simulated sale patterns are compared with those of commercial survey as shown in Table 18.5. Similar to the case of travel modes, most S-shops located in CA and UPA where the ratios of household agents became higher than those of real city in data conversion process, consequently the market share of S-shops is lower than those of other areas as shown in Table 18.5. However, as shown in Table 18.6, the sale patterns are unstable when different proportions of household agents are used for simulation.

Further Research and Discussion

In this chapter we present a tool for integrating GIS dataset with multi-agent system platform, namely Shopsim-MAS. This is to support impact analysis of various planning policies in response to the pressing problem of inner city decline.

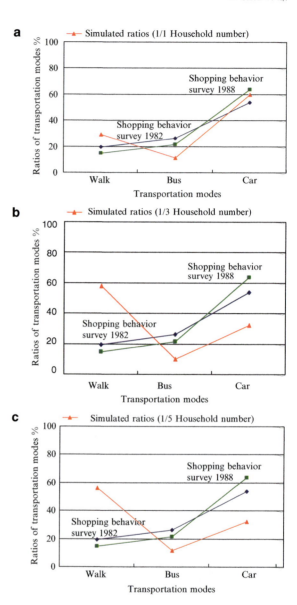

Fig. 18.13 Simulated ratios of transportation modes: (**a**) the case of 1/1 household number; (**b**) the case of 1/3 household numbers; (**c**) the case of 1/5 household numbers

Table 18.4 Accuracy of allocating household agents in the virtual city (the simulation space)

	CA		UPA		UCA		Total	
Real city	16,901	12.0%	103,018	73.0%	21,186	15.0%	141,105	100.0%
Virtual city	16,886	12.6%	101,598	76.0%	15,135	11.3%	133,619	100.0%
Error	−15	0.2%	−1,420	19.0%	−6,051	80.8%	−7,486	100.0%
Accuracy rate	99.9%		98.6%		71.4%		94.7%	

18 Integration of MAS and GIS Using Netlogo 387

Table 18.5 Comparison sale patterns of B-shops and S-shops in real city and virtual city

Grid cell	Retail stores		Value of sales in real city or total times of shopping in virtual city	Sale patterns (%)
Real city (1 km^2)	S-shop		27,263,237	62
	B-shop		16,361,276	38
Simulation urban space in Netlogo (0.25 km^2)	Tick 1	S-shop	667,850	76
		B-shop	209,440	24
	Tick 2	S-shop	1,129,960	72
		B-shop	432,742	28
	Tick 10	S-shop	12,911,270	77
		B-shop	3,780,424	23
	Tick 30	S-shop	42,932,912	77
		B-shop	12,508,927	23

Table 18.6 The proportion of household agents and sale patterns

Proportion (household number)	Retail stores	Total times of shopping in virtual city in Tick 2	Sale patterns (%)
1/2(56,416)	S-shop	839,178	82
	B-shop	181,858	18
1/3(24,244)	S-shop	186,408	70
	B-shop	78,474	30
1/5(13,914)	S-shop	116,792	72
	B-shop	45,064	28
1/10(2,328)	S-shop	1,548	17
	B-shop	7,596	83

Specifically, the project attempts to identify the potential impacts of various policy scenarios on the renewal of inner city commercial environments. This simulation tool can be employed to explore the agents' self-organization processes of shopping decision-making while reflecting utility of shopping behaviours and interactions between households. We suggest that this model allows policy decision makers to use an interactive dynamic tool to examine impacts of planning policies on urban systems through computers experiments.

For the immediate future, we wish to perform accuracy evaluations of the developed tools and of the simulation results. To evaluate the simulation tool and simulation results, one can compare it with real city data. Although it is most often impossible to compare that of individual agents due to data unavailability, the aggregated survey data are available for the comparison. Firstly, through checking the consistency between the prepared households in the simulation environment and those of the real city in each statistic unit of census survey, the accuracy of the conversion tool can be evaluated. Secondly, numbers of different types of shops can also be verified in the same way using the commercial survey data. Thirdly, the market shares of the two types of shops can be aggregated at the level of 500-m mesh for final comparison with real commercial survey data.

Acknowledgements This research is supported by Grants-in-Aid for Scientific Research of Japan Society for the Promotion of Science (JSPS), No. 19656147.

References

Akira, Namatame (1998), Multi-Agents and Complex Theory, Morikita Publisher, in Japanese.

Arentze, T., Borgers, A., Timmermans, H. (2000), A knowledge-based system for developing retail location strategies, Computers, Environment and Urban Systems, Volume 24, 489–508.

Benenson, I., and Torrens, P.M. (2004), Geosimulation: Automata-based modelling of urban phenomena, John Wiley & Sons, London.

Brown, D.G., Riolo, R., Robinson, D.T., North, M. and Rand,W. (2005), Spatial process and data models:Toward integration of agent-based models and GIS, Journal of Geographical Systems, 7 (1), 25–47.

Crooks, A., Castle C., and Batty, M. (2008), Key challenges in agent-based modelling for geospatial simulation, Computers, Environment and Urban Systems,Vol. 32, Issue 6, 417–430.

Crooks, A. (2008b), Constructing and Implementing an Agent-Based Model of Residential Segregation through Vector GIS, http://www.casa.ucl.ac.uk/publications/workingPaperDetail.asp?ID.133, CASA working paper, No. 131

Chen P., Shen Z., Kawakami, M. (2006), Study on Development and Application of MAS for Impact Analysis of Large-scale Shopping Center Development, Journal of the City Planning Institute of Japan, Vol. 413, 271–276.

Ma, L., Arenze, T., Borgers A., and Timmermans H. (2004), Recent Advances in Design & Decision Support Systems in Architecture and Urban Planning. Kluwer Academic Publishers, The Netherlands. 129–144.

Maguire, D.J. (2005), Towards a GIS platform for spatial analysis and modelling. In: D.J. Maguire, M.Batty and M.F. Goodchild, Editors, GIS, spatial analysis and modelling, ESRI Press, Redlands, California, CA, 19–39.

Moreno, N., Ménard, A., Marceau, D. J. (2008), VecGCA: a vector-based geographic cellular automata model allowing geometric transformations of objects, Environment and Planning B: Planning and Design 35(4), 647–665

Sembolonia, F., Assfalgb, J., Armenib,S., Gianassib, R., and Marsonic F. (2004): CityDev, an interactive multi-agents urban model on the web, Computers, Environment and Urban Systems, Volume 28, 45–64.

Shen, Z., Kawakami, M., Chen, P. (2006), Study on decision support system for land use planning of large-scale shopping center Location Using Multi-agent System. In Progress in Design & Decision Support Systems in Architecture and Urban Planning, Endihoven University Press, 169–184

Shen, Z., Kawakami, M., Kawamura, I. (2009), Geosimulation model using geographic automata for simulating land-use patterns in urban partitions, Environment and Planning B: Planning and Design 36(5), 802 – 823

Vancheri, A., Giordano, P., Andrey, D., Albeverio, S. (2008a), Urban growth processes joining cellular automata and multiagent systems. Part 1: theory and models, Environment and Planning B: Planning and Design 35(4), 723–739

Vancheri, A., Giordano, P., Andrey, D., Albeverio, S. (2008b), Urban growth processes joining cellular automata and multiagent systems. Part 2: computer simulations, Environment and Planning B: Planning and Design 35(5), 863–880

Wu, Fulong (1996), A linguistic cellular automata simulation approach for sustainable land development in a fast growing region, Computers, Environment and Urban Systems, Volume 20, Issue 6, 367–387.

Wu, F., and Webster, Christopher J. (1998), Simulation of natural land use zoning under free – Market and incremental development control regimes, Computers, Environment and Urban Systems, Volume 22, Pages 241–256.

Index

A

ABM. *See* Agent-based modeling
Absolute control (AC), 290
Active server pages (ASP), 211
Adjacent parcels, 346
Advisors, 233–234
Agent-based modeling (ABM), 74
Agent interaction, 58–59
Ancient paintings, 242
Animal activities, 332
Annual returns, 352
ArcGIS, 269
ArcToolBox, 355, 359
Ask price, 113
ASP. *See* Active server pages
Asset management, 345
Assumption, 34
Attribute-based modeling technique, 136
AutoCad, 179
Automatic generation, 266, 269–270
Avatar, 229

B

Base map, 326
Base patch, 325
Beijing city master plan (BCMP), 13
Beijing metropolitan area, 285
Bid price, 112–113
Budgetary burden, 108
Building
    covered area, 352
    elevations, 232
    model, 163–166
    types, 345, 359
Build-up process, 343
Built-up urban space, 95

C

CA. *See* Cellular automata
CA transition rules, 349
CAUFN. *See* Cellular automata for urban form
    of neighborhoods
CAUFN tool, 355–359
CCA. *See* Constraint cellular automata
Cellular automata (CA), 3
Cellular automata for urban form of
    neighborhoods (CAUFN), 343
Central core patches, 326
CFs. *See* Control factors
CGI. *See* Common gateway interface
Changsha city, 61
Chat room deliberations, 197
Chuandong area, 31
Color binary function, 331
Common gateway interface (CGI), 332
Communication, 226
Constraint cellular automata (CCA), 5, 27
Construction cost, 352
Construction investment, 352
Consulting company, 219, 221
Control factors (CFs), 290
Control points, 164
Control zones, 290
Conventional commission meetings, 188
Conventional meeting, 199, 201
Cooperative design, 205
    tool, 210
Cooperative equilibrium, 375–377
Coordinating, 230–231
Core patch, 325
Corridors, 325, 326
Courtyard house, 312
Cultural heritage, 155
Cultural property, 315

D

Database processing, 357, 359
Decision making process, 186
Decision tree, 311, 313
Deliberation process, 220
Demand side, 110
DENP. *See* District ecological network plan
Design
    alternatives, 229, 236–237, 277–280
    cooperation, 205
    coordination, 206
    elements, 231
    guidelines, 228, 231–237, 257
    proposal, 206
    schemes, 219
Design review board, 237
Developing land unit (DLU), 288
DHTML. *See* Dynamic Hyper Text Markup
    Language
Digital Beijing VR system, 145
Digital simulation, 170
Digital terrain model (DTM), 157
Digital urban planning, 150–151
District ecological network plan (DENP), 324
DLU. *See* Developing land unit
Documents, 226
3D representations, 236
3ds Max, 179
DTM. *See* Digital terrain model
Dynamic Hyper Text Markup Language
    (DHTML), 215
Dynamic learning process, 233–234

E

Eco-city, 185
Ecological networks, 185, 323
    planning, 323
Economic and social development plan, 28
EDM. *See* Electromagnetic distance
    measurement
Electromagnetic distance measurement
    (EDM), 156
Electronic visualization laboratory (EVL), 133
EventIn, 230
EventOut, 230
EVL. *See* Electronic visualization laboratory
External 3D (X3D), 133

F

Facilitators, 188, 199
Farmer agents, 56–57

Farmland conversion, 49, 56
    quotas, 49
FFD. *See* Free-form deformation
Form scenario analysis (FSA), 3
Fort San Domingo, 161–162
Free-form deformation (FFD), 172
FSA. *See* Form scenario analysis

G

Garbage can model, 77–78
Geodatabase, 354
Geographic information system (GIS), 370
Geomatics models, 155, 166, 177
Geo-referenced 3D models, 165
GIS. *See* Geographic information system
Google 3D Warehouse, 243
Google Earth, 241
Google sketchup, 242
Government agents, 51–52

H

Hirosaka Green Corridor, 188–190
Historical district, 225
Historical urban forms, 9
Households, 30
    agents, 379, 381
    agents attitudes, 376
    numbers, 384
    policy attitude, 115
    satisfaction, 112
Household water consumption simulation
    (HWCSim), 107
HTML, 191
HWCSim. *See* Household water consumption
    simulation
Hypothetical urban space, 118

I

Icon game, 210
Image-based measurement method, 159–160
Imagery classification, 327
Individual design proposals, 218, 220
Industrial enterprise agents, 54–56
Information delivery, 373
Inner polygon, 273–275
Institutional constraints, 8
Integration modeling method, 137
Interaction, 57–58
Interactions among agents, 373
Internet environment, 207–208
Interview survey, 220

J

JSHAPE, 332

K

Kanazawa city, 188
Kanazawa Machiya, 227
KAPPA, 93, 20
Knowledge-based rules, 337

L

Land cover, 330
Land readjustment project, 343
Land-use
    simulation, 345–350, 357
    suitability, 8–9
    types, 328, 345
    zonings, 381
Learning tool, 226
Learning process, 188, 194–195
Learning tool, 226, 233
Local community, 220
Local water resources, 110
Locational constraints, 7
Log analysis, 197–199

M

Machiya elevations, 228–229
MAS. *See* Multi-agent system
Mask data set, 329
Merchant townhouses, 251–252
Messages, 198
Model behavior, 122
Model integration, 171
Multi-agent system (MAS), 50
Multidisciplinary approach, 308
Multimedia, 185
Multiple bounded polygons, 275–277

N

Nagoya city, 266, 277–280
Natural land cover, 330
Negotiation process, 97
Neighbor blocks, 346
Neighborhood constraint, 8
Neighborhood residents, 209
Neighbor parcels, 346
Netlogo, 117, 370
Net participation, 188

Neural network (NN), 350
    learning process, 350
NN. *See* Neural network

O

ODBC. *See* Open database connectivity
Officers, 219
Online deliberation, 185
Online Expo 2010 Shanghai, 145
Online learning, 230–231
Online multimedia, 191, 200–201
    tool, 188, 201
Open database connectivity (ODBC), 212
Orientation, 171
Orthogonal building polygons, 266
Orthogonal polygons, 270–272
Ownership-based plan, 310

P

Paper drawings, 226
Parcels, 346
Participants, 198
Participatory planning and design, 209
Partitioning scheme, 270–272
Patterns of farmland conversion, 49
Payoff matrix, 376
Photogrammetry, 159
Plan compilation, 301
Planned urban form, 94
Planners, 219
Planning
    activities, 73
    behaviors, 74–79
    information, 187
    meetings, 186
    practice, 228
    process, 205–206
    regulations, 309–310
    requirements, 186
    scheme, 301–303
    solution, 186
Planning alternatives, 19, 315
    using VRML, 317
Planning support system (PSS), 285
Policy attitude, 377
Policy parameter identification (PPI), 20, 92
Policy parameters, 11
Polygon expression, 270–272
Population density, 41
PPI. *See* Policy parameter identification

392

Prisoner's dilemma games, 84
Probabilities, 353
Professional planners, 225
Property ownership, 309–310
PSS. *See* Planning support system
Psychological expectations, 57
Public
    good, 113
    involvement, 311, 336–339
    learning, 188
    opinions, 186
    participation, 256

Q

Questionnaire, 194

R

Random utility model (RUM), 372
RC. *See* Relative control
Real estate management, 313, 359
Real estate management simulation,
    352–353
Reality-based virtual conditions, 378
Reconstruction, 315
Relative control (RC), 290
Resident agents, 52–54
Residential environment improvement, 315
Resource use, 34
Restoration, 315
    targets, 316
Review process, 195–197
RUM. *See* Random utility model

S

Samurai houses, 248
Satellite image, 328
Scale, 171
Scenarios, 65
    configuration, 355–356
Server–client model, 226
ShareEvent, 230
Shop agents, 379, 381
Shop-choice
    behaviour, 371
    decisions, 374
    simulation, 375–377
Shopsim-MAS, 370
Simulated historical landscape, 252–253
Simulated urban form, 92

Simulation
    of building types, 350
    space, 378
Social experiment, 193
Space evolution, 83–86
Spatial
    constraints, 95–97, 99–100
    factors, 289–293
    patterns, farmland conversion, 49
    processes, 80–81
    processing, 356–357
Spatial strategic plan support system
    (SSP-SS), 27
Spatial–temporal limitations, 209
Spatio-temporal allocation, 49–69
SSP-SS. *See* Spatial strategic plan support
    system
Stakeholders, 97, 209, 229
States of land use, 348
State transition, 116
Straight Skelton, 273–275
Strategic choices, 376
Strategic spatial plan, 29
Street park, 207–208
Subsidy regulations, 257
Subsidy system, 312
Supply side, 110
System functions, 228

T

Teramachi temple area, 247
Terrestrial laser scanned building models, 166
Terrestrial laser scanning (TLS), 157
Terrestrial surveying, 156–157
TLS. *See* Terrestrial laser scanning
TNDC. *See* Total number of developed cells
Total amount control, 114
Total number of developed cells (TNDC), 97,
    102–105
Total station instruments, 156–157
Townscape, 228, 237
Townscape design guidelines, 225
Traditional temple area, 241
Training data set, 329
Training program, 235
Transaction costs, 352
Transition rules, 8
Transportation mode, 374

U

UAZ. *See* Uniform analysis zone
UCA. *See* Urban control area

Index 393

UGB. *See* Urban growth boundaries
UGCP. *See* Urban growth control planning
UGC-PSS. *See* Urban growth control planning
    support system
Uniform analysis zone (UAZ), 288
UPA. *See* Urban promote area
Urban
    center development, 40
    conservation, 307
    containment, 287
    form, 91
    furniture, 211
    master plans, 18
    plans, 28
Urban control area (UCA), 381
Urban Greenery Technology Development of
    Japan, 324
Urban growth
    control, 285
    scenario, 40
    simulation, 16–18
Urban growth boundaries (UGB), 3
Urban growth control planning (UGCP), 285
Urban growth control planning support system
    (UGC-PSS), 286
Urbanization, 49
Urban planning and managing organization, 140
Urban promote area (UPA), 381

V
Virtual
    communication, 126–127, 229, 235
    3D city model, 265, 266
    globe, 241
    meetings, 199–200, 201
    representation, 205–206, 256

Virtual planning committee meetings,
    190, 193
Virtual reality (VR), 131
Virtual reality modeling language
    (VRML), 133
Visual classroom, 228
Visualization tool, 228
VR. *See* Virtual reality
VR application, 139–140
VRML. *See* Virtual reality modeling language
VR world, 235

W
Waste discharge, 34
Water
    consumption, 111
    market, 110
    price, 114–117
    resources, 108
    supply capacity, 113
Web3D, 133
Web database, 207–208
WebGIS, 332
Willingness to accept (WTA), 110
Willingness to pay (WTP), 109

Y
Yamanoue Street Park, 209